The Doppler Method for the Detection of Exoplanets

AAS | IOP Astronomy

AAS Editor in Chief

Ethan Vishniac, John Hopkins University, Maryland, US

About the program:

AAS-IOP Astronomy ebooks is the official book program of the American Astronomical Society (AAS), and aims to share in depth the most fascinating areas of astronomy, astrophysics, solar physics and planetary science. The program includes publications in the following topics:

Books in the program range in level from short introductory texts on fast-moving areas, graduate and upper-level undergraduate textbooks, research monographs and practical handbooks.

For a complete list of published and forthcoming titles, please visit iopscience.org/books/aas.

About The American Astronomical Society

The American Astronomical Society (aas.org), established 1899, is the major organization of professional astronomers in North America. The membership (~7,000) also includes physicists, mathematicians, geologists, engineers and others whose research interests lie within the broad spectrum of subjects now comprising the contemporary astronomical sciences. The mission of the Society is to enhance and share humanity's scientific understanding of the universe.

The Doppler Method for the Detection of Exoplanets

A P Hatzes
Thüringen Landessternwarte Tautenburg, Sternwarte 5, D-07778 Tautenburg, Germany

IOP Publishing, Bristol, UK

ISBN 978-0-7503-1689-7 (ebook)
ISBN 978-0-7503-1687-3 (print)
ISBN 978-0-7503-1774-0 (myPrint)
ISBN 978-0-7503-1688-0 (mobi)

DOI 10.1088/2514-3433/ab46a3

Version: 20191201

AAS–IOP Astronomy
ISSN 2514-3433 (online)
ISSN 2515-141X (print)

British Library Cataloguing-in-Publication Data: A catalogue record for this book is available from the British Library.

Published by IOP Publishing, wholly owned by The Institute of Physics, London

IOP Publishing, Temple Circus, Temple Way, Bristol, BS1 6HG, UK

US Office: IOP Publishing, Inc., 190 North Independence Mall West, Suite 601, Philadelphia, PA 19106, USA

To Gordon Walker and Bill Cochran, early pioneers in the radial velocity detection of exoplanets who inspired this work.

Contents

Preface

The first radial velocity (RV) measurements on a star using the Doppler method were made late in the 19th century. The discovery of the first exoplanet, found using the Doppler method, was made near the end of the 20th century. Why did the discovery of exoplanets take nearly one hundred years? In short, it is due to the spectacular improvement in the measurement precision—a 10,000-fold increase over the past 100 years, with most improvements occurring in the past few decades. This is the subject of this book.

The field of exoplanets has developed into one of the most exciting and vibrant fields in astronomy, and that is all owed to the Doppler method. By discovering the first exoplanets, it essentially created the field. Although the detection efficiency of exoplanets using the Doppler method has been surpassed by the photometric transit method, the Doppler method still plays a vital role in confirming transit discoveries and giving a mass, one of the most fundamental parameters of a planet. It is one of the few methods (along with astrometry) that gives you a "dynamical" mass—dynamical in the sense that one derives the mass using the laws of Kepler and Newton, rather than statistics and theoretical models. The Doppler method still ranks as one of the most important exoplanet detection methods in use today.

The pioneering transit-search space missions *COnvection ROtation and planetary Transits* (*CoRoT*) and *Kepler* have produced a treasure chest of transiting planets. As of this writing, NASA's *Transiting Exoplanet Survey Satellite* (*TESS*) is performing a transit survey among the brightest stars in the sky, and the RV community has its hands full determining the mass for candidate transiting planets. For all of these missions, ground-based spectroscopic measurements, in particular RV measurements, have played a vital role in characterizing the planet discoveries. Within a decade, the *PLAnetary Transits and Oscillations of stars* (*PLATO*) mission of the European Space Agency will also search for transiting planets around bright stars, but with the goal of finding Earth-like planets in the habitable zone of stars. So, the Doppler method is poised to continue its important role in exoplanet studies well into the 2030s.

Although almost 1000 exoplanets have been discovered with the Doppler method, this book will not focus on the results from the various RV planet search programs. This can be gleaned from the literature or from Perryman's *The Exoplanet Handbook*. Rather, this work will focus purely on the method. This includes how one can achieve a high RV measurement precision as well as the challenges, limitations, and potentials of this technique. It will include other aspects of the method, such as instrumentation, wavelength calibration, finding periodic signals in RV time series, interpreting the signals that you find, and Keplerian orbits. An important aspect is stellar variability, which has been known to trick more than a few astronomers (this author included) into thinking that they have discovered an exoplanet. In short, it will cover every aspect needed for one to detect exoplanets with the RV method, a sort of "handbook" for the Doppler method.

If the reader wants to purse RV follow-up of transiting planets from space missions or wants to perform exoplanet RV surveys, this book should be useful. When it comes to exoplanet discoveries, it is easy to fall into traps, to be misled, or to arrive at erroneous conclusions. As the physicist Richard Feynman once famously said, "Science is a way of trying not to fool yourself. The principle is that you must not fool yourself, and you are the easiest person to fool." This is especially true in the field of exoplanets. I hope that this book will ease the path of those embarking on the use of the Doppler method for the detection and characterization of exoplanets and hopefully, to avoid pitfalls.

Acknowledgments

It is a pleasure to thank all of the scientists and students who helped in the preparation of this book. It would not have been possible without them.

I thank the Tautenburg Observing School: Jaime Avalos, Clark Baker, Dugasa Belay Zeleke, Richard Bischoff, Martin Blazek, Sireesha Chamarthi, Michael Debus, Jana Dvorakova, Vanessa Fahrenschon, Andreea Gornea, Sascha Grziwa, Engin Keles, Hannah Kellermann, Sarah-Jane Köntges, Oliver Lux, Priscilla Muheki, Eva Plávalová Jan Subjak, Jerusalem Tamirat, Fabian Wunderlich, and Jiri Zak. They were kind enough to give up observing time for some crucial tests that are presented in this book.

Silvia Sabotta provided me with an RV time series made with the iodine cell. Priyanka Chaturvedi produced the synthetic stellar spectra used in this work. Figures and data highlighting results from CARMENES were provided by Ansgar Reiners, Ignas Ribas, Guillem Anglada-Escudé, Mathias Zechmeister, and of course, the entire CARMENES consortium. Ulf Seeman provided valuable figures and input for the CRIRES$^+$ gas absorption cell.

A special thanks goes to Michael Hartmann who provided me with results from his PhD. His analysis of two roAp stars nicely demonstrated how the use of different templates for calculating the RV can produce conflicting results. He also provided me with the typesetting of the GLS equations and a short description of the GLS periodograms. This saved me some work.

I thank Patrick Lenz who developed Period04 which was an indispensible tool for the preparation of this book.

I also thank my wife, Ingrid Schutzmann, who tolerated more than a few "book writing" weekends and helped clear time for me.

Finally, a sincere and heartfelt thank you to two very important people: Gordon Walker, who showed us all how to perform precise stellar RVs and whose work was an inspiration to me, and finally, William Cochran, who got me into the business of detecting exoplanets with the RV method. This book is dedicated to them.

Author biography

A. P. Hatzes

Photo credit: Christian Högner

Artie Hatzes is one of the pioneers in searching for extrasolar planets and he brings over 30 years experience in the use of precise stellar radial velocity measurements. Besides searching for extrasolar planets, he has also extended the use of these types of measurements to the study of stellar oscillations in magnetic A-type and K giant stars. Hatzes received his Bachelor of Science with Honor from the California Institute of Technology and his Master of Science and PhD from the University of California in Santa Cruz. In 1988 he joined Bill Cochran at the University of Texas at Austin for the start of the McDonald Observatory Planet Search Program. He has been working exoplanets ever since. Hatzes is currently director of the Thuringian State Observatory and Professor of Physics and Astronomy at the Friedrich-Schiller-University in Jena, Germany.

AAS | IOP Astronomy

The Doppler Method for the Detection of Exoplanets

A P Hatzes

Chapter 1

Introduction

1.1 The Dawn of Doppler Measurements

In 1842, the Austrian physicist, Christian Doppler, published his treatise *Über das farbige Licht der Doppelsterne und einiger andere Gesterne des Himmels* (*On the Colored Light of Binary stars and Some Other Stars in the Heavens*). Doppler postulated that because the pitch of a sound wave depended on the relative speed between the source and the observer that the color of light of a moving star should also change. Doppler thought that this phenomenon could explain the colors of binary stars. Although wrong about the colors of stars, his hypothesis about the change in the frequency of waves relative to a moving source—and that the effect can be used to measure the velocity of stars—proved true. It was shown to be experimentally correct for sound waves and had an easy theoretical explanation, but not so for electromagnetic waves.

At the same time, Armand Hippolyte Louis Fizeau also became involved with aspects of the discovery of the Doppler effect (known as the Doppler–Fizeau effect in France). He focused his work on understanding the effect as applied to light rather than sound and developed the mathematical formalism underlying the principle. He was the first to predict the redshift of electromagnetic waves.

Remarkably, there was early debate among physicists as to whether Doppler's principle could even be applied to light waves (Vogel 1900). An early and elegant demonstration of the Doppler effect applied in astronomy was made by James E. Keeler in his seminal paper "A Spectroscopic Proof of the Meteoritic Constitution of Saturn's Ring" (Keeler 1895, p. 416). It is obvious now, but at that time it was not known whether the rings were solid or consisted of small particles in orbit around Saturn. He also could foresee the power of the Doppler method: "I have recently obtained a spectroscopic proof of the meteoritic constitution of the ring, which is of interest because it is the first direct proof of the correctness of the accepted hypothesis, and because it illustrates in a very beautiful manner

doi:10.1088/2514-3433/ab46a3ch1

(as I think) the fruitfulness of Doppler's principle, and the value of the spectroscope as an instrument for the measurement of celestial motion."

His results, published in the first issue of *The Astrophysical Journal* (Figure 1.1), clearly show the solid body motion of the planet and the Keplerian motion of the ring. Coincidentally, Keeler was eager to apply Doppler measurements for spectroscopic measurement of the velocity of galaxies (Osterbrock 2002). Unfortunately, he died tragically in 1900 at the young age of 42, and the discovery of the expanding universe had to await the work of Edwin Hubble.

Doppler's principle could indeed be applied to light with fruitful results, and it has produced some of the most fundamental discoveries in astronomy. Some examples include

- Hubble's relationship between the distance of a galaxy and its redshift (velocity). This established the fact that the universe was expanding as one of the fundamental principles of cosmology (Hubble 1929).
- The flat rotation curves of galaxies, which was one of the first evidence of dark matter (Rubin et al. 1978).

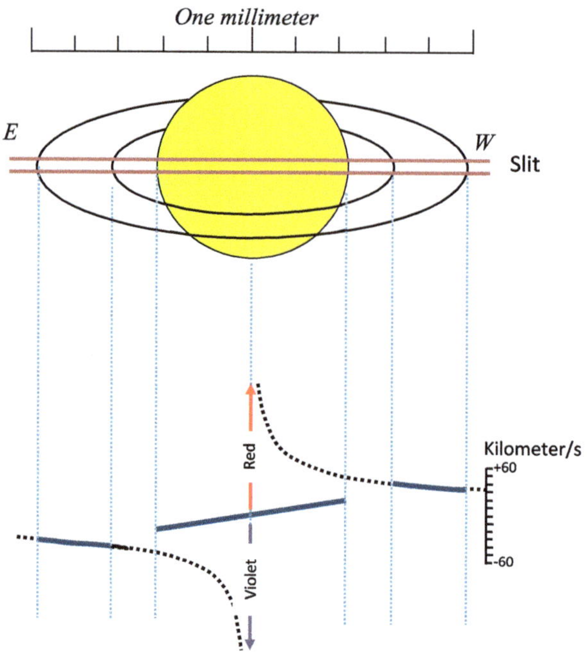

Figure 1.1. A reproduction of the figure from Keeler (1895) showing the Doppler velocity along Saturn using long-slit spectroscopy of Saturn. The spectrum crossing the planet's disk shows the Doppler motion of solid body rotation. The spectrum from the rings show Doppler motion consistent with Keplerian motion.

- The rotation of stars and the rotation break at spectral type mid-F stars. This was early evidence that magnetic activity is responsible for stellar angular momentum loss while a star is on the main sequence (Kraft 1967).
- The discovery of exoplanets (Mayor & Queloz 1995).

This book is devoted to the last item—the use of stellar radial velocity (RV) measurements for the detection and study of exoplanets. Over the past two decades, the field of exoplanets has developed into one of the most vibrant fields of astrophysics. As of this writing, thousands of planets have been discovered orbiting other stars. This exciting field owes its existence to Doppler's method, through which the first exoplanets were discovered.

1.2 Early Work on Stellar Radial Velocity Measurements

The first attempts to measure stellar RV measurements date to the 19th century, when Huggins (1868) visually observed the displacement of stellar hydrogen Balmer lines with respect to those from a hydrogen discharge tube. The "founder" of modern stellar RVmeasurement arguably falls on the German astronomer Herman Carl Vogel, who systematically applied photography in stellar RV measurements. He studied astronomy at the German universities of Leipzig and Jena, and his accomplishments were pioneering. Vogel was the first to measure the rotation of the Sun using Doppler shifts of the approaching and receding limbs. His RV measurements first detected an unseen stellar companion to an eclipsing binary (Vogel 1890, p. 27) using the Doppler method, noting that "...before a minimum Algol was moving away from the Sun, and after a minimum it was moving toward it."

Because the RV precision was of the order of several km s^{-1}, early work focused largely on the study of binary stars. Figure 1.2 shows the velocity curve of the spectroscopic binary star HD 36954 taken with the 36 inch refractor at Lick Observatory in the mid-1930s (Neubauer 1936). The RV measurements have a scatter of 6.9 km s^{-1}, typical of the RV precision of that era.

The first astronomer to recognize that stellar RV measurements could be used to detect exoplanets was Otto Struve. He was a Russian-born astronomer who did most of his astronomical work in the United States. Struve served as director of the Yerkes, McDonald, and National Radio Astronomy Observatories. As a director, he could recognize talent, having hired Subrahmanyan Chandrasekhar and Gerhard Herzberg, two future Nobel Prize winners.

Struve was also a visionary. His remarkable paper "Proposal for a Project of High-precision Stellar Radial Velocity Work" (Struve 1952) was the first to propose using Doppler measurements to search for exoplanets. The discovery of 51 Peg b in 1995—a giant planet in a 4.2 day orbit—was foreseen by Struve. In his paper, p. 200, he argued that "we know that stellar companions can exist at very small distances. It is not unreasonable that a planet might exist at a distance of 1/50 of an astronomical unit. Such short-period planets could be detected by precise radial velocity measurements." His predictive powers did not stop there. He goes on to say that "there would, of course, also be eclipses ··· and the loss of light in stellar magnitudes is

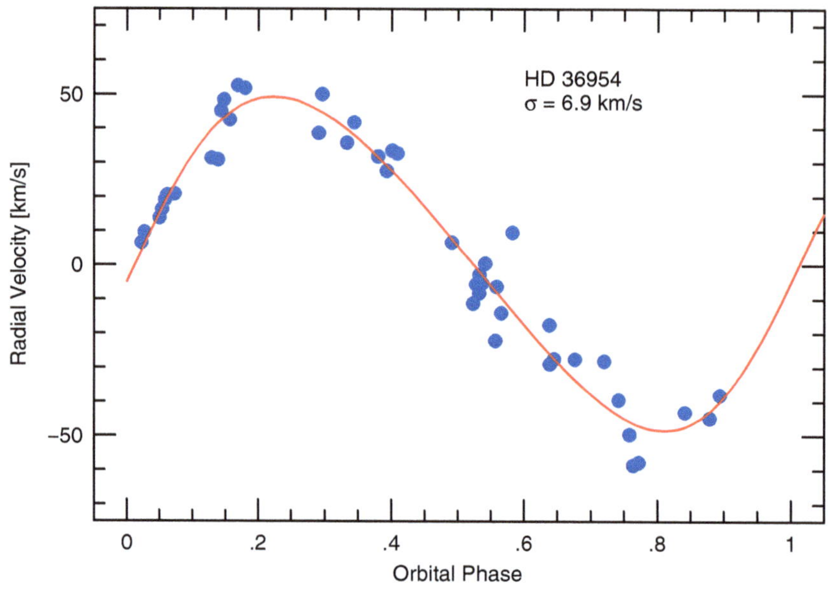

Figure 1.2. Orbital motion of HD 36954 (curve) calculated from radial velocity measurements taken from 1932–1935 using the 36 inch refractor at Lick Observatory (Neubauer 1936). The scatter about the orbital solution is 6.9 km s^{-1}.

about 0.02." Struve not only foresaw the possibility of short-period Jupiter-mass planets, but the use of the transit method to characterize the density. His proposal did not result in the "powerful" spectrograph he advocated, which only shows that science has its own "prophets" who are often ignored. The discovery of exoplanets still had to wait another half century.

1.3 Toward Precise Stellar Radial Velocity Measurements

With 150 years of stellar RV measurements and even proposals from the mid-20th century to build spectrographs capable of such precise measurements, why did it take until the end of the 20th century to discover the first exoplanets? The short answer: a lack of precision.

The Doppler shift of a star due to the presence of planetary companions is small. We can use *Kepler*'s third law to get an estimate of the RV precision needed to detect the reflex motion of star due to the presence of a planetary companion:

$$P^2 = \frac{4\pi^2 a^3}{G(M_s + M_p)}, \tag{1.1}$$

where M_s is the mass of the star, M_p is the mass of the planet, P the orbital period, and a the semimajor axis.

For planets, we are in the regime where $M_s \gg M_p$. If we assume circular orbits and the fact that $M_p \times a_p = M_s \times a_s$, where a_s and a_p are the semimajor axes of the star and planet, respectively, it is trivial to derive

$$V\,[\mathrm{m\,s^{-1}}] = 28.4\left(\frac{P}{1\ \mathrm{yr}}\right)^{-1/3}\left(\frac{M_\mathrm{p}\sin i}{M_\mathrm{Jup}}\right)\left(\frac{M_\mathrm{s}}{M_\odot}\right)^{-2/3}, \tag{1.2}$$

where i is the inclination of the orbital axis to the line of sight. Remember, the Doppler method only measures velocities along the line of sight.

The reflex motion of a 1 $M_\odot$ star due to various planets at different orbital radii calculated with Equation (1.2) is shown Figure 1.3 and a Jupiter analog (1 M_Jup orbiting at a distance of 5.2 au) will induce a 11.2 m $\mathrm{s^{-1}}$ reflex motion of a Sun-like host star with an orbital period of 12 years. To detect such a planet, you would need an RV measurement precision of at least 10 m $\mathrm{s^{-1}}$, which would have to be maintained for over a decade.

Moving this planet to the semimajor axis of a "Struve planet" (0.02 au) would result in a reflex motion 10 times higher. This eases your required measurement precision to a more comfortable 100 m $\mathrm{s^{-1}}$ maintained over several days. If you are bold and you want to detect the first Earth analog (planet at 1 au), you would need a more challenging measurement precision of better than 10 cm $\mathrm{s^{-1}}$. For a "lava" Earth-mass planet orbiting at 0.05 au from the star, the stellar Doppler amplitude would be a more reasonable 1 m $\mathrm{s^{-1}}$, a precision, as we will soon see, that is achieved by modern instruments. This figure shows that to detect planets with the RV method, one needs exquisite precision coupled with long-term stability.

The early exoplanet surveys had a search strategy that was driven by the only example of a planetary system—our own. We thus expected giant planets to all lie at approximately 5 au from the star. The meant we required an instrument capable of

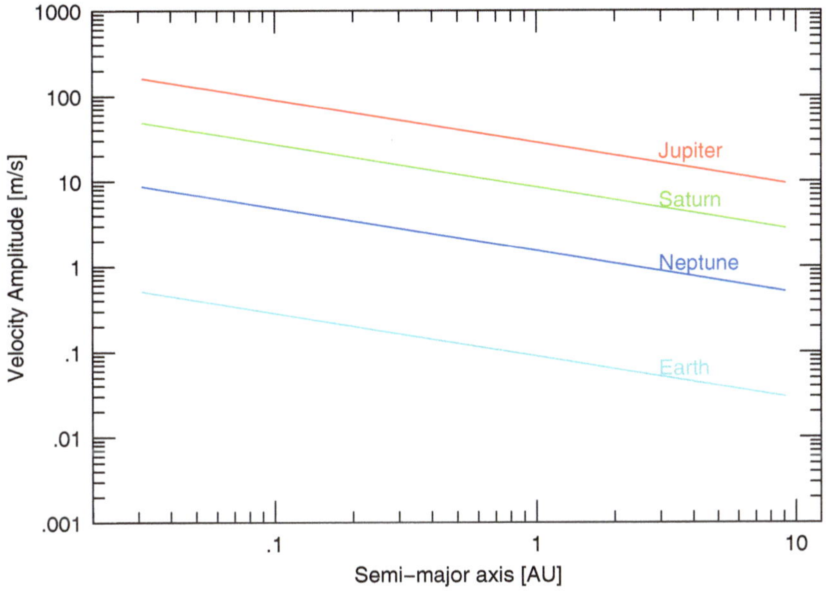

Figure 1.3. The amplitude of the barycentric radial velocity variations for a 1 solar mass star orbited by an Earth, Neptune, Saturn, or Jupiter at various orbital distances.

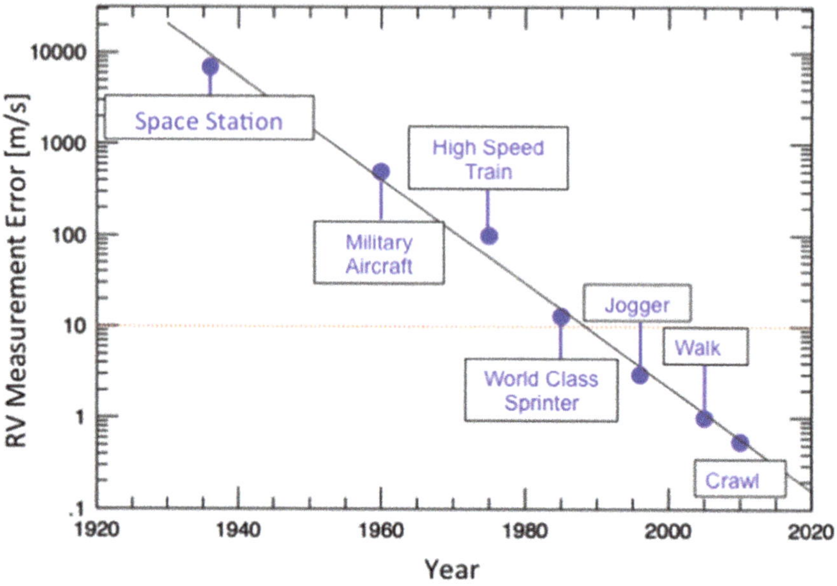

Figure 1.4. The evolution of the radial velocity measurement error as a function of time. The horizontal line marks the reflex motion of a solar mass star with a Jupiter analog.

achieving an RV measurement precision of at least 10 m s^{-1} and with a high premium on long-term stability.

Figure 1.4 demonstrates just why it took 150 years to discover exoplanets with the Doppler method. It shows the approximate RV precision as a function of time. The decrease in measurement error follows a power-law fit (solid line). In the 1960s, we could measure a star moving at the speed of an SR-71 military aircraft, or about a km s^{-1}. Currently, we can measure a stellar RV under 1 m s^{-1}, or the speed of a very leisure walk or a rapidly crawling baby. If the power law holds, we should achieve an RV precision of 10 cm s^{-1} by the mid-2020s. The "magic" precision of 10 m s^{-1}, shown by horizontal dashed line, was only achieved in the mid-1980s. Coincidentally, this was about the time the first exoplanets were discovered.

1.4 The Early Hints of Exoplanets

Although the discovery of 51 Peg b (Mayor & Queloz 1995) is considered as the discovery of the first exoplanet around a Sun-like star, there were hints of discoveries before 1995[1]. Campbell et al. (1988; hereafter CWY) monitored the brighter component of the spectroscopic binary γ Cep using a hydrogen fluoride absorption cell (see Chapter 4). The top panel of Figure 1.5 shows the RV measurements of γ Cep A from CWY. The long-term trend due to the binary motion is obvious, but one can see extra "wiggles." Removing the linear trend shows clear variations, due to a

[1] In 1992 Wolszczan & Frail discovered Earth-mass exoplanets around the pulsar PSR1257 + 12 (Wolszczan & Frail 1992). These were discovered using timing variations.

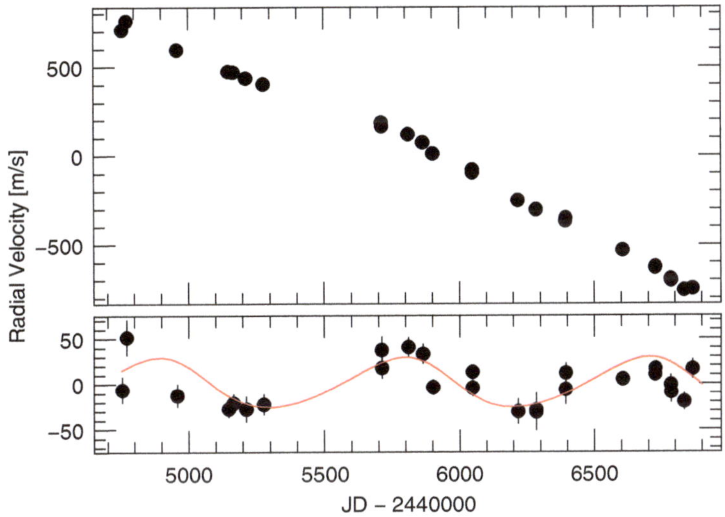

Figure 1.5. (Top) Early RV measurements of γ Cep A made in the 1980s using an H–F absorption cell (Campbell et al. 1988). The trend is due to orbital motion of the binary companion. (Bottom) RV measurements after removing the long-term trend. The RV motion due to the planetary companion can easily be seen (red curve: orbital solution of Hatzes et al. 2003).

possible planetary companion. CWY commented that these variations represented a possible third body in the system with a period of ≈3 yr that might be planetary in nature. Unfortunately, Walker et al. (1992) later attributed these variations to rotational modulation, largely because theoretical work could not produce giant planets in short period orbits (G. A. H. W. Walker, 2013, private communication). Hatzes et al. (2003) later demonstrated that these residual RV variations were indeed due to a 1.7 $M_{\rm Jup}$ giant planet in a 2.48 yr orbit.

Latham et al. (1989) found a possible giant planet with a minimum mass of 11 $M_{\rm Jup}$ orbiting HD 114762 with an orbital period of 83.8 days (Figure 1.6). With an eccentricity of $e = 0.38$, HD 114762 b was the prototype of the so-called massive eccentric planets. These are massive planets ($M \approx 10\ M_{\rm Jup}$) in eccentric ($e \gtrsim 0.3$) orbits. The RV measurements were made with "traditional" methods, i.e., without simultaneous wavelength calibration (see Chapter 4), so the measurement error is $\sigma \approx 400$ m s^{-1}. This is comparable to the RV amplitude of ≈600 m s^{-1} and demonstrates that with sufficient measurements, one can detect a planet with RV amplitude comparable to the measurement error. However, the true nature of the companion is unknown until the orbital inclination is measured. Preliminary results from the GAIA astrometric space mission indicate an orbital inclination of 6.2 degrees which yields a companion mass of $107^{30}_{-27}\ M_{\rm Jup}$, or in the M dwarf star regime (Kiefer 2019)[2].

Inspired by the early work of Walker et al. (1989), who found RV variations in a sample of K giant stars, Hatzes & Cochran (1993) monitored several of these stars

[2] As of this writing, this paper was submitted but not yet accepted.

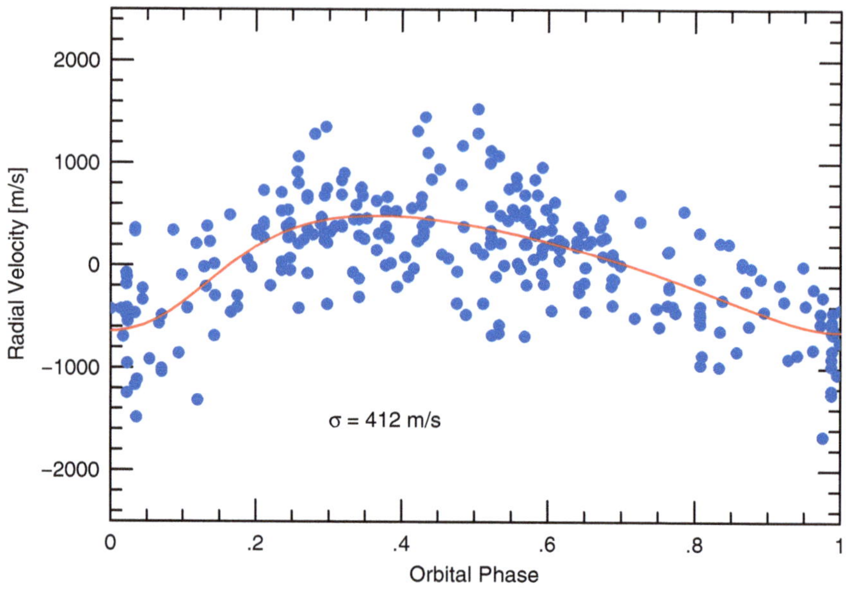

Figure 1.6. The RV variations of HD 114762 due to a planetary candidate companion with an orbital period of 83.8 days and a minimum mass of 11 $M_{\rm Jup}$ (Latham et al. 1989). The rms scatter about the orbital solution (red curve) is 412 m s^{-1} and was typical for high-quality RV measurements before the use of simultaneous wavelength calibration.

and found long-period RV variations in α Tau, α Boo, and β Gem. They hypothesized that these could be due to giant-planet companions. Indeed, it was later shown that early RV measurements for β Gem were due to a giant planet with $M = 2.7\ M_{\rm Jup}$ in a 590 day orbit (Hatzes et al. 2006; Reffert et al. 2006).

1.5 The 51 Peg Revolution

The discovery of 51 Peg b (Mayor & Queloz 1995) clearly marked an explosion of the field[3]. This giant planet ($M \approx 0.5\ M_{\rm Jup}$) in a 4.2 day orbit shocked astronomers, except maybe for the ghost of Otto Struve. It also demonstrated that RV surveys were probing the wrong parameter space (orbital distances of 5 au rather than 0.05 au)—the dangers of planning surveys based on one example, our solar system. The RV amplitude of ≈50 m s^{-1} (Figure 1.7) clearly benefited from the increased precision of Doppler measurements. Figure 1.8 shows the discovery rate of exoplanets found using the Doppler method. The sharp increase after the discovery of 51 Peg is for three reasons. First, once astronomers realized that giant planets could occur in short period orbits, they changed their observing strategies so that these short-period planets could be discovered rather quickly. Second, using the Doppler method to detect exoplanets became quite fashionable, with many groups "jumping on the bandwagon." Prior to 1995, there were only a handful of

[3] In recognition of this discovery Mayer and Queloz were awarded the 2019 Nobel Prize in physics.

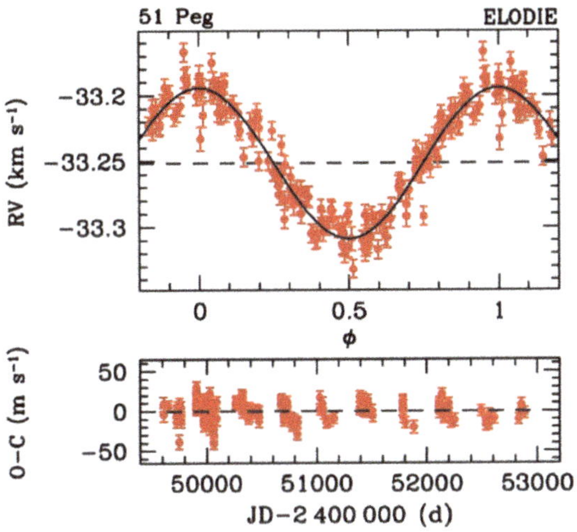

Figure 1.7. The discovery of 51 Peg b. The RV variations have an amplitude of ≈50 m s^{-1} and are phased to the orbital period of 4.2 days. The typical measurement error is ≈15 m s^{-1}. (Adapted by permission from Macmillan Publishers Ltd: Mayor & Queloz 1995.)

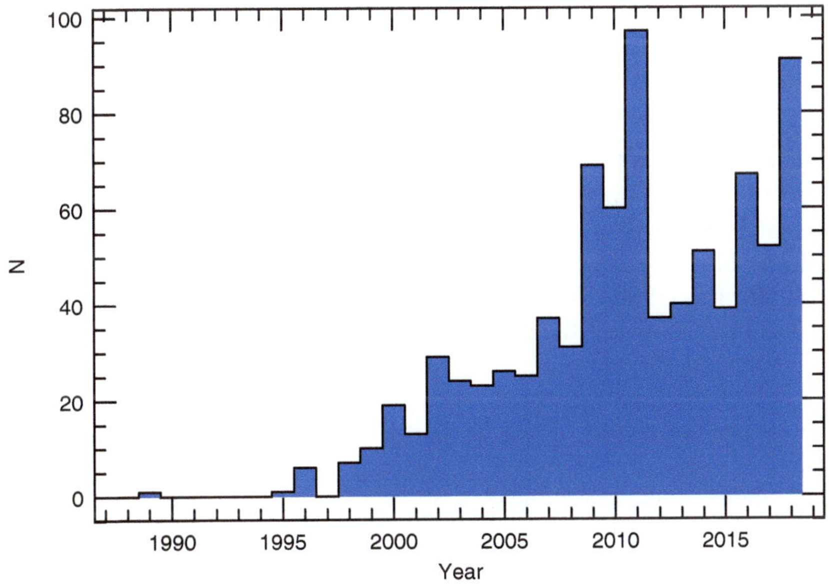

Figure 1.8. The rate of exoplanet discovered planets using the Doppler method (transit discoveries are not included; data from http://www.exoplanet.eu).

groups using precise stellar RVs to search for exoplanets. Now, the number of such groups is in the dozens. Currently, approximately a hundred exoplanets per year are discovered with the Doppler method. Note that RV measurements also play a key role in the confirmation and mass determination of discoveries made by the transit method.

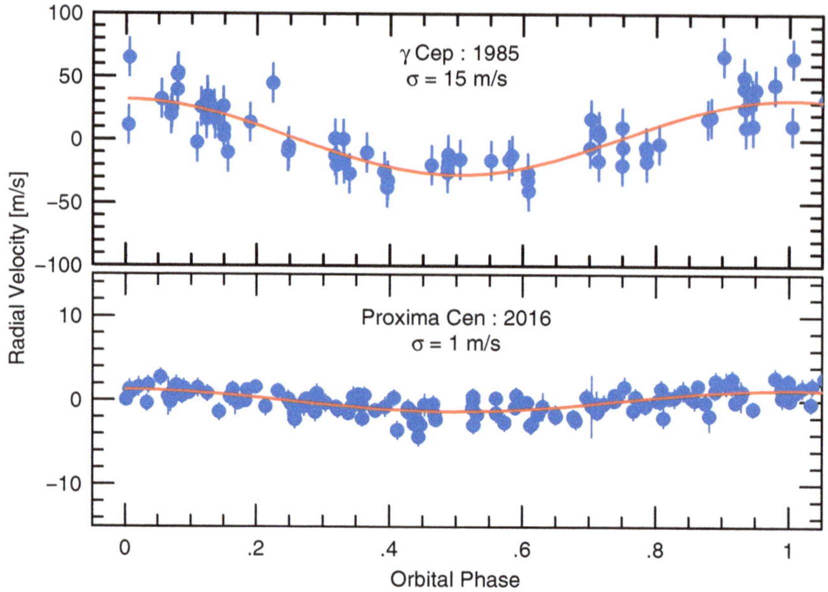

Figure 1.9. (Top) RV measurements of γ Cep phased to the orbital motion using Doppler measurements taken in the mid-1980s. (Bottom) RV measurements of Proxima Centauri phased to the orbital motion as measured in 2016. The box size for this panel is the same as the measurement error for the earlier measurements.

Finally, the increased detection rate of exoplanets using Doppler measurement also benefited from the dramatic increase in RV precision over the past 30 years. The top panel of Figure 1.9 shows precise RV measurements of γ Cep (Walker et al. 1992). These show a scatter of about 15 m s^{-1}—the measurement error they could achieve in the mid-1980s. The lower panel shows modern RV measurements of Proxima Centauri showing the variations of the Earth-mass companion in an 11.8 day orbit (Anglada-Escudé et al. 2016). These have a scatter of a mere 1 m s^{-1}. Note that the scale of the y-axis in this lower panel is the size of the error bar in the top panel. An important aspect of this book is to show how this dramatic increase was achieved.

The parameter space in the mass versus semimajor axis of exoplanets discovered with the Doppler method is shown in Figure 1.10. These are only planets discovered through RV measurements and not those from transit discoveries, although Doppler measurements were important for confirming the nature of these discoveries and measuring the companion mass.

1.6 The Doppler Method

This book is written primarily for astronomers who wish to use the Doppler method for the detection of exoplanets. Given the ubiquity of Doppler measurements in astronomy, its utility is not restricted to this narrow field. Measurement precision is only one aspect of the process.

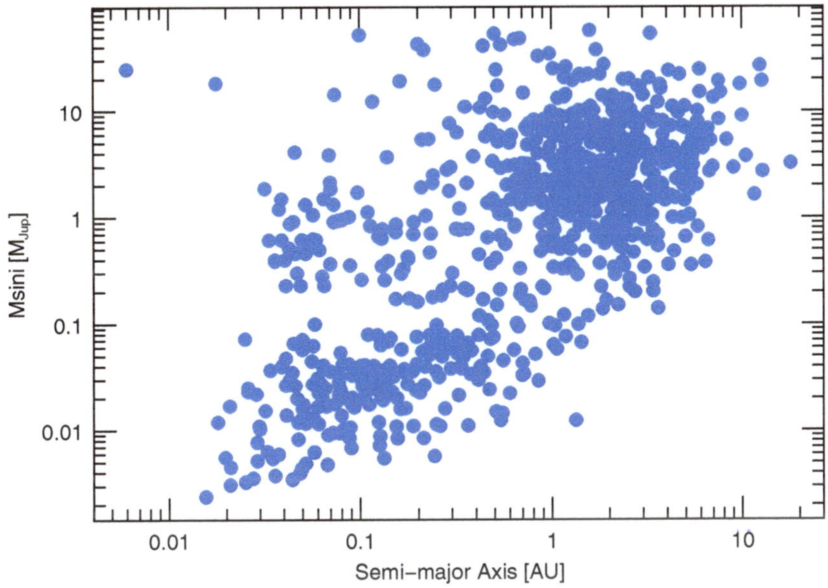

Figure 1.10. The detection parameter space for planets found using the Doppler method. This does not include transit discoveries. The method only measures the mass multiplied by the sine of the orbital inclination, i ($M \sin i$).

The redshift z of an object is defined as

$$z = \left(\frac{\lambda_{\text{meas}}}{\lambda_{\text{emit}}} - 1 \right), \tag{1.3}$$

where λ_{meas} is the measured wavelength of the spectral feature and λ_{emit} is the emitted wavelength.

We are interested in the nonrelativistic Doppler shift, and if the change in wavelength is $\Delta\lambda = \lambda_{\text{meas}} - \lambda_{\text{emit}}$, then the Doppler shift can be converted to an RV by

$$\frac{\Delta\lambda}{\lambda} = \frac{v}{c}. \tag{1.4}$$

So, in principle, the method is quite simple. You measure a position of a spectral feature, compare it to its rest wavelength, and convert that to a Doppler shift in velocity using Equation (1.4). Once you have sufficient velocity measurements, you can fit a Keplerian orbit and derive the companion mass. Simple in theory, but as we shall soon see, challenging in practice.

The Doppler method we will deal with in this book is strictly a *relative*, as opposed to *absolute*, velocity measurement of stars. To detect companions to a star, we just need the relative Doppler shift with respect to a fiducial spectrum, either a standard star or the target star itself. In the latter case, we measure a shift with respect to spectrum of the star taken at time t_0. Measuring relative Doppler shifts is considerably much easier as zeroth-order effects due to wavelength calibration and systematic errors largely cancel out. This is one reason why relative Doppler measurements today can be made to a precision of $\approx$1 m s^{-1}. If you wanted to make an absolute Doppler measurement, your accuracy would be a factor of 10–100 worse.

The detection of exoplanets with the Doppler method is a multistep process:

1. Build or use a high-precision RV instrument.
2. Make sufficient RV measurements.
3. Search your RV time series for periodic signals.
4. Understand the nature of the signals that you find.
5. Fit orbits to your data.
6. Publish your results.

This book will cover most aspects of this process to help the reader in these various steps. These include

- instrumentation for Doppler measurements,
- reducing instrumental errors,
- calculating RVs,
- the frequency analysis of time series data,
- Keplerian orbits, and
- avoiding false planets.

References

Anglada-Escudé, G., Amado, P. J., Barnes, J., et al. 2016, Natur, 536, 437

Campbell, B., Walker, G. A. H., & Yang, S. 1988, ApJ, 331, 902

Hatzes, A. P., & Cochran, W. D. 1993, ApJ, 413, 339

Hatzes, A. P., Cochran, W. D., Endl, M., et al. 2003, ApJ, 599, 1383

Hatzes, A. P., Cochran, W. D., Endl, M., et al. 2006, A&A, 457, 335

Hubble, E. 1929, CMWCI, 3, 23

Huggins, W. 1868, RSPT, 158, 529

Keeler, J. E. 1895, ApJ, 1, 416

Kiefer, F. 2019, arXiv: 1910.07835

Kraft, R. P. 1967, ApJ, 150, 551

Latham, D. W., Mazeh, T., Stefanik, R. P., Mayor, M., & Burki, G. 1989, Natur, 339, 38

Mayor, M., & Queloz, D. 1995, Natur, 378, 355

Neubauer, F. J. 1936, LicOB, 481, 185

Osterbrock, D. E. 2002, James E. Keeler: Pioneer American Astrophysicist (Cambridge: Cambridge Univ. Press)

Reffert, S., Quirrenbach, A., Mitchell, D. S., et al. 2006, ApJ, 652, 661

Rubin, V. C., Ford, W. K. J., & Thonnard, N. 1978, ApJ, 225, L107

Struve, O. 1952, Obs, 72, 199

Vogel, H. C. 1890, PASP, 2, 27

Vogel, H. C. 1900, PASP, 12, 223

Walker, G. A. H., Yang, S., Campbell, B., & Irwin, A. W. 1989, ApJ, 343, L21

Walker, G. A. H., Bohlender, D. A., Walker, A. R., et al. 1992, ApJ, 396, L91

Wolszczan, A., & Frail, D. A. 1992, Natur, 355, 145

AAS | IOP Astronomy

The Doppler Method for the Detection of Exoplanets

A P Hatzes

Chapter 2

The Instruments for Doppler Measurements

The measurement of stellar radial velocities (RVs) requires spectrographs that break the light into its component wavelengths and detectors to record the resulting spectrum. In this chapter, we give a brief overview of the essential equipment needed for stellar RV measurements, namely spectrographs and detectors. Entire books are required to do the subject justice, and an excellent source is the text *Spectroscopic Instrumentation: Fundamentals and Guidelines for Astronomers* by Eversberg & Vollman (2015). Rather, in this chapter, we will cover just the basics principles of instrumentation needed for RV measurements.

2.1 Echelle Spectrographs

The left panel of Figure 2.1 shows the classic layout of a high-resolution spectrograph. Light from the telescope comes to a focus at the slit and then diverges. A collimator having the same focal ratio[1] as the telescope then converts this diverging beam into a parallel one that strikes a dispersing element. This is generally a reflection grating (see below) that breaks the light up into its component wavelengths. The dispersed light is then focused onto the detector by a camera, which is either a reflective, Schmidt-type camera (mirror plus corrective lens), or a refractive, lens-based system. Reflective cameras are generally used for spectrographs with slits, while lens cameras are the choice for spectrographs fed by a fiber optic.

The cross-disperser that is shown is an optical element that disperses the light in the direction perpendicular to the grating dispersion in order to separate the spectral orders. This will be discussed in more detail below. Classic spectrographs from 40 years ago typically used finely ruled gratings at low spectral order. Filters had to be inserted into the light path to block out light from unwanted orders. Modern spectrographs use echelle gratings at high orders coupled with a cross-disperser.

[1] The focal ratio is defined as the focal length of the telescope, f, divided by its diameter, D.

doi:10.1088/2514-3433/ab46a3ch2

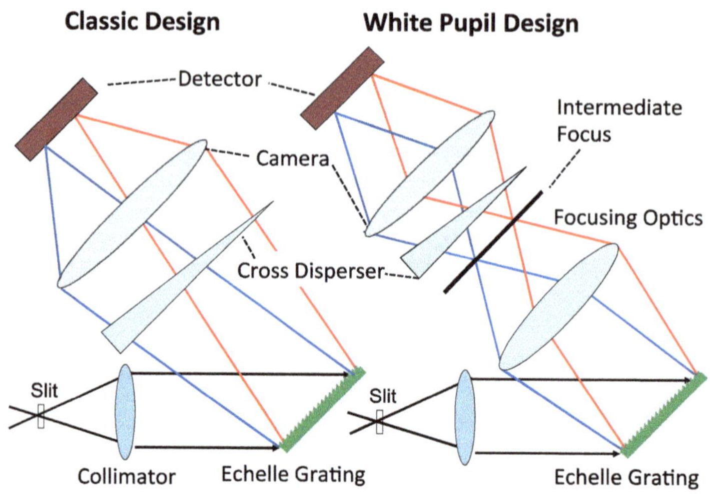

Figure 2.1. The layout of a classic spectrograph (left) and a white-pupil design spectrograph (right). Spectrographs using fine-ruled gratings do not have the cross-dispersing element.

So, before the advent of cross-dispersing elements, the layout looks the same, except for the cross-dispersing element.

It is important to note that a spectrograph is merely a camera (you can consider it a telescope as it is also bringing a parallel wavefront, like starlight, to a focus). The only difference is the presence of the echelle grating to disperse the light, and in this case, the cross-dispersing element. Remove these and what you would see at the detector is a white-light image of your entrance slit. Reinsert the grating, and the spectrograph now produces a dispersed image of your slit at the detector.

Many modern echelle spectrographs are designed after the white-pupil concept (Baranne 1972) shown in the right panel of Figure 2.1. In this design, an additional optical element is used to produce an intermediate focus between the grating and the cross-disperser at a position where the various spectral orders created by the grating are not yet separated. Here, there is a superposition of all wavelengths, and an intermediate white-light image of the slit is formed.

Naively, one may think that adding an additional optical element is unwise as it reduces the efficiency of your spectrograph, due to the extra optical element, but there are two good reasons to do this. First, because of the intermediate focus, all subsequent apertures, like the camera, can be made smaller. As a general rule, smaller optics translates into reduced costs. Furthermore, smaller optics can make for a more compact spectrograph, which is easier to stabilize thermally and mechanically. As we shall see, this is important for precise RV measurements.

Second, a spatial filter at the intermediate focus can reduce stray light, which is not desired in your spectrograph. Overall, the advantages of the white-pupil design outweighs the disadvantage of the small loss of light due to the extra optical element.

2.1.1 Gratings

The key part of any spectrograph is a dispersing element that breaks the light up into its component wavelengths. For high-resolution astronomical spectrographs, this is almost always a reflecting grating, a schematic that is shown Figure 2.2. The grating is ruled with a groove spacing, σ. Each groove, or facet, has a tilt at the so-called blaze angle, ϕ, with respect to the grating normal. This blaze angle diffracts most of the light into higher orders, m, rather than the $m = 0$ order, which is white light with no wavelength information.

Light hitting the grating at an angle α is diffracted at an angle β_b. and satisfies the grating equation:

$$\frac{m\lambda}{\sigma} = \sin\alpha + \sin\beta. \tag{2.1}$$

Note that at a given λ, the right-hand side of Equation (2.1) is $\propto m/\sigma$. This means that the grating equation has the same solution for small m and small σ (finely grooved), or alternatively, for large m and large σ (coarsely grooved).

One can compute the angular dispersion $d\lambda/d\beta$ by taking the derivative of the grating equation:

$$\frac{d\beta}{d\lambda} = \frac{m}{\sigma\cos\beta}. \tag{2.2}$$

Thus, a higher dispersion can be achieved by using higher spectral orders.

The grating equation (Equation (2.1)) can be used to eliminate the spectral order number and obtain

$$\frac{d\beta}{d\lambda} = \frac{\sin\alpha + \sin\beta}{\lambda\cos\beta}. \tag{2.3}$$

If we chose the blaze angle, θ_B such that $\alpha = \beta = \theta_B$, we get

$$\frac{d\beta}{d\lambda} = \frac{2}{\lambda}\tan\theta_B. \tag{2.4}$$

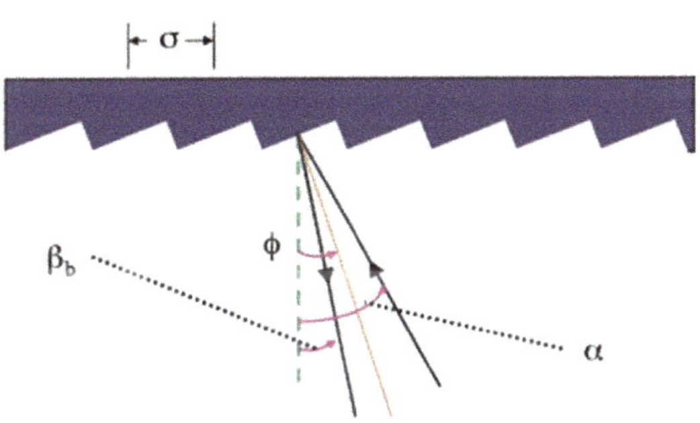

Figure 2.2. Schematic of an echelle grating. Each groove facet has a width σ and is blazed at an angle θ_B with respect to the grating normal (dashed line). Light strikes the grating at an angle α and is diffracted at an angle β, both measured with respect to the grating normal.

In other words, large angular dispersions require large blaze angles. The blaze angles of echelle gratings are typically 63.4° or 75.9°. These are often called "R2 grating" or "R4 grating," due to the fact that the tangent of 63.4° or 75.9° is 2 and 4, respectively. Because of the tangent, the angular dispersion is a steeply increasing function of the blaze angle. Note that the angular dispersion of an R4 grating is a factor of two larger than that of an R4. Increasing the blaze angle by just another 7° would result in another factor of two increase in the dispersive power.

Now let us consider the case of a constant diffraction angle, $\beta = \beta_c$. The right side of Equation (2.1) is now a constant, and the associated central wavelength of an order, λ_c, is inversely proportional to the order number m:

$$\lambda_c(m) = \frac{\sigma}{m}(\sin\alpha + \sin\beta_c). \qquad (2.5)$$

If we differentiate with respect to m, we get the change in central wavelength with respect to m, $d\lambda/dm \sim 1/m^2$. Thus, the distances between the central wavelengths of spectral orders will decrease as $1/m^2$. This means at high orders, the wavelength intervals will overlap (Figure 2.3).

Most modern, high-resolution echelle gratings operate at high spectral orders of $m \sim 100$, such that these occur spatially at the same location (same β) as the detector. To eliminate the undesired light from the other orders, one has to resort to blocking filters to isolate the light from the desired wavelength range. But why waste light? The elegant solution is to use an additional optical element to disperse the light perpendicular to the grating dispersion (i.e., the so-called "cross-disperser"). The use of two-dimensional detectors such as charge-coupled devices (CCDs) means that the separated spectral orders can now all be recorded at the same time.

The advent of echelle spectrographs, which provide a large wavelength coverage and high quantum efficiency of two-dimensional digital detectors (as opposed to photographic plates), was the main driver for the dramatic increase in RV precision in the past couple of decades. Figure 2.4 shows a spectrum of sunlight taken with an echelle spectrograph, and it nicely shows the large wavelength coverage one can achieve with an echelle spectrograph.

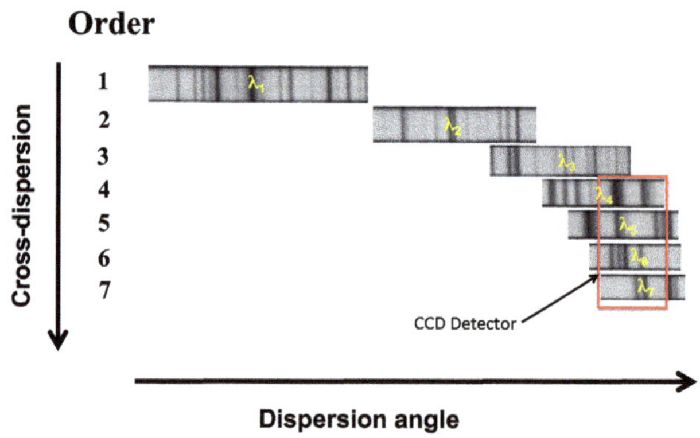

Figure 2.3. Schematic of the wavelength intervals in different orders from an echelle grating. For higher orders, the central wavelength of the order λ_m has approximately the same dispersion angle.

Figure 2.4. A spectrum of sunlight taken with a prism cross-dispersed echelle spectrograph.

2.1.2 Cross-dispersers

The cross-disperser element currently comes primarily in three forms: prisms, gratings, and grisms. Each provides its own wavelength dependence on the dispersive power in the cross-dispersion direction, y.

Prisms provide a separation of spectral orders that is inversely proportional to the central wavelength of the order

$$\Delta y \propto \lambda^{-1} \text{ (prism)}. \tag{2.6}$$

Thus, for prism cross-dispersers, the separation between spectral orders increases as one goes to bluer wavelengths. There are two advantages to using prisms. First, they have a high throughput and tend to be more efficient than other options such as gratings. Second, they provide an efficient packing of spectral orders and thus effectively use the real estate of the CCD detector. Prisms are often the choice for echelle spectrographs built for 2–3 m class telescopes (Vogt 1987; Tull et al. 1995; Kaufer & Pasquini 1998; Vogt et al. 2014)

There are two main disadvantages to using prisms as cross-dispersers. First, the packing of orders at red wavelengths becomes too tight, making it more difficult to determine interorder scattered light (see below). Second, for large diameter telescopes prisms do not provide enough dispersive power for good order separation. One must use several prisms in a chain or, preferably, another type of cross-disperser, such as a grating.

Grating cross-dispersers provide a separation of spectral orders that is proportional to the wavelength squared:

$$\Delta y \propto \lambda^{2} \text{ (grating)}. \tag{2.7}$$

In contrast to prisms, gratings have a spectral order separation that increases toward redder wavelengths. They have the advantage in that they provide a much larger separation of orders which can be set by the designer by choosing a grating with the appropriate groove spacing. This makes it much easier to remove the interorder scattered light. On the other hand, gratings have an efficiency of ≈70%, which is generally less than that for prisms. Because they are more effective dispersing devices, they are the choice of cross-disperser on echelle spectrographs built for large telescopes (Vogt et al. 1994; Dekker et al. 2000; Strassmeier et al. 2015).

Another option is a grism cross-disperser, which is a combination of prisms and gratings. This produces an order separation that should roughly be proportional to wavelength,

$$\Delta y \propto \lambda^{-1} \text{ (prism)} \times \lambda^{2} \text{ (grating)} \propto \lambda \text{ (grism)}. \tag{2.8}$$

The advantage of grisms is that they provide a more uniform spacing of orders from blue to red wavelengths. However, they tend to have even lower throughput than either prisms or grating cross-dispersers. The Tautenburg Echelle Spectrograph is one example of an echelle spectrograph that uses grisms as cross-dispersers. Figure 2.5 shows the spectral format of the various types of cross-dispersing elements.

2.1.3 Dispersion and Spectral Resolution

The spectral resolution is one of the most important parameters of your spectrograph for precise stellar RV measurements. It fundamentally determines the smallest shift of a spectral line that you can measure.

The spectral resolution, $\delta\lambda$, is defined as the difference in wavelength of two monochromatic beams that can just be resolved by your spectrograph. The spectral resolving power, R, of a spectrograph is defined by the ratio

$$R = \frac{\lambda}{\delta\lambda}. \tag{2.9}$$

It is a common mistake for astronomers to interchange the term "spectral resolution" with "spectral resolving power." The two are related, but technically not the same.

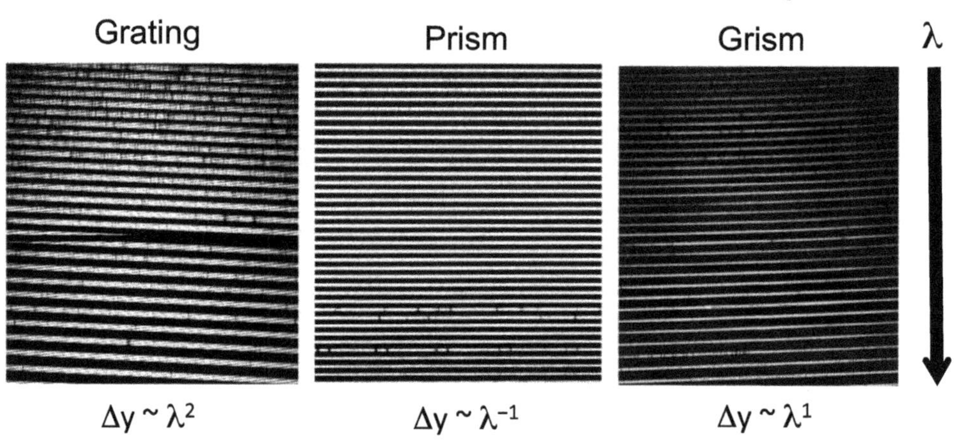

Figure 2.5. The various types of cross-dispersers. The wavelength of the central order increases toward the bottom. (Left) A grating cross-disperser produces an order separation, $\Delta\lambda$, that increases rapidly toward longer wavelengths ($\Delta\lambda \propto \lambda^2$). (Center) A prism cross-disperser produces orders that are more tightly packed at redder wavelengths ($\Delta\lambda \propto \lambda^{-1}$). A grism produces a more uniform spacing of orders, but with a linear increase toward the red ($\Delta\lambda \propto \lambda$).

Spectral resolution, $\delta\lambda$, has the dimensions of length, typically angstroms, and the smaller it is, the easier it is to distinguish fine spectral details. On the other hand, R is a dimensionless quantity, and the larger it is, the higher the resolution (equivalent to small $\delta\lambda$).

Before we derive the spectral resolution of a spectrograph, it is useful to consider the basic geometry, parameters, and angles that are involved in the telescope plus spectrograph system. Figure 2.6 shows the layout of the telescope and spectrograph. For a detailed discussion see Schroeder (1987). The important parameters are as follows:

D : telescope diameter;
f: telescope focal length;
d_1: collimator diameter;
f_1: collimator focal length;
d_2: camera diameter;
f_2: collimator focal length;
A: dispersing element (e.g. the echelle grating);
w, h: slit width and height (for fibers, the diameter);
w', h': projected slit width and height at detector.

The slit width subtends an angle $\phi = w/f$ on the sky. At the detector, it subtends an angle of $\phi' = w/f_1$. For the highest efficiency, the focal ratio of the collimator (f_1/d_1) should be the same as the telescope (f/D). In general, because of dispersion, the diameter of the camera is larger than that of the collimator so we can define an anamorphic magnification, $r = d_1/d_2$. We can also define a "focal ratio" of the camera as $F_2 = f_2/d_1$. The projected slit width at the detector is

$$w' = rw\left(\frac{f_2}{f_1}\right) = r\phi DF_2. \tag{2.10}$$

This expression is important for matching the slit width to the detector. A typical pixel size, Δ, for a CCD detector is $\Delta = 15$ μm. To adhere to the Nyquist sampling (see Chapter 7) of a 2 pixel projection of the slit requires $2\Delta = r\phi DF_2$.

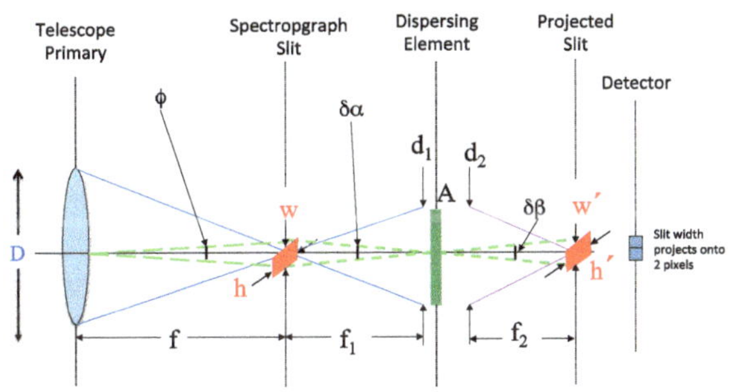

Figure 2.6. A schematic showing the geometry and angles of a spectrograph. See text for the definition of the various elements.

Equation (2.10) represents one of the major problems facing designers of high-resolution spectrographs. To match the slit width to the detector, you either have to use a narrow slit (smaller ϕ) or build a faster camera, i.e., a smaller F_2. In optical design, a faster camera translates into higher costs, or problems in its manufacture.

The problem becomes worse for large telescopes. Suppose you have a 4 m telescope at a site that has a median seeing of 1″. This requires a camera with $F_2 \approx 1.5$. If you build a spectrograph for a 10 m telescope at a better site (0.5″ median seeing), you require a faster camera of $F_2 \approx 1.2$.

What is important for RV measurements is not the angular dispersion but the linear dispersion at the detector as this determines the displacement of a spectral line in detector pixels for a given Doppler shift in wavelength. The change in the dispersion angle, $\delta\beta$, can be converted to a displacement at the detector of $dx = f_2\, d\beta$, which, along with Equation (2.3), yields the dispersion at the detector:

$$\frac{d\lambda}{dx} = \frac{d\lambda}{dx}\frac{d\lambda}{d\beta} = \frac{1}{f_2}\frac{1}{d\beta/d\lambda} \quad (\text{Å mm}^{-1}). \tag{2.11}$$

This dispersion is in units of Å mm^{-1}, which can be converted to the more useful Å/pixel simply by multiplying by the CCD pixel size in mm per pixel.

Using Equation (2.11), a displacement in wavelength by $\delta\lambda$ results in a displacement (in millimeters) at the detector:

$$dx = f_2\frac{d\beta}{d\lambda}\delta\lambda. \tag{2.12}$$

Setting this equal to the projected slit width,

$$f_2\frac{d\beta}{d\lambda}\delta\lambda = r\phi D F_2, \tag{2.13}$$

yields a spectral resolution of the spectrograph of

$$\delta\lambda = \frac{r\phi}{A}\frac{D}{d_1}, \tag{2.14}$$

where we have used $F_2 = f_2/d_1$ and $A = d\beta/d\lambda$.

Converting to spectral resolving power results in

$$R = \frac{\lambda}{\delta\lambda} = \frac{\lambda A}{r}\frac{1}{\phi}\frac{d_1}{D}. \tag{2.15}$$

It is instructive to reflect for a moment on this equation and the difficulties involved in designing high-resolution spectrographs for very large telescopes. The resolving power depends on the ratio of the collimator to telescope diameter, and inversely on the projected slit width. Suppose you want to build a spectrograph with a fixed resolving power, say $R = 100{,}000$, on a telescope with a large D. To do this, you either have to increase the collimator diameter (which is more expensive), or you have to use a smaller slit width, which results in loss of light. The latter defeats the purpose of using a large diameter telescope!

Table 2.1. Telescope (D) and Collimator (d_1) Diameters for Various Image Sizes (ϕ) for R = 100,000

D (m)	ϕ (arcsec)	d_1 (cm)
2	1.0	10
4	1.0	20
10	1.0	52
10	0.5	26
2	0.5	77
2	0.25	38

Table 2.1 summarizes the various combinations of telescope diameters and median seeing, and the required collimator diameter needed to achieve a resolving power of R = 100,000 using a dispersive power $A = 1.7 \times 10^{-3}$. One can see that as one designs larger diameter telescopes, the collimator diameter, and thus the overall size of your optical system, becomes larger. This can drive up the cost of your spectrograph dramatically and is the main reason that an R = 100,000 spectrograph may cost approximately one million Euros for a 3 m telescope, 10 million Euros for an 8 m class telescope, and 50 million Euros for a 30 m class telescope. Furthermore, a larger spectrograph means that it is more difficult to stabilize, and it will be more susceptible to instrumental shifts (see Chapter 4).

It is of interest to consider Equation (2.15) in the context of the diffraction limit of a telescope:

$$\theta = \frac{1.22\lambda}{D}. \tag{2.16}$$

Adaptive optics (AO) is the discipline where you correct the image for atmospheric distortions (see *Principles of Adaptive Optics* by Tyson 1987). The performance of an AO system is measured by how close the image comes to achieving the diffraction limit of the telescope. Although in practice this is rarely achieved, let's imagine that we have the perfect AO system, one that indeed produces diffraction-limited images. In this case, the image quality is not determined by the atmospheric seeing, but rather by the diffraction limit of the telescope. If we substitute the diffraction limit for ϕ in Equation (2.15), we find that $R \propto d_1$. In other words, the resolving power depends only on the diameter of your collimator and is independent of the telescope diameter. In principle, an R = 1,000,000 spectrograph coupled to a telescope with the perfect AO system can be built with a collimator diameter of only 6 cm!

AO systems are often used to improve the image quality for imaging measurements, but the improvement that they provide for spectroscopic measurements is often overlooked. Even if an AO system is imperfect and does not achieve the diffraction limit, it can

1. improve the efficiency of your spectrograph by allowing more starlight to enter your slit or fiber.

2. allow you to build a spectrograph with smaller optical components. This decreases construction costs as well as enables you to better stabilize your spectrograph mechanically and thermally.
3. stabilize the image at your slit or fiber (see Chapter 12).

2.1.4 The Blaze Function

One rather annoying feature of spectral data taken with echelle gratings is the so-called "blaze function." The blaze function results from the interference pattern of the single grooves of the grating. Each facet is a slit, so the interference pattern is described by a sinc function.

Recall from optical interference that a smaller slit results in a broader sinc function.[2] So, finely grooved gratings will produce a much wider blaze function. For a grating with rulings of ≈1000 grooves/mm like the ones employed decades ago, the groove spacing is small, so the blaze (sinc) function is rather broad and was hardly noticeable, or at least easily removed. However, for echelle gratings that have wide facets, the blaze function becomes much narrower and the wavelength coverage larger such that the blaze function can become an issue in data reductions.

Figure 2.7 shows an extracted spectral order from an echelle spectrograph that has a strong blaze function. This should be removed for RV measurements. The blaze function can have an influence on your RV measurement in two ways. First, if

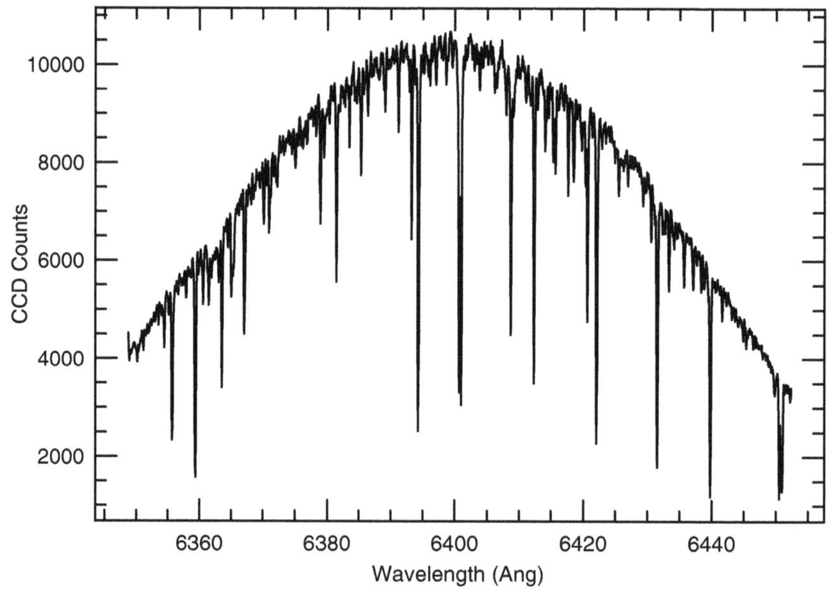

Figure 2.7. A spectral order from an echelle spectrograph that has not been corrected for the blaze function. The blaze function is the interference pattern of a single groove and has the shape of a sinc function.

[2] The interference pattern is merely the Fourier transform of the aperture, and as we will see in Chapter 7, there is an inverse relationship between the spatial domain and the Fourier frequency domain.

you use the cross-correlation method (see Chapter 5), the blaze should be removed (flattened) from all extracted spectral orders before calculating RVs. As we shall soon see, the cross-correlation method matches two signals, and in this case it will simply be matching the blaze functions rather than the location of spectral lines.

The second way the blaze function can influence the RVs is by altering the photon statistics after removal of the blaze. Note in Figure 2.7 that the number of counts at the edges of the spectrum are about half that at the center of the spectrum. This means that the signal-to-noise ratio (S/N) is a function of position along the spectrum, being a factor of 1.4 lower at the edges. If you normalize the spectrum in order to remove the blaze, the count rate in the continuum is no longer related to the true photon count. This will alter your statistics if you want to weight your RV measurement according to the S/N ratio.

Typically, the blaze function is not removed via the normal flat-fielding process (see below), so one needs to take an extra step in the reduction process to remove it. There are two approaches to doing this. One way to remove the blaze function is by dividing it with a low order polynomial fit to the continuum points in your spectrum. This can be tricky as it depends on the choice of continuum points. As you see in Figure 2.7, it is sometimes difficult to identify the true continuum. This is particularly true for late-type stars with an abundance of spectral lines.

Alternatively, you can divide the blaze function with the spectrum of a rapidly rotating hot star with no spectral features. This has the advantage in that a hot star spectrum was taken through the same optical path as the science target. The disadvantage is that you are introducing additional noise, and for high-precision RV work, you not want to avoid increasing the noise level if possible. Also, there may still be stellar lines present in the hot star spectrum even though rapid rotation will make these shallow. These will alter the shape of the continuum.

2.1.5 Scattered Light

Scattered light comes from photons that are redirected into the optical path. Because the light has not properly traversed the optical path of the spectrograph, it is recorded on the detector at a position that has no relationship to its original wavelength. In short, it is light appearing on your detector where it should not be. It primarily comes from dust on the grating, optics, light leaks, etc., as well as light reflected off (in an undesired direction) optical surfaces.

Figure 2.8 shows a cut perpendicular to spectral orders of an unreduced stellar spectrum. The peaks represent the spectral orders where most of the light should be, but one can see that the intensity in between the orders does not reach zero (the bias level has been subtracted; see below), which would be the case if there were no scattered light.

Improper scattered-light subtraction will alter the true depths of the spectral lines. This is definitely important if you want to perform stellar abundance studies, which depend on the line strength, but it can also have an influence on the RV measurement. This is primarily through slight changes in the line depth (see Chapter 3), but

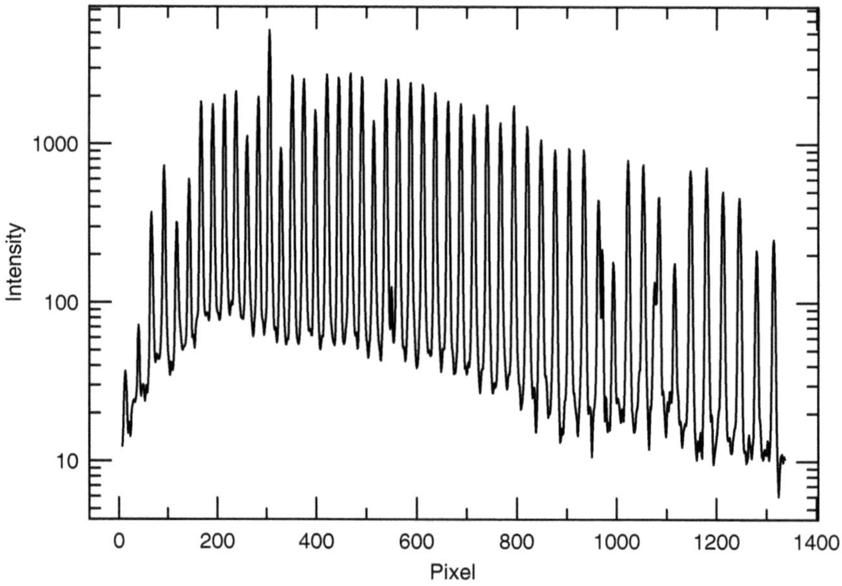

Figure 2.8. A cut along CCD detector columns of an unreduced echelle spectrum. The peaks are the spectral orders. Note the nonzero intensity in the interorders due to scattered light.

also because of the slight mismatch between the stellar observation and the template used to calculate the RV (see Chapter 5).

Proper scattered-light subtraction requires a good separation between orders, which should be considered when designing a spectrograph.

2.2 Fourier Transform Spectrometers

Most stellar RV measurements are made with spectrometers using an echelle grating as a dispersing element. Another form of spectrometer is a Fourier transform spectrometer (FTS). Unlike a dispersing spectrometer which measures the intensity over a narrow range of wavelengths at a time, an FTS simultaneously collects high spectral resolution data over a wide spectral range. The FTS is not directly used for RV measurements, but it plays an important role in providing a reference spectrum for the gas absorption cell method (Chapter 6).

An FTS is a Michelson interferometer with a movable mirror (Figure 2.9). Light from a source is split into two beams using a beam splitter. One beam is reflected off a fixed mirror while the other is reflected off a movable mirror, which can vary the optical path length. The combined beams interfere and are recorded at the detector. By measuring the temporal coherence of the light at a different optical path difference, one can convert the time domain signal into a spatial coordinate.

If d is the optical path difference of the combined beams at the detector, then for a monochromatic beam of frequency $\nu_0(=\lambda^{-1})$, the intensity is

$$I(d) = \frac{I_0}{2}[1 + \cos(2\pi\nu_0 d)]. \tag{2.17}$$

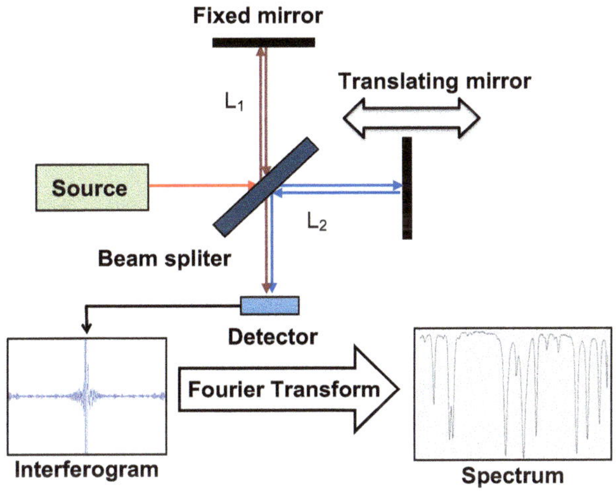

Figure 2.9. The Fourier transform spectrometer. The light source goes through a beam splitter, which divides the light equally into two light paths, L_1 and L_2. The optical path length L_1 remains fixed, but a translating movable mirror changes the optical path length L_2. The combined interference beam is recorded at the detector. Moving the translating mirror over a fixed range produces an interferogram in frequency space. A Fourier transform converts this into a spectra in the spatial domain.

For a polychromatic source, $S(\nu)$, only at $d = 0$ will all the light waves add constructively to produce a signal. The output signal is thus

$$I(d) = 1/2 \int_0^{\infty} S(\nu)[1 + \cos(2\pi\nu)d]d\nu. \tag{2.18}$$

If $\bar{I}(\nu) = 2I(d) - I(0)$, then

$$\bar{I}(\nu) = \int_0^{\infty} S(\nu)\cos(2\pi\nu d)d\nu. \tag{2.19}$$

The function $\bar{I}$ is called the interferogram, and it is merely the Fourier transform of the source, which can be recovered by the Fourier integral theorem:

$$S(\nu) = \int_0^{\infty} \bar{I}(d)\cos(2\pi\nu d)d\nu. \tag{2.20}$$

Hence, as the name FTS implies, it is a device that records the Fourier transform of your spectrum.

The resolution of an FTS is determined by the maximum distance that the translating mirror moves during the observation. Thus, it is relatively easy for an FTS to achieve a very high resolving power, typically up to $R = 500{,}000 - 1{,}000{,}000$. If you want lower resolution, simply move the translating mirror a shorter distance.

There are no reported instances of an FTS being used for precise RV measurements. These instruments, however, play an important role in the absorption cell method (see Chapter 4). The top panel of Figure 2.10 shows a portion of the spectrum of molecular iodine, I_2, recorded using an echelle spectrograph with reasonably high resolution, $R = 67{,}000$. The lower panel shows the same wavelength

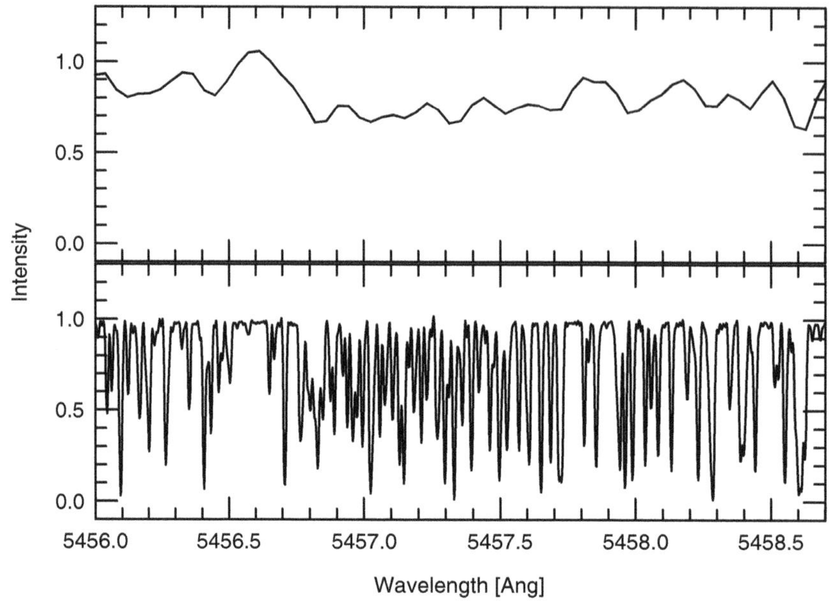

Figure 2.10. (Top) A portion of a spectrum of molecular iodine taken using an echelle grating spectrometer with resolving power $R = 67{,}000$. (Bottom) The same wavelength segment of molecular iodine taken with an FTS with resolving power $R \approx 1{,}000{,}000$.

region recorded with the McMath FTS (see Deming & Plymate 1994). The resolving power is $R \approx 1{,}000{,}000$, and one can see that the broad and shallow spectral features seen at lower resolving power is really a myriad of very narrow and deep lines.

2.3 Charge-coupled Device Detectors

After a spectrograph has dispersed the light into its component wavelengths, you now need a detector to record the photons. CCD detectors are universally used on spectrographs working at optical wavelengths (≈3000–10,000 Å). There are three important advantages to using CCD detectors: (1) they have high quantum efficiency; (2) they are two-dimensional devices, ideal for recording the two-dimensional format of an echelle spectrum;[3] and (3) the output is in digital form, making it easier to process the data. Here we will cover just the basics of CCD detectors.

The use of photographic plates at the turn of the last century certainly revolutionized both imaging and spectroscopic astronomical observations. Spectroscopists could not only record their observations, which is certainly easier than using your eye to detect Doppler shifts (like Huggins!), but photographic plates also enabled astronomers to integrate for long exposures, thus allowing the observations of fainter objects. In spite of this important development, photographic plates could rarely achieve S/N ~ 20 for high-resolution spectra. If we define the quantum efficiency as

[3] In the 1980s, my fellow graduate student G. Donald Penrod once made the poetic remark, "There is a glorious match between CCD detectors and echelle spectrographs." So true!

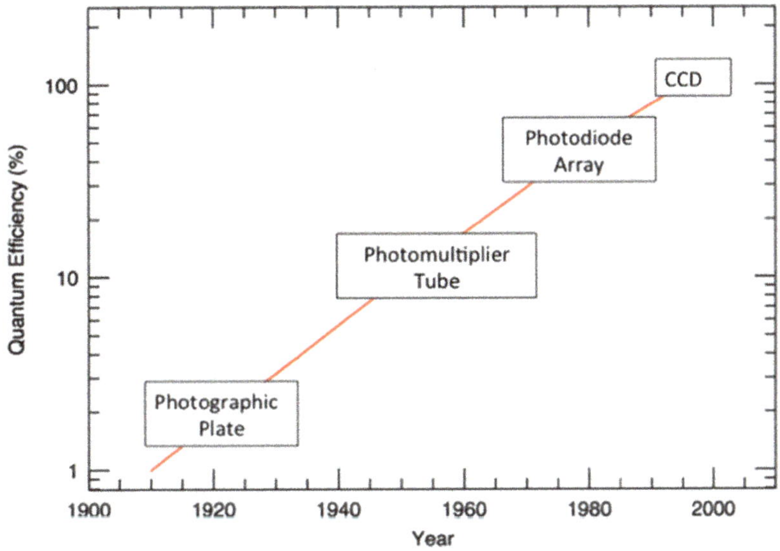

Figure 2.11. The increase in the QE of astronomical detectors as a function of time.

the ratio of incoming to detected photons, photographic plates had a rather low quantum efficiency (QE) of 1%–3%.

The advent of electronic detectors played the first key role in improving stellar RV measurements. The first major developments, starting in the 1940s, came with photomultipliers, which provided a QE of about 10%, or a factor of 10 increase over photographic plates. By the early 1980s, photodiodes improved the QE to ~50%. Currently, virtually all astronomical instruments employ CCDs. These devices saw routine use in the 1990s, and a modern CCD detector can have a QE of ~80%–100%.

Figure 2.11 summarizes the evolution in quantum efficiency of detectors over the past decades and is the companion for detectors to Figure 1.4. Although the trend seems linear in a logarithm plot, the development came in quantum leaps (pun intended). For example, with photographic plates there was no improvement in the QE for many decades before photomultiplier tubes were developed. Since then, the developments have come much more rapidly, but still in steps.

It is interesting to reflect on how much CCDs have revolutionized observational astronomy. In the 1920s Edwin Hubble discovered the expansion of the universe using the Mt. Wilson 2.5 m telescope taking spectra of galaxies with photographic planets, often with exposure times of several days duration (and without autoguiding!). With CCDs, he could have achieved the same result with a 40 cm telescope!

2.3.1 The Structure and Operation of a CCD

Figure 2.12 shows the basic structure of single CCD element or pixel ("pixel element"). It has two layers of silicon substrates that are in contact. One is doped such that it has an excess of electrons (n-Si type) and the other doped such that it has an excess of positive charge (p-Si type), often referred to as "holes." On top of the Si

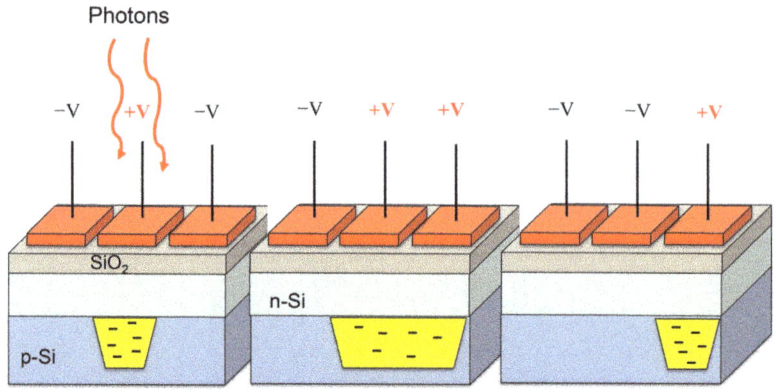

Figure 2.12. The structure of a CCD pixel and the charge readout. The pixel has three gates with an applied voltage. Below these is an insulating layer of SiO_2. The silicon substrate is doped to have excess (n-Si) or depleted (p-Si) electrons. (Left) Photons striking the substrate collect under the potential well of the central gate. (Center) A positive voltage is applied to the left gate, causing the potential well to reside under the center and right gate. (Right) After applying a negative voltage to the central gate, the charge has been completely transferred to the potential well under the right gate.

substrate is an insulating material of silicon dioxide, and finally polysilicate gates for controlling the voltages.

At the p–n junction, electrons from the n-type material diffuse into the p-type and fill the extra holes. This diffusion results in the n-Si being positively charged and the p-Si negatively charged. This creates an electric field that ultimately slows and stops the diffusion process.

The boundary where the holes and electrons have combined is called the depletion zone, and this is the photosensitive part of the CCD pixel. By applying a voltage, one can increase or decrease the electric field across the depletion zone, and this controls the flow of current between the two Si substrates. Photons striking the depletion zone dislocate electrons from Si atoms via the photoelectric effect. Because of the applied voltage ($+V$) of the polysilicate gate, these remain trapped in the potential well of the depletion zone.

After exposing the CCD and accumulating charge, one needs to transfer and record the charge packet. This is done by phasing (clocking) the voltages of the adjacent gates, the process shown in Figure 2.12. Applying a $+V$ voltage to the right gate allows the charge packet to straddle two of the gates. By changing the central voltage to $-V$, we have now transferred the charge packet in the pixel to the right gate. In this way, the charge packet can be transferred from pixel to pixel in the so-called "bucket brigade."

A schematic of the basic operation of the CCD is shown in Figure 2.13. Charge packets are first transferred along a parallel register (along columns). On the final row, charge is shifted along a serial register and an amplifier, where the charge is read, converted to analog-to-digital units (ADUs) and stored on a computer.

The readout time of a CCD can range from approximately 10 s to several minutes, depending on the size of the CCD and the control electronics. Most of the time is spent along the serial register, especially the last one, where the charge is read

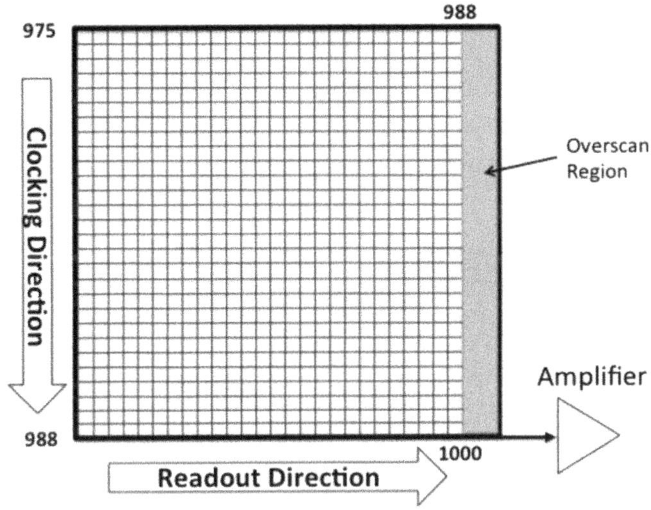

Figure 2.13. Schematic of the operation of a CCD. Charge packets are first transferred along columns in the clocking direction. The last row is then read in the serial direction and into the amplifier. The overscan region is masked off so as not to record light. The numbers on each corner represent the number of recorded photons for 1000 detected photons. Losses are due to charge transfer inefficiency.

out. The transfer along columns and then along rows is relatively fast (of order milliseconds) compared to the charge transfer along rows.

In order to speed up the readout process, some CCDs have amplifiers at each corner of the array. This will decrease the readout time by a factor of 4. In these cases, the user has the option of using either one or four amplifiers in the readout. This speed, however, comes at a price. In this case, each amplifier and readout electronics has its own characteristics, read noise, etc. You have essentially converted your single CCD chip into four independent detectors so the data from each quadrant will have to be treated as four independent data sets. The factor of four in time savings by the reduced readout time of the CCD now translates into at least a fourfold increase in the time needed to reduce the data. For spectroscopy, this can cause problems as a spectral order now straddles two quadrants. For RV work, it is advisable not to use multiple amplifiers in the readout.

2.3.2 Quantum Efficiency

The QE of a modern CCD depends on whether the CCD is front- or back-side illuminated. The front side of the CCD is where the polysilicon wires that control the charge transfer are attached. These absorb all the UV (or blue) photons before they can reach the photosensitive layers (left panel of Figure 2.14).

To improve the efficiency, one can turn over the CCD to avoid photons striking the polysilicon wires (right panel of Figure 2.14). One can also apply an antireflection (AR) coating, which is not possible with the polysilicon wires present. However, now the CCD p-Si layer is too thick for photons to reach the depletion zone. This is remedied by thinning the p-Si layer from 625 μm to about 15 μm. This

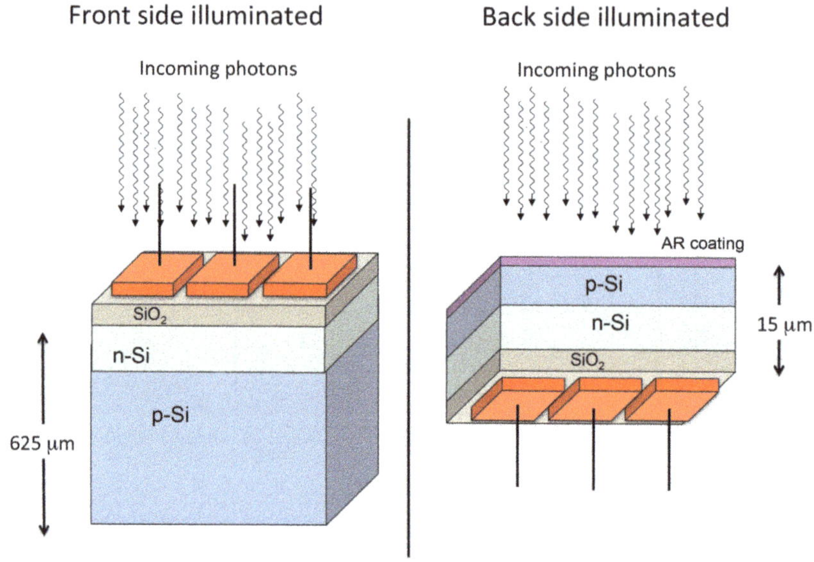

Figure 2.14. Comparison of a front-side-illuminated CCD (left) to a back-side-illuminated CCD (right).

is a nonstandard process that decreases the yield of CCD chips and thus increases the costs. Devices are often destroyed during the thinning process. However, the increase in QE, often by a factor of 2–3, is well worth the risk.

One disadvantage of thin devices is that the CCD is now more transparent to near-infrared photons, making the red response of the CCD poorer. The gain in the blue, however, more than offsets the loss at redder wavelengths. Virtually all modern CCDs in use at astronomical observatories are thin, front-side-illuminated devices.

Modern CCDs can be optimized for better performance in either the red or blue wavelength regions. Figure 2.15 shows the QE curves for red- and blue-optimized CCDs from MIT Lincoln Labs. The red-optimized CCD reaches a QE of nearly 100% in the wavelength range 7000–8000 Å. However, the response below 4000 Å is quite low at approximately 20%. The blue-optimized CCD now has a much improved QE of approximately 90%, but with only a slight loss at red wavelengths. The performance of all CCD detectors dies off at approximately 10,000 Å, due to the fact that photons no longer have sufficient energy to dislodge electrons from the silicon atoms.

2.3.3 Bias Level

The bias level of a CCD is an electronic offset that is artificially applied to the data at the time the CCD is read out. Its purpose is to ensure that the analog-to-digital converter (ADC) always receives a positive signal. This bias level must be removed from the data values if these are to reflect the true number of counts (photons) recorded by each pixel.

All astronomical CCD detectors have a so-called overscan region of the chip that is masked so that it receives no light during an exposure (Figure 2.13). The overscan region can be easily found by taking an exposure of a lamp source and plotting along

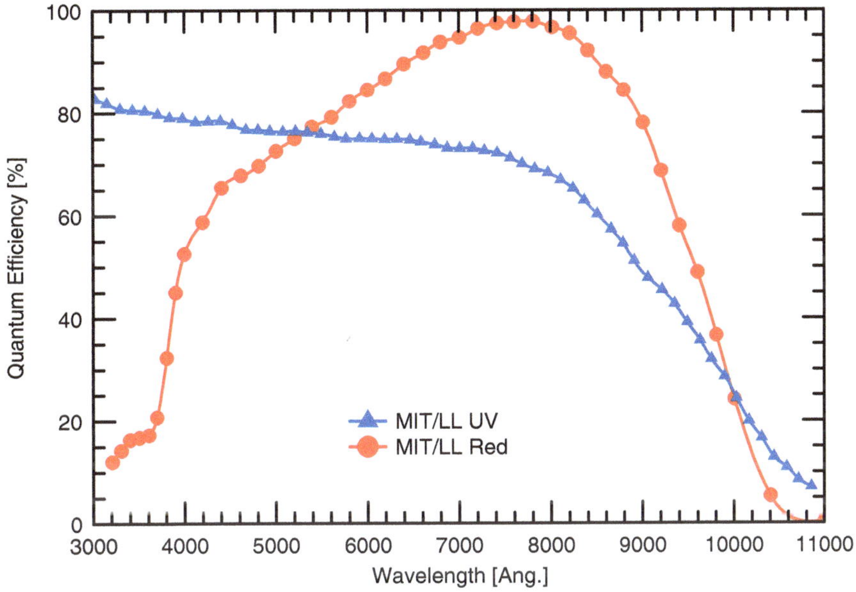

Figure 2.15. The quantum efficiency versus wavelength for an MIT/LL CCD. Such curves are typical for modern CCD detectors. (Data courtesy of ESO.)

a row or column to see where the intensity sharply drops (seen as a step function). Typical bias levels are a few hundred to about a thousand ADUs.

The first step of any CCD reduction process is to remove the bias level. The bias frame is simply an observation with the CCD shutter closed (dark frame) with a zero exposure time. The bias levels can be removed using several methods:

1. Measure the count level in the overscan region and subtract this value from each frame. The disadvantage of this method is that it will not remove any structure in the bias level if some are present across the CCD.
2. Take a series of several (>10) frames and compute a "master bias," consisting of a frame having the average (or better yet, the median) value for each pixel. This master bias is then subtracted from each science frame. The disadvantage is that this will introduce some noise into the data frames.
3. Take a single bias frame, or the median of several, and fit a surface to it using a low-order, two-dimensional polynomial. Subtracting this fit will not introduce noise into your data frame. This can be done on the master bias and is the preferred method.

Most CCDs in use today are relatively stable so that one can take the bias frames at the beginning or end of the night. The bias level is the first indication of problems with the CCD, so it is a good idea to monitor the stability of the CCD by checking it regularly. This is easily done by looking at the counts in the overscan region. The author once had the experience where the bias level was seen to change throughout the night, but this only became evident during the reduction process after the observing run. Fortunately, the overscan region provided a "real time" measurement of the bias level. We will see in Chapter 12 that a significant change in the bias level will be an early indication that something is amiss in your detector.

2.3.4 Gain

For photon statistics, the uncertainty in the detected photons, N_p, is $\sigma = \sqrt{N_p}$. The S/N is simply S/N = $N_p/\sigma = N_p\sqrt{N_p} = \sqrt{N_p}$. So, for astronomical observations, the important quantity is the number of detected photons. However, CCD pixel values are ADUs, which are related to the actual number of photons through the gain factor (G), $N_p = GN_p$. When using a CCD for the first time, it is wise to check the gain factor and not merely rely on documentation or information in the headers of the observations as these may be outdated. The CCD gain can be easily measured using either of two methods.

Method 1
For photon statistics, $\sigma^2 = N_p = GN_p$. A series of exposures of a white-light source at increasing intensity levels (exposure times) is made, the bias level from each frame is subtracted, and the standard deviation on a region of the detector is calculated. The slope of the σ^2 versus the intensity (ADU) level (after subtracting the bias level) yields the gain (Figure 2.16).

Method 2
The gain can also be calculated from just a couple of flat and dark frames:

1. Two frames are taken, f_1 and f_2, at reasonably high intensity levels (~10,000) as well as two bias frames, b_1 and b_2 (zero exposure times).

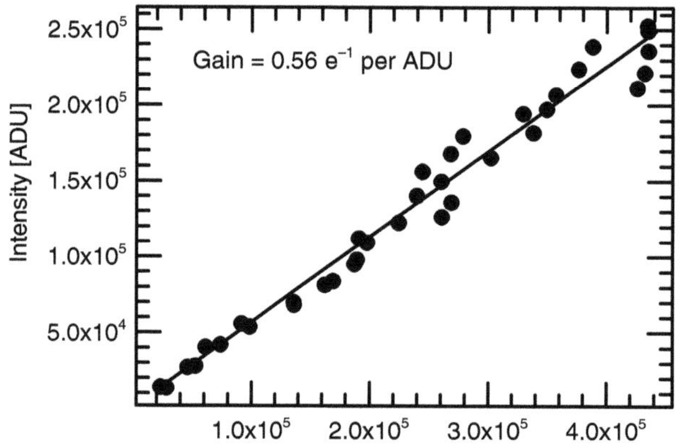

Figure 2.16. The standard deviation of ADU values from a CCD as a function of the intensity (ADU) level. The slope indicates a CCD gain of 0.56 electrons per ADU.

2. Difference frames are produced from the bias and flats: $b_{12} = b_1 - b_2$ and $f_{12} = f_1 - f_2$. The use of differences removes any underlying structure that may be present.
3. One then calculates the mean values of the bias frames, m_{b1} and m_{b2}, the flat frames, m_{f1} and m_{f2}, as well as the standard deviations of the difference frames, σ_{f12} and σ_{b12}.
4. The gain is given by

$$G = \frac{m_{f1} + m_{f2} - m_{b1} - m_{b2}}{\sigma_{f12}^2 - \sigma_{b12}^2}. \tag{2.21}$$

2.3.5 Readout Noise and Dark Current

The readout noise results from the conversion of the electrons in each pixel to a voltage on the on-chip amplifier. The readout is the ultimate noise limit of a CCD. Figure 2.17 shows the S/N that is achieved as a function of the detected photons and readout noise. With no readout noise, this follows photon statistics, with S/N merely the square root of the number of detected photons. However, with nonzero readout noise, this deviates from the ideal expression for low signal levels. For a rather high readout noise of 10 e^{-1}, the performance of the CCD is seriously affected and one can never achieve an S/N below about 3.

When CCDs first came into regular use by astronomers in the 1980–1990s, CCDs with readout noise of 10 e^{-1} were common. Modern CCDs now have a readout noise of typically 2–3 e^{-1}. Most precision RV measurements are taken at S/N > 10,

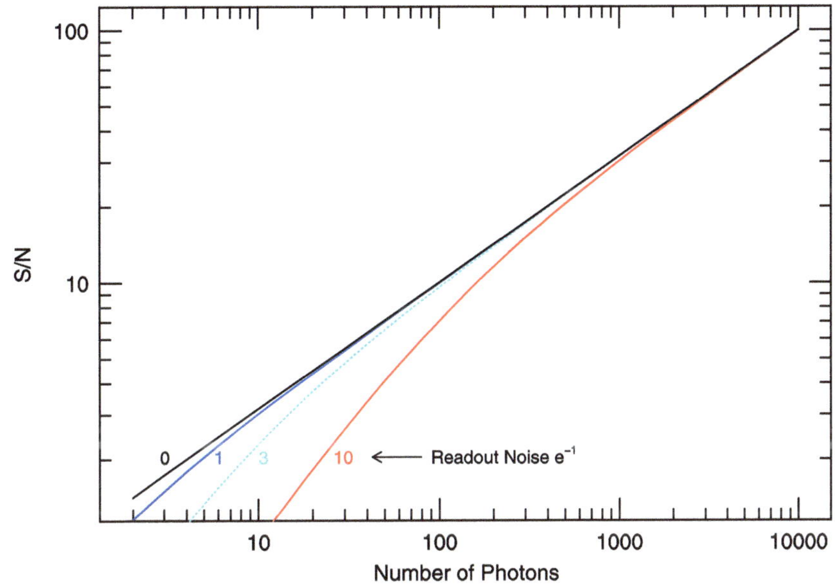

Figure 2.17. The signal-to-noise ratio (S/N) as a function of the number of detected photons for different levels of readout noise (0, 1, 3, and 10 e^{-1}). The black line is for the ideal case of photon noise (no readout noise).

so the readout noise is never an issue influencing the uncertainty in the RV measurement.

The readout noise of a CCD can easily be measured by taking these steps:

1. Take a large number (≈10) of bias frames.
2. Create a master bias by taking the median of the bias frames.
3. Subtract the master bias from each bias frame.
4. Take the standard deviation of these images and multiply by the gain because you want the readout noise in electrons and not ADUs. This value is your readout noise.

2.3.6 Charge Transfer Efficiency

The charge transfer efficiency (CTE) is a measure of the fraction of the charge that is lost when moving from pixel to pixel. A poor CTE can affect the shapes of your instrumental profile (see Chapter 6 for a discussion on the instrumental profile). Early CCD devices often had poor CTE, but a modern CCD device has a CTE that is typically about 99.9997%, so for the most part this should not be a concern.

The amount of charge lost depends on the number of times the packet has been transferred, and this of course depends on the initial location of the charge on the CCD array. The numbers in the corners of the CCD shown in Figure 2.13 show how many photons are actually recorded for 1000 photons striking the CCD. Charge packets near the readout amplifier suffer little loss, but a charge packet in the upper right will suffer a loss of 2.5% because it undergoes the maximum number of transfers: 2048 times along columns followed by another 2048 times along the final row (for a 2048 × 2048 pixel array).

The CTE may become important as astronomers use ever larger CCD detectors. For example, if you have a 10,000 × 10,000 CCD a charge packet at the left corner would have lost ~6% of its charge. Although these are relatively low losses, especially for high signal-to-noise data, this can become more important for low light level observations, or if one wants to push the RV precision down to the cm s^{-1} precision. In Chapter 12, we will discuss further possible sources of error due to CTE.

2.3.7 Linearity

An important characteristic of a CCD is its linearity in response to the incident light. The observed count should be linearly proportional to the intensity of the light—expose for twice as long and you should get detect twice as many photons.

Most modern CCD detectors generally have excellent linearity (Figure 2.18). The linearity of a CCD is trivial to check. Observe a white-light source at different exposure levels and plot the total counts (minus the bias level!) as a function of exposure time. If one sees a deviation from a linear behavior (red dashed line in Figure 2.18), then one should avoid exposure times which produce counts above this level. For Figure 2.18 where we have marked a hypothetical nonlinearity with a red line, this is at approximately 170,000 ADUs.

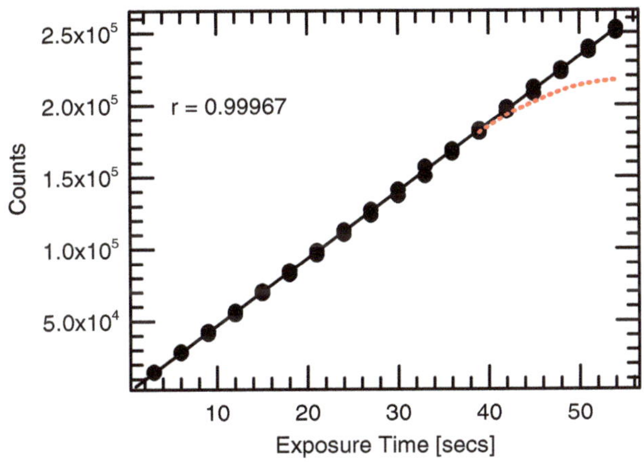

Figure 2.18. The CCD linearity as indicated by the number of detected photons as a function of the exposure time. The red dashed line shown would indicate a strong non-linearity at high count rates.

2.3.8 Flat Fielding

CCDs have slight variations in the QE from pixel to pixel. The recorded CCD frame may also have ghost images, reflections, or other artifacts from the spectrograph. If these variations are not properly removed, this will affect the quality of your spectrum and ultimately, your RV precision.

The process of "flat fielding" consists of taking an observation (spectrum or image) of a white-light source commonly referred to as the "flat lamp." Divide your observation with your flat-field exposure. Flat-field errors can arise when the variations are not taken out completely. This can be the case because flat lamps are usually inserted in the light path just before the entrance slit of the spectrograph and thus do not follow the same optical path as your observation of the star. For this reason, it is a common practice to take a "dome flat." The telescope is pointed to a white screen mounted on the interior of the dome, which is then illuminated by the white-light source. The light from the flat field thus follows, more or less, the same optical path through the telescope as the starlight.

Figure 2.19 shows an example of the flat-fielding process as applied to imaging observations where you can better see the artifacts and their removal. The top-left image is a raw frame taken with the Schmidt Camera of the Tautenburg 2 m telescope. The top-right image shows an observation of the flat lamp, where one can see the structure of the CCD, as well as an image of the telescope pupil caused by reflections. The lower image shows the observed image after dividing by the flat lamp observation. Most of the artifacts and intensity variations have been removed by the division.

2.3.9 Saturation and Blooming

The voltage potential well in each pixel can only hold a fixed number of electrons. Above this value, called the "full-well capacity," the pixels have saturated, and additional electrons then spill over into adjacent pixels along columns in an effect

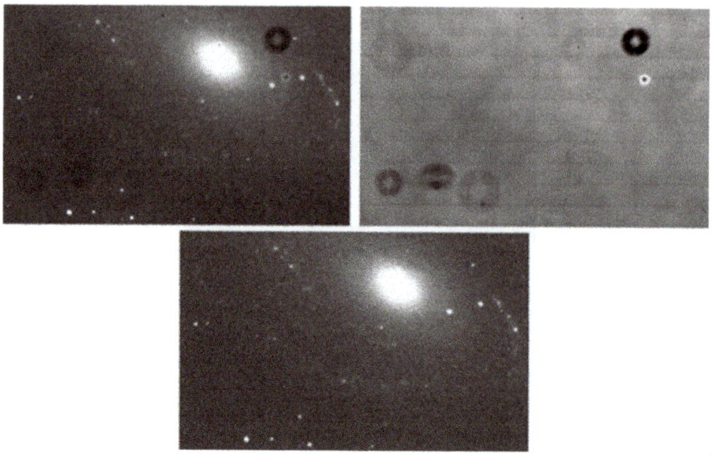

Figure 2.19. The flat-fielding process for CCD reductions. (Left, top) A raw image taken with a CCD detector. (Right, top) An image taken of a white-light source (flat field) that shows the CCD structure and optical artifacts. (Bottom) The original image after dividing by the flat field.

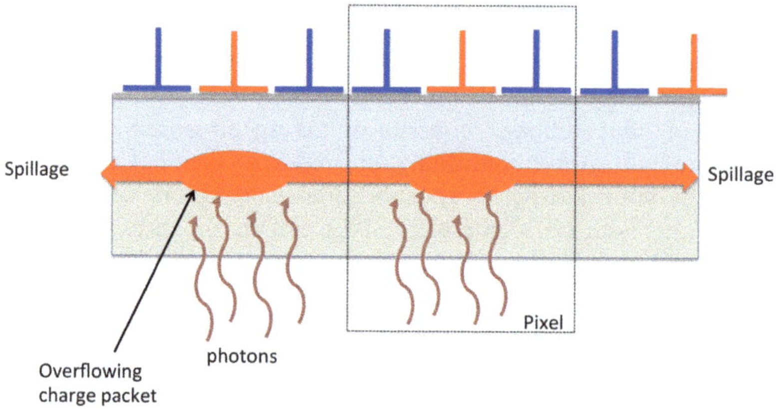

Figure 2.20. Blooming in CCD. When the amount of charge exceeds the potential well of the pixel, it starts to spill over into the direction of readout, i.e., columns.

called "blooming" (Figure 2.20). The saturated pixels appear as a hot column with values of the full well (Figure 2.21).

Modern CCDs have a full-well capacity of ≈100,000–200,000 e^{-1}. The full-well capacity defines the maximum S/N you can achieve in a single exposure. For example, a full well of 100,000 e^{-1} means you cannot achieve an S/N higher than about 316 in a single exposure.

Antiblooming CCDs can eliminate the effects of saturation. This is done with additional gates that bleed off the overflow due to saturation. The disadvantage of this is that these "bleed off" gates cover about 30% of the pixel. This results in

Figure 2.21. A CCD image of the Pelican Nebula (north is to the right). The vertical streaks are due to blooming of saturated pixels. Image credit: Thüringer Landessternwarte Tautenburg.

reduced sensitivity, smaller well depth, and lower resolution, due to the increased size of gaps between pixels.

2.3.10 Fringing

Fringing is another problem with CCDs that is caused by the small thickness of the CCD. It occurs because of the interference between the incident light and the light that is internally reflected at the interfaces of the CCD. Figure 2.22 shows a spectrum of a white-light source taken with an echelle spectrograph (see below). Red wavelengths are at the lower part of the figure where one can clearly see the fringe pattern. This pattern is not present in the orders at the top, which are at blue wavelengths.

CCD fringing is mostly a problem at wavelengths longer than about 6500 Å. For RV measurements made with the iodine technique (Chapter 6), this is generally not a concern because these cover the wavelength range 5000–6000 Å. However, the simultaneous Th–Ar method (Chapter 4) can be extended to longer wavelengths where improper fringe removal may be an issue.

In principle, the pixel-to-pixel variations of the CCD and the fringe pattern should be removed by the flat-field process, but again, this may not be perfect, and this can introduce RV errors.

2.3.11 Persistence

Suppose you have recorded an image with a CCD that had high intensity values (in particular saturated pixels). If you take a subsequent dark exposure, you may find that instead of being completely dark, the frame has a memory, or residual image, of

Figure 2.22. Fringing in a CCD. The exposure is of a white-light source taken with an echelle spectrograph. Blue orders (≈5000 Å) are at the top, and red orders (≈7000 Å) are at the bottom. Fringing becomes more pronounced at longer wavelengths.

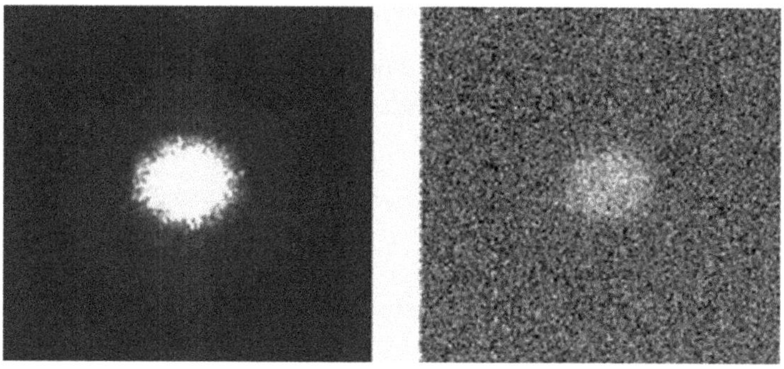

Figure 2.23. Residual images in a CCD. (Left) An image of a star with a high count level. (Right) An image of the CCD after reading out the previous exposure. There is a low-level (a few counts) image of the star remaining on the CCD.

the previous exposure. This effect is called persistence and is an extraneous signal that is not removed by flushing (reading out) the CCD. It is caused by electrons generated in the previous exposure that are trapped at impurity sites on the detector. The electrons that are released in subsequent exposures appear as a residual image (Figure 2.23). The time it takes for the trapped electrons to be released depends on the device.

For infrared (IR) arrays, the persistence effect can be larger than that for CCDs. Figure 2.24 shows the persistence measured for a Hawaii 4 IR array used by the European Southern Observatory. For an exposure level of 100,000 detected electrons, after 30 s, the residual image has approximately 250 electrons. This is not so much of a problem if the next exposure is also at a high intensity level (a

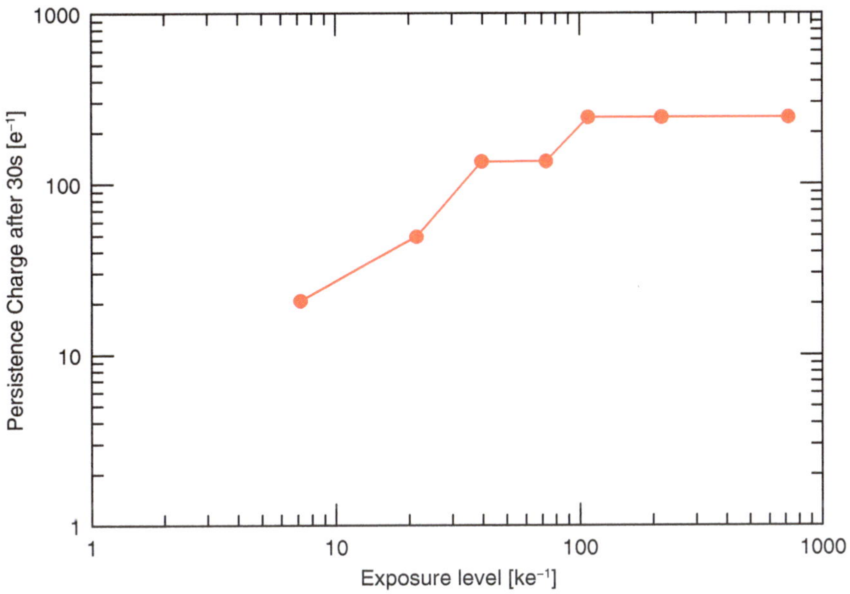

Figure 2.24. Residual images in a CCD. (Left) An image of a star with a high count level. Persistence as measured in a Hawaii 4 infrared detector. (Data courtesy of ESO.)

0.25% effect). However, a subsequent exposure with, say, 10,000 detected electrons (S/N = 100) means that it will be contaminated by the spectrum of the previous star at the 2.5% level. This contamination will influence your RV measurement.

Persistence in optical CCDs is generally small, and this author, for one, has never had to worry about it. This is particularly true if you are only achieving a modest RV precision of greater than, say, 10 m s^{-1}. However, it you really want to measure ultraprecise RVs below 1 m s^{-1} or in the cm s^{-1}, then persistence in your device should be measured to assess if this is a problem.

Persistence can be handled in three ways:

1. Limiting the intensity levels of the observation.
2. Flushing the CCD with dark bias frames a number of times to reduce the intensity of the residual image. However, this reduces the efficiency of your observations.
3. Treating the contamination from the previous observations in your data reduction pipeline.

References

Baranne, A. 1972, Auxiliary Instrumentation for Large Telescopes, ed. S. Laustsen, & A. Reiz (Geneva: ESO/CERN), 227–39

Dekker, H., D'Odorico, S., Kaufer, A., Delabre, B., & Kotzlowski, H. 2000, Proc. SPIE, 4008, 534–45

Deming, D., & Plymate, C. 1994, ApJ, 426, 382

Eversberg, T., & Vollman, K. 2015, Spectroscopic Instrumentation (Berlin: Springer)
Kaufer, A., & Pasquini, L. 1998, Proc. SPIE, 3355, 844
Schroeder, D. J. 1987, Astronomical Optics (New York: Academic)
Strassmeier, K. G., Ilyin, I., Järvinen, A., et al. 2015, AN, 336, 324
Tull, R. G., MacQueen, P. J., Sneden, C., & Lambert, D. L. 1995, PASP, 107, 251
Tyson, R. K. 1987, Principles of Adaptive Optics (New York: Academic)
Vogt, S. S. 1987, PASP, 99, 1214
Vogt, S. S., Allen, S. L., Bigelow, B. C., et al. 1994, Proc. SPIE, 2198, 362
Vogt, S. S., Radovan, M., Kibrick, R., et al. 2014, PASP, 126, 359

IOP Astronomy

The Doppler Method for the Detection of Exoplanets

A P Hatzes

Chapter 3

Factors Influencing the Radial Velocity Measurement

The radial velocity (RV) precision that one can achieve depends on many factors: the performance of the spectrograph, the properties of the star, sources of instrumental errors, and finally, stellar variability. Some of these you have control over. For instance, you can take efforts to minimize the systematic errors of your instruments, or you can get more observations of a star in order to "beat down" the noise due to stellar variability. Other factors you cannot control, and these depend on the basic properties of the spectrograph, which are fixed at the design level (e.g., wavelength coverage, resolution) while others depend on the type of star you observe.

So, we can divide these "uncontrollable" factors into two broad categories:

1. Factors due to the instrumental characteristics.

   These primarily include the spectral resolution, wavelength coverage, and signal-to-noise ratio (S/N) of your spectrum. There are of course other factors that can introduce errors such as instrumental shifts, variations in the spectrograph, improper barycentric corrections, etc. These topics will be covered in subsequent chapters. What we will address here are the basic properties of the spectrograph and the RV precision you would achieve if it worked as a perfect instrument.
2. Factors due to the properties of the star. This is largely dominated by the spectral type of the star (spectral information) and its rotational velocity.

In other words, the only control the researcher has is in the choice of an appropriate target star. Stars can also influence the RV measurement through its intrinsic variability in the form of pulsations or stellar activity. The "stellar noise" aspects will be addressed in Chapters 9 and 10. What we will discuss here are the properties of the star if it were a boring rotating sphere of gas that shows no variability.

doi:10.1088/2514-3433/ab46a3ch3

Several authors have done detailed studies on the theoretical uncertainties from RV measurements (Beatty & Gaudi 2015; Bouchy et al. 2001; Bottom et al. 2013). It is not our intention to go into a deep mathematical investigation into the uncertainties of RV measurements for two reasons. First, it is more important to have a general feel as to how a spectrograph or a star of given characteristics will influence your RV precision. These will be done through simple numerical simulations. Second, as we shall soon see, systematic instrumental errors and intrinsic stellar variability will introduce errors in your RV measurement that will completely overwhelm the theoretical errors due to photon statistics. You rarely get perfect results.

3.1 Instrumental Characteristics

The fundamental characteristics of a spectrograph that can influence the RV precision are

1. the wavelength coverage,
2. the resolving power of your spectrograph, and
3. the S/N of your data.

An increased wavelength coverage usually results in more spectral lines to measure your Doppler shift. All other factors constant (i.e., the properties of the star), the spectral resolution determines the recorded width of your spectral line and how easy it is to measure its centroid. The S/N naturally depends on the brightness of the star, the exposure time, and the observing conditions. If all these factors are equal, then the S/N also depends on the efficiency of the spectrograph. We therefore include it among the instrumental characteristics.

3.1.1 Wavelength Coverage

Every spectral line gives you a Doppler measurement of the star with a certain error. Use a second line and your measurement error should decrease by the square root of two. Indeed, a simple simulation using spectral lines that all have the same strength confirms that if you achieve an RV error of σ for a single line, the total RV error decreases by $\sigma_{\mathrm{RV}} = \sigma\sqrt{(N)}$, where N is the number of spectral lines (left panel of Figure 3.1). So naively, if we have a wavelength coverage of $\Delta\lambda$ (not to be confused with spectral resolution, $\delta\lambda$) then we expect the RV error to scale as

$$\sigma_{\mathrm{RV}} \propto (\Delta\lambda)^{-0.5}. \tag{3.1}$$

Reality is a bit more complicated. The right panel of Figure 3.1 shows the behavior for real RV measurements on a solar-type star taken with the Tautenburg Coudé Echelle spectrograph (TCES). The method employed was the iodine absorption cell method, which will be covered in more detail in Chapters 4 and 6. The RV was calculated using an increasingly larger wavelength region.

The value of σ_{RV} indeed decreases with wavelength coverage, $\Delta\lambda$, but for a short wavelength coverage, it is markedly worse than the predicted behavior. However, it quickly falls in line with the predicted behavior. Beyond a wavelength coverage of

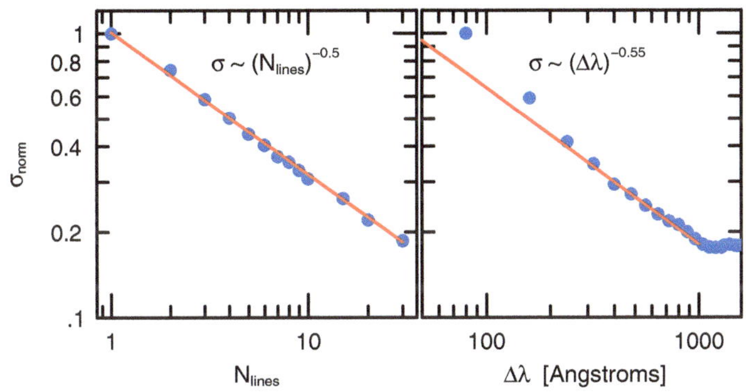

Figure 3.1. (Left) Simulations (points) showing the normalized RV error $\sigma_{\rm RV}$ as a function of the number of spectral lines, $N_{\rm lines}$, used in the RV measurement. The fit (line) shows that $\sigma_{\rm RV} \propto N_{\rm lines}^{-0.5}$. (Right) The normalized RV uncertainty as a function of wavelength coverage for real data using the spectrum of a Sun-like star. The line represents a fit with $\sigma_{\rm RV} \propto N_{\rm lines}^{-0.55}$.

about 1000 Å, the measurement error is flat, that is, increasing the wavelength coverage results in no substantial reduction in the measurement error. A fit to the data results in $\sigma \propto (\Delta\lambda)^{-0.55}$, close to the theoretical expectation.

There are three reasons for these deviations with the $\Delta\lambda^{-0.5}$ law in Figure 3.1. First, an increase in wavelength coverage is not always followed by an increase in the number of useful spectral lines for a Doppler measurement. Depending on the effective temperature of the star, there can be spectral regions where the number density of lines is sparse. This is especially true of early-type stars. Furthermore, not all spectral lines have the same strength, and as we shall shortly see, this will affect the RV precision.

Second, once your wavelength coverage extends beyond about 6000 Å, telluric lines start to become prevalent (see Chapter 12). These features are not tied to the Doppler motion of the star, and they only serve to decrease the RV error precision.

Finally, as we shall soon see, the molecular iodine lines that were used to compute the relative RV shifts in Figure 3.1 become weak beyond about 5800 Å, and this results in a degradation of the RV precision. This, along with the presence of telluric lines, largely explains the flattening of the RV precision.

In practice, when performing precise RV measurements on a star, it is useful to explore how the measurement error behaves as one uses various spectral regions for the Doppler calculation. It is best to avoid those regions where there is no noticeable improvement in the error, or at least weight these accordingly. However, when performing RV measurements for solar-type stars in the optical region, we can expect $\sigma_{\rm RV} \propto (\Delta\lambda)^{-0.5}$.

3.1.2 Signal-to-noise Ratio

In the absence of other sources of noise, the error in your spectral data, $\sigma_{\rm p}$, is determined purely by the number of photons you detect, $N_{\rm p}$, and according to photon statistics, this is simply $\sigma_{\rm p} = \sqrt{N_{\rm p}}$. The photon noise is the theoretical limit to

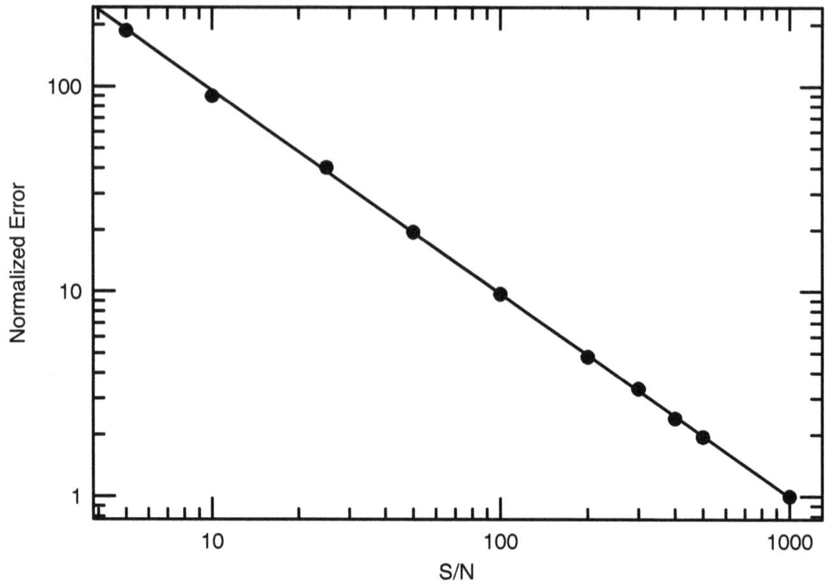

Figure 3.2. Simulations (points) showing the normalized RV measurement error as a function of signal-to-noise ratio, S/N. The error has been normalized to unity for S/N = 1000. The behavior shows that $\sigma_{RV} \propto (S/N)^{-1}$.

the RV precision you can achieve for a given exposure. If you do this, then you have a perfect spectrograph and have done an excellent job of eliminating systematic errors, or at least to a level where they are not important.

Figure 3.2 shows the result of simulations where increasing levels of noise were added to a spectral line and the relative Doppler shift was calculated with respect to a noise-free line. The RV uncertainty follows the simple power law

$$\sigma \propto (S/N)^{-1}, \tag{3.2}$$

At face value, it would seem that simply increasing the S/N is a quick way of achieving a lower RV uncertainty. However, as with most things in life, an increase in quality comes at a price, and in this case, it is paid with exposure time. Suppose you achieve an RV error of $\sigma = 10$ m s^{-1} for S/N = 50 using a 15 minute exposure. Decreasing this error to 1 m s^{-1} requires a 10-fold increase in the exposure time to about 2.5 hr. This would not be an effective strategy if you want to survey a large number of stars, or if you have limited telescope resources. For bright stars, it is wise to limit your the S/N ratio to no higher than about 200. Above this, there is relatively little gain in the RV precision. Furthermore, using a higher S/N to achieve RV precisions of 1 m s^{-1} is probably an inefficient use of telescope time because other factors, such as systematic errors or intrinsic stellar variability, will determine your precision before you hit the photon noise limit.

3.1.3 Resolving Power

Spectral resolution is the spectrograph characteristic that largely determines the basic RV precision you can achieve. Clearly, if you have more sampling points across a stellar line profile, the more accurately you can determine the centroid and thus Doppler shift of the line. Most spectrographs are designed to satisfy the Nyquist criterion, i.e., two detector pixels cover the spectral resolution, $\delta\lambda$. So, if a spectrograph has a resolving power of $R = 100{,}000$, this will produce a dispersion of 0.055 Å per pixel, or a velocity resolution of 3 km s^{-1} at a wavelength of 5500 Å.

Table 3.1 lists the expected velocity resolution for spectrographs over a wide range of resolving powers, $R = \delta\lambda/\lambda$. Also listed is the expected shift in millimeter at the detector assuming CCD pixels of 15 μm size a wavelength of 5000 Å and a Doppler shift of 1 m s^{-1}. If you can measure the position of a spectral line to a fixed fraction of a CCD pixel, it make sense to have a higher resolving power. Clearly, one should avoid low-resolution spectrographs for precision RV measurements. For example, if you want to get an RV precision of 1 m s^{-1} with an $R = 1000$ spectrograph, this would result in a shift at the detector of only 6.7×10^{-6} pixels, or equivalently only 10^{-4} μm (= 1 Å!).

Several works have investigated the dependence of the RV uncertainty on spectral resolution, or resolving power. Bouchy et al. (2001) found that $\sigma_{RV} \propto R^{-1}$. This result is consistent with simple simulations using a single spectral line generated at different resolutions of spectral lines generated using model atmospheres (Figure 3.3). Bottom et al. (2013) found that the RV uncertainty behaved as $\sigma_{RV} \propto R^{-1.2}$. Earlier work by Hatzes & Cochran (1992) reported a steeper variation of the uncertainty with resolving power, $\sigma_{RV} \propto R^{-1.5}$, which is consistent with what Beatty & Gaudi (2015) reported. Why the discrepancies?

One possibility is how the Doppler shifts were calculated. The simulations of Hatzes & Cochran (1992) mimicked the data that were taken with an iodine gas absorption cell. The method will be discussed at length in the next chapter, but basically Doppler shifts are calculated with respect to molecular iodine absorption lines which are unresolved even at a resolving power $R = 100{,}000$. The shape of these lines are thus dominated by the instrumental profile. Figure 3.4 summarizes the

Table 3.1. The Doppler Shift of 1 m s^{-1} for Different Resolving Powers

Resolving Power	Dispersion (Å/pixel)	Velocity Resolution (m s^{-1} $pixel^{-1}$)	Shift in Pixels	Shift at Detector (mm)
1000	2.5	150,000	6.7×10^{-6}	10^{-7}
5000	0.5	30,000	3.3×10^{-5}	5.0×10^{-7}
10,000	0.25	15,000	6.7×10^{-5}	1.0×10^{-6}
25,000	0.10	6000	1.7×10^{-4}	2.5×10^{-6}
50,000	0.05	3000	3.3×10^{-4}	5.0×10^{-6}
100,000	0.025	1500	6.7×10^{-4}	1.0×10^{-5}
200,000	0.0125	750	1.4×10^{-3}	2.0×10^{-5}
500,000	0.005	300	3.3×10^{-3}	5.0×10^{-5}

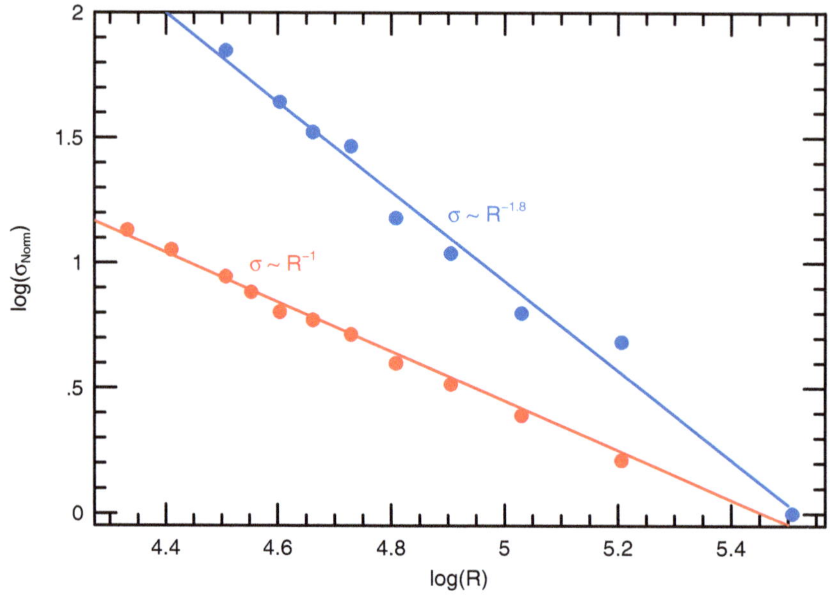

Figure 3.3. (Red circles) Simulations of the normalized RV measurement error (σ_{RV} as a function of resolving power R, normalized for the error at R = 120,000). These simulations used synthetic spectral lines with a rotational velocity v sin i = 1 km s^{-1}. The red dashed line shows that $\sigma \propto R^{-1}$. (Blue squares) Simulations of the RV measurement error as a function of R, but this time using a δ-function as the "spectral line," i.e., a feature that is unresolved at R = 100,000. The broadening of this feature is dictated by the instrumental profile. In this case, $\sigma \propto R^{-1.8}$.

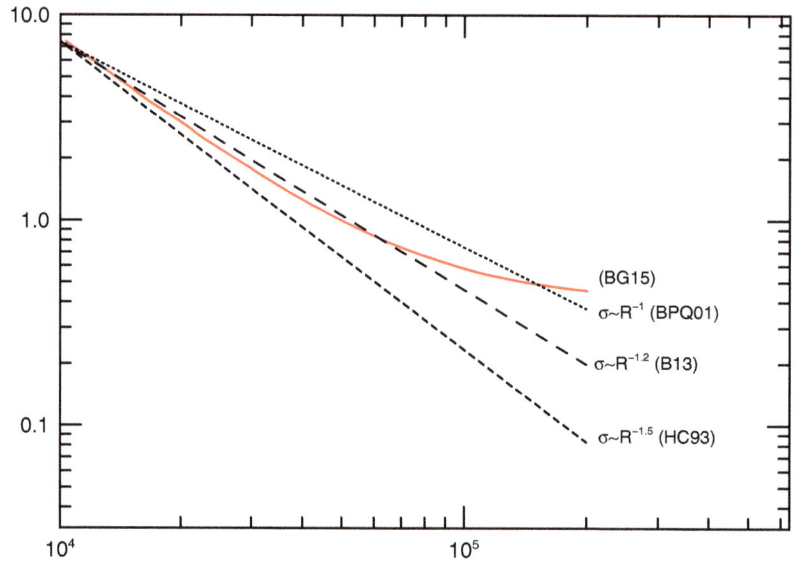

Figure 3.4. The behavior of the RV uncertainty as a function of spectral resolving power from various studies: BG15 (red line): Beatty & Gaudi (2015), BPQ01: Bouchy et al. (2001); B13: Bottom et al. (2013); HC93: Hatzes & Cochran (1992).

different power dependencies of the RV uncertainty on resolving power found by various investigations. Although Beatty & Gaudi (2015) reported an $R^{-1.5}$ dependence in the RV uncertainty, a close inspection of their figures shows that the dependence follows the red line shown in the figure.

Interestingly, when one computes the relative shift of an unresolved profile (δ-function) that is convolved with a Gaussian instrumental profile, the RV uncertainty has a much steeper dependence on R, namely $\sigma_{\mathrm{RV}} \propto R^{-1.8}$ (Figure 3.3). This is because the uncertainty depends on the line depth (see below), and for lower resolution, this decreases more rapidly for lower resolution than for resolved lines.

From the various studies, we expect the RV uncertainty to follow a power law, $\sigma_{\mathrm{RV}} \propto R^{-\alpha}$ with $\alpha = 1$–1.5. For a good approximation of the dependence of σ_{RV} on R, it is sufficient to take the average of the extreme values for α, namely $\sigma_{\mathrm{RV}} \propto R^{-1.2}$. In subsequent discussions, we will adopt this as the dependence of the RV uncertainty with spectral resolving power.

If you were building a high RV precision spectrograph, you might naively assume that it should have the highest resolving power possible. However, there are trade-offs to consider. First, high-resolution spectrographs are much more expensive to build. Everything is bigger: collimator optics, gratings, cameras, etc. Bigger implies more expensive. If you are designing a spectrograph, budget constraints may be the factor that dictates resolving power.

Second, at higher resolving powers, you are dispersing the light over more CCD pixels. This will decrease the count rate and thus S/N of your spectrum for a given star and exposure time. Double your resolving power, and you have just decreased the S/N and thus RV precision by a factor of 1.4.

Finally, at higher resolving power for a fixed-sized detector, you have a decreased wavelength coverage, and this scales as $\Delta\lambda \propto R^{-1}$. If you have a CCD detector with a certain size and you install it on a spectrograph with twice the resolving power, you will have decreased the wavelength coverage by a factor of 2. The RV uncertainty for a fixed S/N has just increased by the square root of 2, due to the lost wavelength coverage. You can of course try to compensate for this by using more detectors in a mosaic configuration, but again, that comes with increased cost for the spectrograph.

It is easy to calculate how the RV changes if the same detector were moved to a spectrograph with higher resolving power, i.e., for a "fixed-size detector." In this case, what you gain in precision from the increased resolving power is partially offset by the loss in precision due to the smaller wavelength coverage. The uncertainty due to resolving power scales as $\sigma_R \propto R^{-\alpha}$. The uncertainty with wavelength coverage, $\Delta\lambda$, scales as $\sigma_{\Delta\lambda} \propto (\Delta\lambda)^{-1/2}$. The wavelength coverage scales as R^{-1}, so substituting into the previous expression, one gets $\sigma_{\Delta\lambda} \propto R^{1/2}$. So, for the case of the fixed-sized detector, the total uncertainty is given by the product of the two, namely $\sigma_{\mathrm{Total}} \propto R^{1/2-\alpha}$.

Figure 3.5 shows the actual RV error determined from solar spectra (day sky observations) using an iodine absorption cell at resolving powers of R = 2300, 15,000, and 200,000. These were taken with the same detector and spectrograph, but different gratings to provide different resolutions. The solid red curve is the function

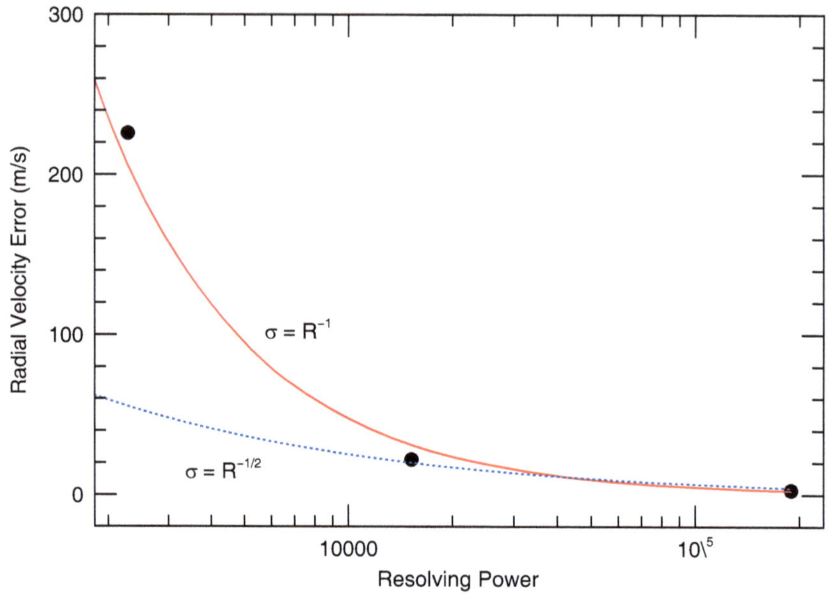

Figure 3.5. (Points) The radial velocity error taken with a spectrograph at different resolving powers. This is the actual data taken of the day sky all with the same S/N values. The solid red line shows a $\sigma \propto R^{-1}$ fit. The dashed black line shows a $\sigma \propto R^{-1/2}$ fit. The detector size is fixed for all data, thus the wavelength coverage is increasing with decreasing resolving power.

$\sigma \propto R^{-1}$, while the dotted curve is $\sigma \propto R^{-1/2}$. At first glance this seems to support $\sigma \propto R^{-1}$. Therefore, $(1/2 - \alpha) = -1$, which implies that $\alpha = 3/2$ as opposed to unity. Keep in mind the caveat that these data were taken with the iodine absorption cell.

3.2 Stellar Characteristics

The properties of the star also have a large influence on your RV precision. The RV uncertainty depends on three major fundamental features of the star (stellar spectrum):

1. The projected rotational velocity of the star, or $v \sin i$.
2. The strength of stellar spectral lines.
3. The number density of stellar lines.

3.2.1 Stellar Rotational Velocity

In the absence of stellar variability, stellar rotation has the largest influence on the RV measurement error. Rotation broadens the width of the stellar lines and makes them shallower, thus making it more difficult to determine the centroid. Figure 3.6 shows the spectral region of two stars, the top of a B9 star and the lower panel of a K5 star. The hot star only has one spectral line in this region, and it is quite broad and shallow, due to the high projected rotation rate of the star, $v \sin i$, where v is the true rotational velocity of the star and i is the inclination of the spin axis to the

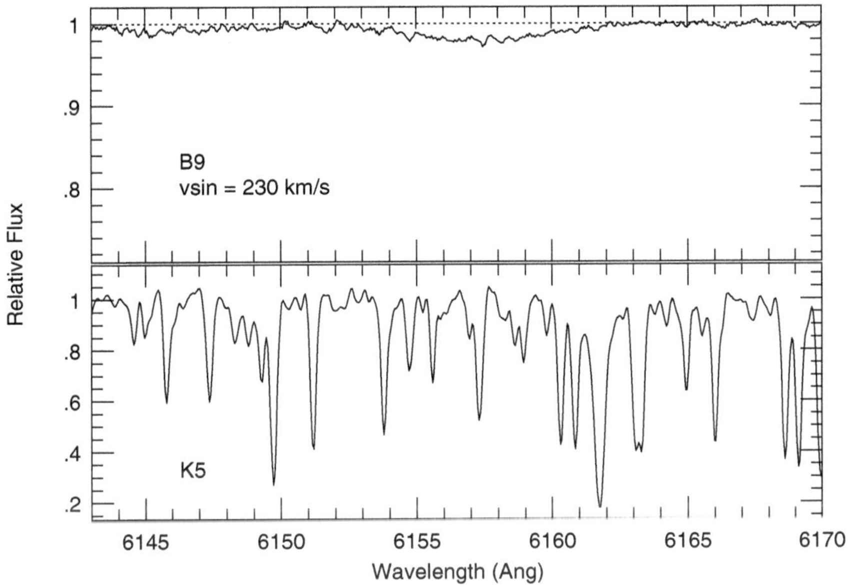

Figure 3.6. (Top) A spectrum of a B9 star rotating at 230 km s^{-1}. (Bottom) The spectrum of a K5 star.

Table 3.2. Median Rotational Velocities of Stars

Spectral Type	$v \sin i$ (km s^{-1})
O4	110
O9	105
B5	108
A0	82
A5	80
F0	44
F5	11
G0	4
G5	3
K0	3
K5	2
M0	10
M4	16
M9	10

observer. Clearly, it would be more difficult to determine the centroid position and thus a Doppler shift of the rotationally broadened spectral line.

Table 3.2 lists the median $v \sin i$ as a function of stellar types for main-sequence stars taken from the Glebocki et al. (2000). Note the sharp drop in rotational velocities at mid-F, which is often called the "rotation break" or "Kraft break," which was first noted by Kraft (1967). This results from the fact that around mid-F, stars start to develop a substantial outer convection zone. This, coupled with rotation, leads to magnetic activity, and the star loses angular momentum through magnetic braking.

In terms of good RV precision, early-type stars are poor targets for RV measurements for two reasons. First, these stars are hot, and as such, they have much fewer spectral lines for RV measurements than for stars at the lower end of the main sequence. This is seen in the lower panel of Figure 3.6, where the K5 star has a higher density of spectral lines. Second, rapid rotation greatly degrades the RV precision.

Figure 3.7 shows the behavior of the RV uncertainty as a function of $v \sin i$ and different resolving powers of the spectrograph. These curves were generated by measuring the relative shift of a single synthetic spectral line at the appropriate resolution and $v \sin i$. Noise at a level of S/N = 50 was added, but the shape of the curves are the same for a fixed S/N. For each curve (fixed R), the uncertainty was normalized to the value of $v \sin i \approx 2$ km s^{-1}. The ordinate thus represents the factor by which the uncertainty scales with the stellar rotational velocity.

Equation (3.3) and the curves in Figure 3.7 should only be used to estimate the RV uncertainty for a star with a certain $v \sin i$ by scaling the known performance of the same spectrograph with a given resolving power. As an example, suppose you use a spectrograph with $R = 60{,}000$ and get an RV precision of 3 m s^{-1} on a slowly rotating star ($v \sin i \approx 2$ km s^{-1}). If you observe the same type of star rotating at 50 km s^{-1} with the same spectrograph and at the same S/N, then your RV uncertainty will be a factor of 13 worse, or about 40 m^{-1}.

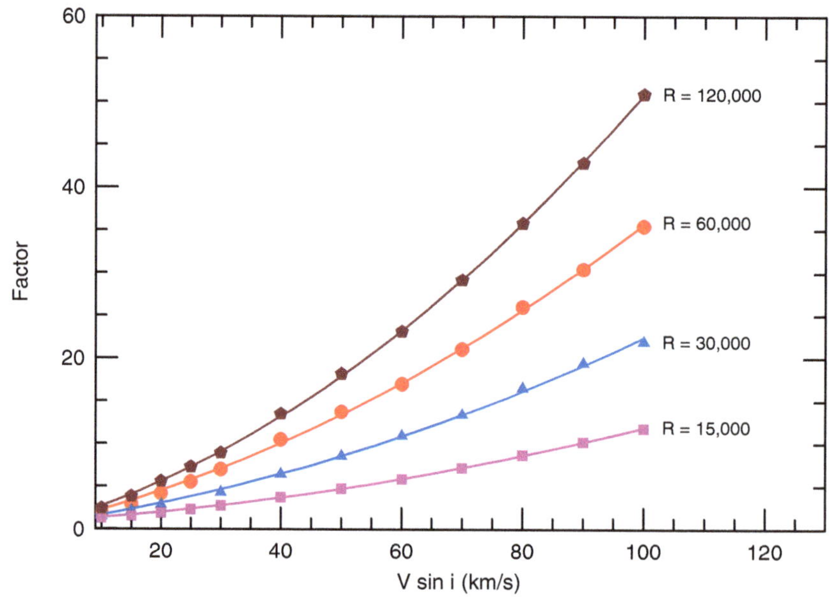

Figure 3.7. The scale factor for the increase in the RV uncertainty as a function of the stellar $v \sin i$ and for several values of the resolving power (R = 15,000–200,000). Each curve has been normalized to the uncertainty for $v \sin i = 2$ km s^{-1}.

Formally, the curves can be fit by the function

$$f(V) \propto 0.62 + (0.21 \log R - 0.86)V + (0.00260 \log R - 0.0103)V^2, \qquad (3.3)$$

where V is the projected rotational velocity in km s^{-1}.

One can also use more simple relationships to scale between stars of different rotational velocities depending on whether you have slow or fast rotating stars.

For typical resolving powers of spectrographs used for precise RV measurements, the RV precision due to stellar rotation scales as:

for $v \sin i < 2$ km s^{-1},

$$\sigma_{\mathrm{RV}} \propto (v \sin i)^{0.2}; \qquad (3.4)$$

for $v \sin i > 10$ km s^{-1},

$$\sigma_{\mathrm{RV}} \propto (v \sin i)^{1.3}. \qquad (3.5)$$

For example, if you have an RV precision of 3 m s^{-1} on a star rotating at 2 km s^{-1}, then you should get an RV precision of approximately 4 m s^{-1} on a star rotating at 5 km s^{-1}. Likewise, a star rotating at 70 km s^{-1} will have an RV a factor of 2 larger than a star rotating at 40 km s^{-1}.

Note that the latter equation has a slightly higher dependence on $v \sin i$ than the rough approximation of $\sigma \propto v \sin i$ given by Hatzes (2016). However, the linear expression probably is sufficient for getting a rough estimate of the uncertainty for rapidly rotating stars with $v \sin i > 5$ km s^{-1}.

3.2.2 Spectral Line Strength

The RV precision depends on the depth of the stellar line. Clearly, if your line is too weak it will get lost in the noise, and an RV measurement will be next to impossible. Let us define the depth, d, as a value from 0 to 1. A line with $d = 1.0$ has a depth that is 100% of the continuum, i.e., zero flux in the core of the line. A weak line with $d < 1.0$ will produce a measurement error that is some factor, F, times the measurement error of the stronger line.

Figure 3.8 shows a simulation of the Doppler measurement error as a function of spectral lines of fixed width, but varying depths. This simulation used an S/N = 50, but the results are insensitive to the exact S/N chosen. It shows that $1/F$ scales linearly with line depth. That is to say, if a line has a depth of one-fifth the continuum, it will have an RV measurement error that is five times greater than that of a line with a depth of 100% of the continuum.

However, real spectral lines do not have a depth that scales linearly with the line strength. Rather, these follow the so-called curve of growth. As one increases the line strength as measured by the equivalent width (EW), the line depth increases, but the width remains fairly constant (Figure 3.9). Once the line starts to saturate, the line depth remains constant, but the wings, and thus the line width, starts to increase. The triangles in Figure 3.8 show how the factor F varies with the EW for real spectral lines. At first, the curve follows the one for lines of fixed width, but after

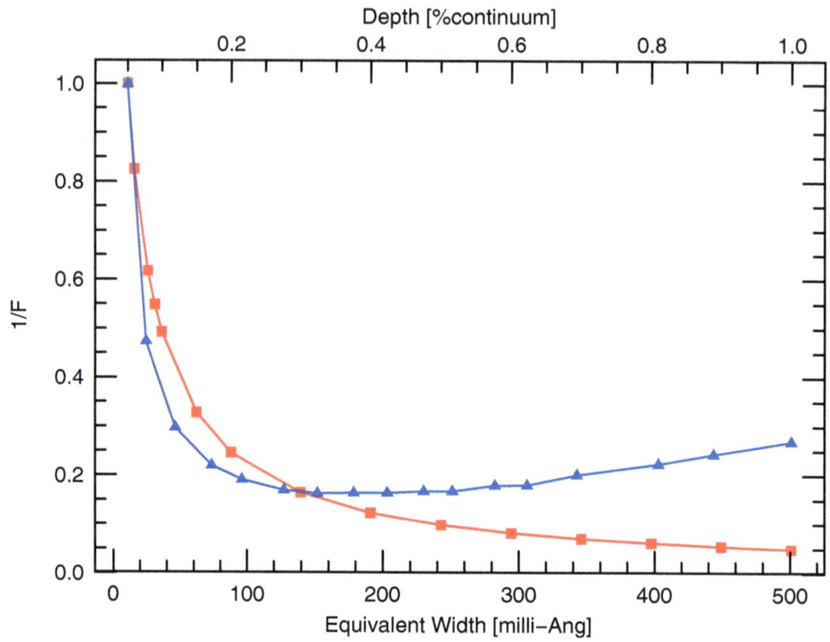

Figure 3.8. The inverse of the multiplicative factor in the RV uncertainty, *F*, as a function of line strength. (Blue triangles) The factor in the uncertainty for spectral lines of constant width, but a depth that is a fraction of the continuum value (top abscissa). (Red squares) The factor in the RV uncertainty as a function of equivalent width (lower abscissa) for a real spectral line. The best RV precision is for strong, unsaturated lines.

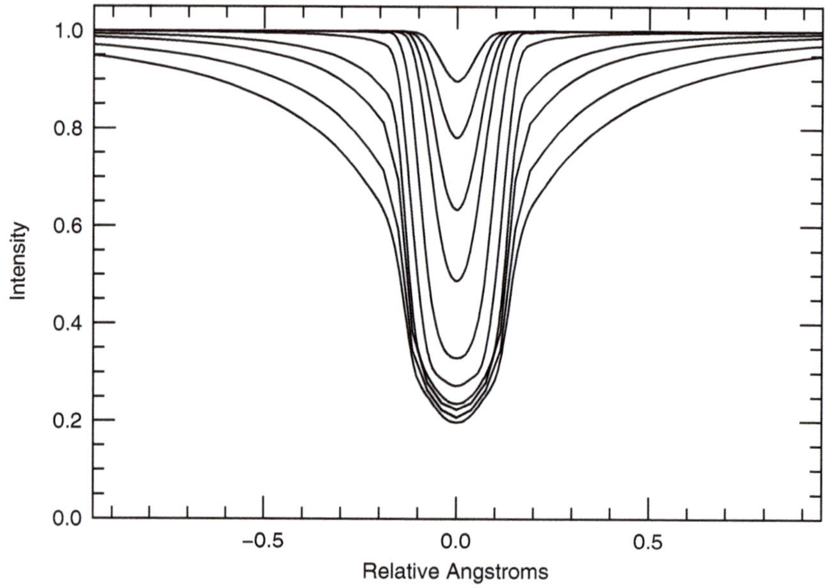

Figure 3.9. The change in spectral line shape as a function of increasing line strength.

EW ≈ 100 mÅ, the curve starts to flatten out. For the strongest lines, the RV uncertainty actually starts to increase with increasing line strength (Figure 3.8).

This behavior can easily be understood in terms of how the RV uncertainty varies with line depth and widths. For weak spectral lines, the increase in EW is due primarily to an increase in the line depth, so the RV uncertainty follows the behavior in our simple simulation (squares in Figure 3.8). Once the line saturates, the line depth increases very slowly, but there is a more rapid increase in the width of the line. It is more difficult to determine the centroid of a broad line as opposed to a narrow line. Whatever gain in precision is achieved by a slightly deeper line is more than offset by the larger line width. In this case, a spectral line with an EW of 100 mÅ yields the same Doppler uncertainty as a line twice as strong. Finally, for the strongest lines, the increase in line width dominates, and the RV uncertainty actually increases with line strength. So, the largest Doppler information is found in strong, yet unsaturated, spectral lines.

3.2.3 Number Density of Spectral Lines

The Doppler precision not only depends on the wavelength of your spectrograph, but also on the number density of stellar absorption lines. The latter of course depends on the effective temperature of the star—hot stars have much fewer absorption features.

The left panel of Figure 3.10 shows the approximate number of strong spectral lines in the wavelength range 4000–7500 Å[1] as a function of the effective temperature of the star. This is for main-sequence stars in the effective temperature range

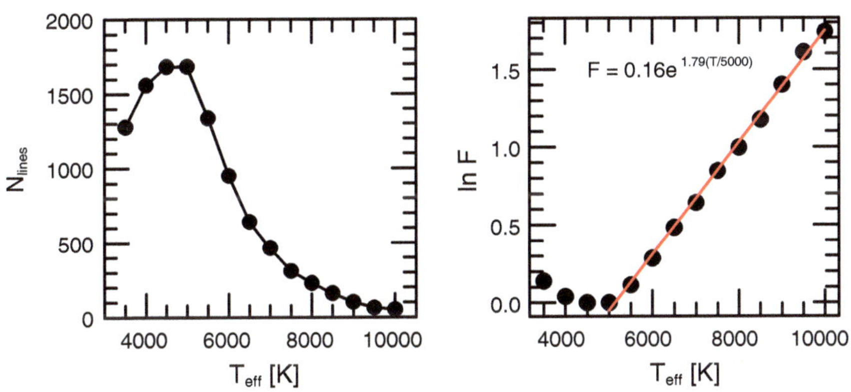

Figure 3.10. (Left) The number of strong lines (50% of the continuum) in the wavelength range 4000–7500 Å as a function of effective temperature, T_{eff}, of the host star. The drop-off at lower temperatures is due to line blending at bluer wavelengths. (Right) The scale factor, F, in the RV uncertainty due to the line density as a function of T_{eff} referenced to $\sigma = 1$ m s^{-1} at $T_{\text{eff}} = 5000$ K. Over the temperature range 5000–10,000 K, this can be well fit by $F = 0.16e^{1.79(T/5000)}$.

[1] We take the typical wavelength coverage of an optical spectrograph used for RV measurements.

$T_{\rm eff}$ = 3500–10,000 K. Here we define a “strong” line has having a depth deeper than 50% of the continuum value.

The number of spectral lines increases sharply as the effective temperature decreases, but surprisingly, this flattens out at cooler temperatures. The reason for this is that for wavelengths less than about 5000 Å, cool stars simply have too many spectral lines. Line blending suppresses the continuum, causing even strong lines to have a relatively small depth. The line blending also results in few clean, isolated lines which provide the higher Doppler information.

The right panel of Figure 3.10 shows the natural logarithm of the scaling factor for the RV uncertainty, F, as a function of effective temperature. Beyond a temperature of 5000 K, this follows a linear trend, so the scale factor for $T_{\rm eff}$ < 5000 K can be well fit by the expression

$$F = 0.16e^{1.79(T_{\rm eff}/5000)}. \tag{3.6}$$

Therefore, a main-sequence star with $T_{\rm eff}$ = 8000 K will have an RV uncertainty 2.8 times higher than a main-sequence star with $T_{\rm eff}$ = 5000 K just from the decreased line density (same S/N and stellar rotational velocity).

3.3 RV Precision across Spectral Types

We can now put together all we have learned to get a rough estimate of the RV error as a function of $T_{\rm eff}$, or of the spectral type for main-sequence stars. Using the mean rotational velocity of stars (Table 3.2) and the mean density of lines as a function of $T_{\rm eff}$ results in Figure 3.11. The horizontal dashed line marks an RV precision of 10 m s^{-1}, the nominal value if you want to detect a Jupiter analog around a solar-type star.

Early RV surveys for planets strove for an initial precision of approximately 10 m s^{-1}, the nominal precision to detect Jupiter analogs. By this criterion, you should not be able to detect planets around stars of early spectral types. Indeed, up until the mid-2000s, the earliest spectral type for which a planet had been detected was about F6. This lack of precision for early stars factored into the biases in the early surveys—investigators simply avoided stars with spectral types earlier than mid-F. It was only in the mid-2000s that RV surveys began to survey more early-type stars (Galland et al. 2005; Hartmann & Hatzes 2015).

Figure 3.11 largely explains the distribution of planet discoveries as a function of spectral type (Figure 3.12). RV surveys largely ignored stars with spectral early than mid-F ($T_{\rm eff} \approx$ 6000 K) due to the poor RV precision. Stars later than about K5 ($T_{\rm eff}$ < 4000 K) were simply too faint. The early RV surveys were largely performed on 2–3 m class telescopes (Cochran & Hatzes 1993; Butler & Marcy 1997; Queloz et al. 1998), so one could not get good S/N ratios for observations on very cool stars.

The distribution of RV discoveries also highlights the bias of the technique when it comes to the mass of the host star. On the main sequence, there is a one-to-one mapping between effective temperature and stellar mass. About two-thirds of the host stars of RV-detected planets have effective temperatures in the range 4500–6500 K, and this translates into the narrow mass range of M = 0.7–1.2 $M_\odot$ for the mass of the host star.

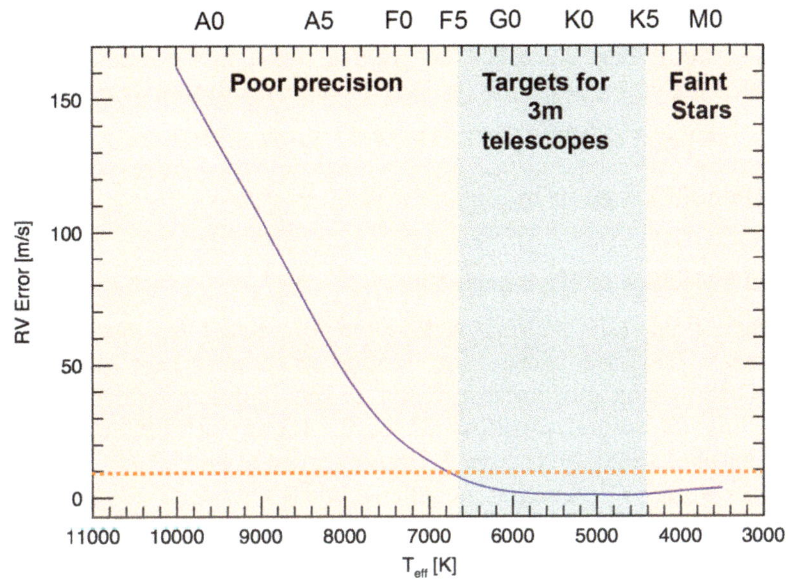

Figure 3.11. The expected radial velocity error as a function of spectral type. This was created using the mean rotational velocity and approximate line density for a star in each spectral type. The horizontal line marks the nominal precision of 10 m s^{-1} needed to detect a Jovian-like exoplanet.

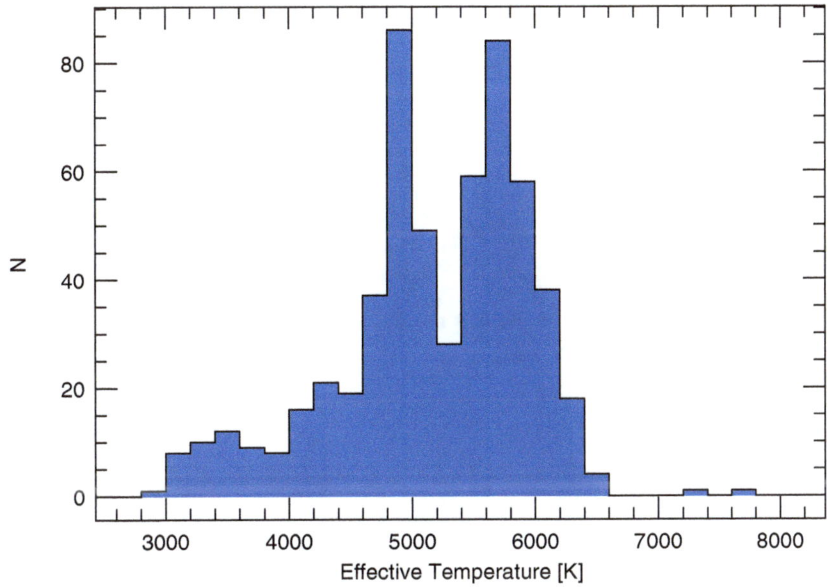

Figure 3.12. The distribution of planet detections as a function of the effective temperature of the host star. The distribution roughly coincides with the green shaded region shown in Figure 3.11.

We can now put together a grand scaling relationship for the expected RV precision that combines the S/N, the spectral resolving power R, the effective temperature of the star T, and the projected stellar rotational velocity V:

$$\sigma\ [\mathrm{m/s}] \propto \Delta\lambda^{-1/2}(S/N)^{-1}R^{-1.2}f(V)(0.16e^{1.79(T/5000)}). \tag{3.7}$$

The function $f(V)$ is given by Equation (3.3).

3.3.1 Radial Velocities of High-mass Stars

Although early-type stars are now well suited for precise RV measurements, that does not mean they are useless for exoplanet studies. Low precision can be compensated by taking more measurements. This is demonstrated by the case of WASP-33. This star hosts a transiting planet in a 1.2 day orbit (Christian et al. 2006; Collier Cameron et al. 2010). It is an A5 main-sequence star ($T_{\mathrm{eff}} = 8100$ K) rotating with a $v \sin i = 90$ km s^{-1}, which is a challenge for RV work. To complicate matters, it is a δ-Scuti star, and the stellar oscillations add an additional noise component (see Chapter 10).

Figure 3.13 shows RV measurements of WASP-33 phased to the orbital period (Lehmann et al. 2015). These measurements have been filtered for the δ-Scuti pulsations. One can clearly see that the orbital variation and the measurements have an rms scatter of 245 m s^{-1} about the orbital curve. By taking many measurements,

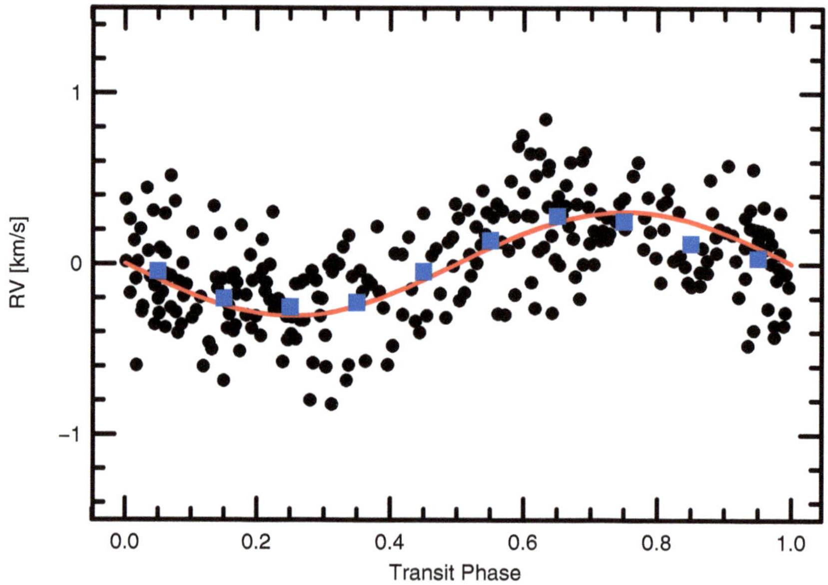

Figure 3.13. RV measurements of WASP-33 phased to the orbital period of the transiting planet (Lehmann et al. 2015). WASP-33 is an A5 star with $T_{\mathrm{eff}} = 8100$ K that exhibits δ-Scuti pulsations. The star has a mass of 1.5 $M_\odot$ and rotates with $v \sin i = 90$ km s^{-1}. The individual measurements have an rms scatter of 245 m s^{-1} while the phase-binned averages (blue squares) have an rms scatter of 23 m s^{-1}. The red curve is the orbital solution.

one can take binned averages to reduce the scatter to 23 m s^{-1} (squares in Figure 3.13).

Let's see if the measurement errors are consistent with the predictions from our scaling relationships. The RV data for WASP-33 were taken with the TCES. This spectrograph has a resolving power of $R \approx 60,000$ and can achieve an RV precision of 3 m s^{-1} on a slowly rotating solar-type star with data having S/N of 100. Figure 3.7 and Equation (3.3) indicate that at this resolving power, a star rotating at 90 km s^{-1} should have an RV uncertainty a factor of 30 times larger than for our slowly rotating "reference" star stemming just from the larger rotational velocity. An A5 type star has approximately 6.5 times fewer spectral lines (Figure 3.10) than a solar-type star. This gives another factor of 2.5 increase in the measurement uncertainty. The RV measurements of WASP-33 have roughly the same S/N as our reference star, so we expect an RV error of ≈230 m s^{-1}, comparable to the actual rms scatter. This means that the RV measurements for WASP-33 have an error fairly close to the expected uncertainty due to photon statistics.

A way to circumvent the low RV precision for early-type stars is to use evolved giant stars as proxies for investigating the frequency of planets around stars more massive than the Sun. As an intermediate-mass star ($M = 1.5$–$3\ M_{\odot}$) evolves off the main sequence, it expands and becomes cooler, thus it shows more spectral lines. More importantly, its rotation rate slows. A $2M_{\odot}$ K giant star has an effective temperature $T_{\rm eff} \approx 4000$ K and rotates at a few km s^{-1}, thus it is highly amenable to precise RV measurements. This fact has inspired a large number of surveys for planets around evolved intermediate-mass stars (Setiawan et al. 2003; Reffert et al. 2006; Johnson et al. 2007; Döllinger et al. 2007; Sato et al. 2008; Niedzielski et al. 2009; Han et al. 2010; Wittenmyer et al. 2011; Quirrenbach et al. 2015).

The K0 III star β Gem was shown to host a giant planet with a mass of $3M_{\rm Jup}$ in a 589 day orbit (Hatzes et al. 2006; Reffert et al. 2006). Figure 3.14 shows RV measurements taken from the McDonald and Tautenburg Observatories phased to the orbital period of the planet. The star has a mass of $1.9M_{\odot}$ (Hatzes & Zechmeister 2007; Hatzes et al. 2012). Because the star is cool and slowly rotating, the rms scatter about the orbit is 13 m s^{-1}, or a factor of 20 less than for WASP-33, a main-sequence star of comparable mass.

However K giant stars may avoid one problem (better RV measurement precision), but it is replaced by others. First, determining the mass of the host star is more problematic. On the main sequence, there is a tight relationship between spectral type and stellar mass. So if you measure the effective temperature of the star, you have a good handle on the stellar mass. For K giant stars, one must rely on evolutionary tracks, so the stellar mass is model dependent. Unfortunately, the main sequence covers a wide range of masses ($\approx 1-4M_{\odot}$) with evolutionary tracks that converge to the same region of the color–magnitude diagram. Placing the star on the correct track requires an accurate effective temperature, luminosity, metal abundance, and of course, good stellar models.

The second problem is that K giant stars show relatively high-amplitude RV variations due to stellar oscillations, and these create excess "noise," which can hinder the detection of low-mass companions (see Chapter 10). However, as the

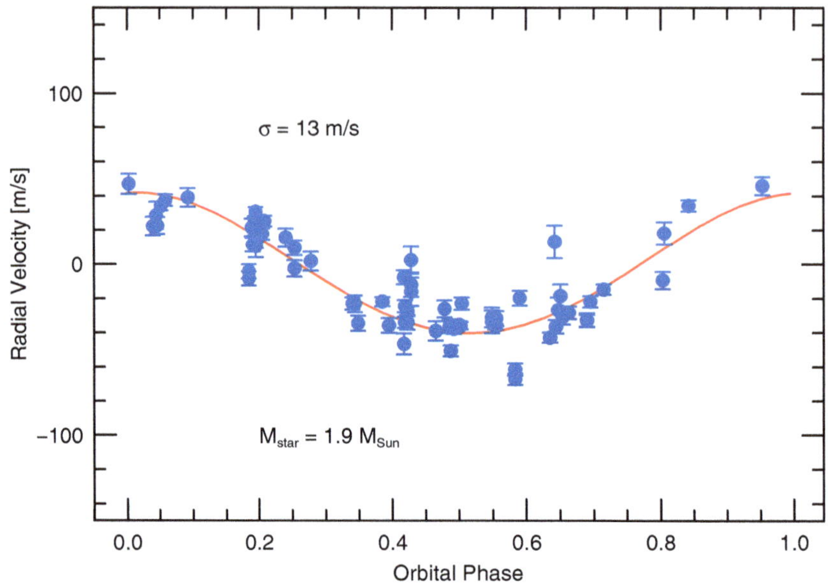

Figure 3.14. RV measurements of β Gem phased to the 585 day orbital period of the planet (Hatzes et al. 2006). This star is a K0 III giant star with a mass of 1.9 $M_{\odot}$. The red curve is the orbital solution. The rms scatter about the orbit is 13 m s^{-1}.

saying goes, "one person's noise is another person's signal," and we can exploit the second problem (more "noise" in the RV measurement) to address the first (poorly known stellar mass). One can use the properties of these stellar oscillations to derive the mass of the host star as was done in the case of β Gem (Hatzes & Zechmeister 2007; Hatzes et al. 2012). In particular, the *Kepler* space mission has provided us with a wealth of data on stellar oscillations in K giant stars, and these can provide a good sample of giant stars with well-determined masses that are suitable for planet searches (Hrudková et al. 2017) In Chapter 10, we will return to the subject of stellar oscillations as a noise component to RV measurements.

3.3.2 Radial Velocities of Low-mass Stars

There are still relatively few exoplanets that have been found around low-mass stars (Figure 3.12). Here, the problem is not a lack of spectral information or high rotation rates, but a lack of photons. M-dwarf stars have low effective temperature, small radii, and thus low luminosities. Even nearby M-dwarf star have visual magnitudes $V < 10$, which makes observations at high spectral resolutions. These often require large telescopes (8–10 m) with efficient spectrographs. However, there is keen interest in RV surveys for these stars as an Earth-mass planet in the habitable zone of these stars can have an RV amplitude of a few m s^{-1}, which is achievable with state-of-the-art instruments.

M-dwarf stars have low effective temperatures, so one way to address the low number of photons from these stars is to make measurements in the near-infrared

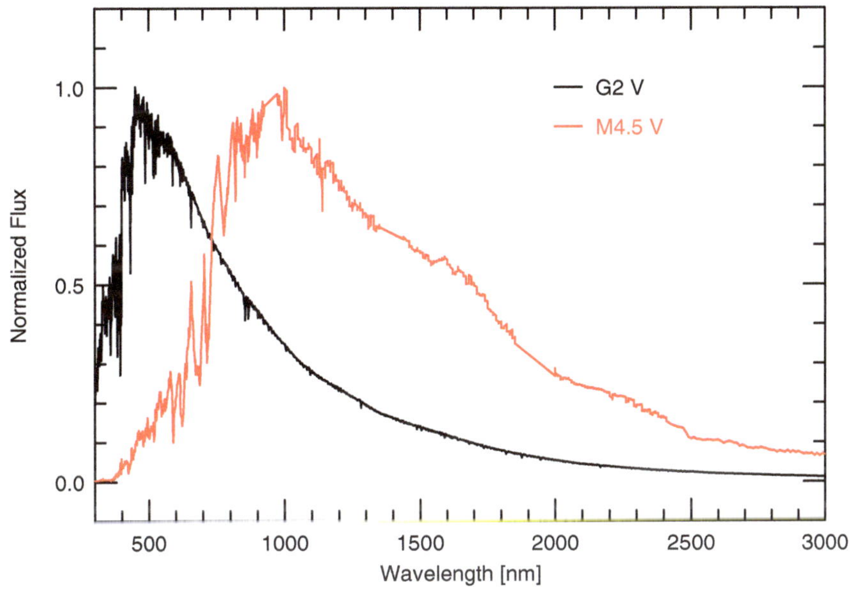

Figure 3.15. The spectral energy distribution of a G2 V star (black curve) compared to that of an M4.5 V star (red curve).

(NIR). Figure 3.15 shows the normalized spectral energy distributions of a G2 V star (the Sun) and an M4.5 V star. The G2 star has a peak in the energy distribution at 500 nm. On the other hand, the flux of the M dwarf peaks at ≈1000 nm. Clearly, if you wanted a spectrograph to detect more photons from the low-mass star, it should cover the NIR wavelengths. For this reason a number of NIR spectrographs have been built for precise RV measurements of cool stars (Mahadevan et al. 2014; Quirrenbach et al. 2018; Reiners et al. 2018).

However, the RV precision one can actually achieve depends on the spectral content of a wavelength region, the number of photons you detect, but also on the presence of telluric features, which become a problem as one observes at longer wavelengths. CARMENES was the first NIR instrument that was dedicated to RV searches around M-dwarf stars (Reiners et al. 2018). It is a dual-arm high-resolution spectrograph with a resolving power $R \approx 80{,}000$. The visual channel (VIS) operates in the wavelength range λ = 520–960 nm, and the NIR channel covers the range λ = 960–1719 nm.

Reiners et al. (2018) investigated the RV precision as a function of wavelength for the full range of M-dwarf spectral types. For M0–M3 stars (left panel of Figure 3.16), the RV precision improves from 520 nm to 700 nm. At 700 nm, the TiO absorption band brings a lot of spectral information. In the range 700–900 nm, the RV error increases, but still produces good RV precision. At 760 nm, the TiO system at 770 nm starts to kick in. There is then a gradual increase in the RV error toward the longest wavelengths. So, for spectral types down to about M5, the RV information content in the VIS channel outperforms that in the NIR channel by a factor of 2.5.

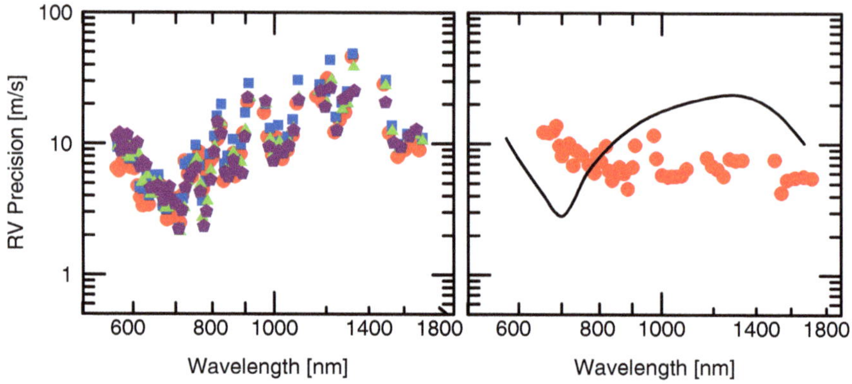

Figure 3.16. The empirical RV precision as a function of wavelength taken from individual spectral orders from the CARMENES spectrograph. (Left) The RV precision for M0 (dots), M1 (squares), M2 (triangles), and M3 (pentagons) dwarfs. (Right) The RV precision for M6+M7 (triangles) and M9 (dots) dwarfs. The curve is a spline fit through the M0–M3 dwarf data for comparison. (Data from Reiners et al. 2018.)

For late-type stars later than M6, the spectral energy distribution starts to play a larger role. RVs can still be computed down to 600 nm, but the poorer S/N due to the decreased flux is more than a factor of 2 worse, this results in $\sigma \approx 20$ m s^{-1}, compared to 3–7 m s^{-1} for the earlier spectral types. Because of the low flux level, the RV cannot be computed for M8 and M9 stars at wavelengths shorter than 600 nm. However, in the wavelength range $\lambda = 1000$–1600 nm, the RV performance is a factor of 2–3 higher than that for the M0–M3 stars ($\approx$6 m s^{-1} compared to 10–20 m s^{-1}). This is due to more spectral content at longer wavelengths but also more relative flux (higher S/N).

The take-home message is that for stars earlier than M5, the "sweet spot" for precise RV measurements is 600–900 nm. Suppose you want to design a spectrograph to measure precise RVs for a sample of M-dwarfs, but you only have access to a 4 m class telescope. Then a good strategy would be to build an optical high-resolution spectrograph that is optimized for the 600–900 nm region. The dwarf stars later than spectral type M5 will simply be too faint for the 4 m telescope. If on the other hand you have access to an 8 m class telescope and want to observe dwarf stars with spectral type later than M5, then a high-resolution NIR spectrograph is the instrument of choice (A. Reiners, 2018, private communication). In fact, the most efficient instrument would be a dual-arm visual and NIR spectrograph like CARMENES. In this way, high-quality RVs can be obtained for the full spectral range M0–M9.

References

Beatty, T. G., & Gaudi, B. S. 2015, PASP, 127, 1240

Bottom, M., Muirhead, P. S., Johnson, J. A., & Blake, C. H. 2013, PASP, 125, 240

Bouchy, F., Pepe, F., & Queloz, D. 2001, A&A, 374, 733

Butler, R. P., & Marcy, G. W. 1997, IAU Colloq. 161, Astronomical and Biochemical Origins and the Search for Life in the Universe, ed. C. B. Cosmovici, S. Bowyer, & D. Werthimer (Bologna: Editrice Compositori), 331

Christian, D. J., Pollacco, D. L., Skillen, I., et al. 2006, MNRAS, 372, 1117

Cochran, W. D., & Hatzes, A. P. 1993, in ASP Conf. Ser. 36, Planets Around Pulsars, ed. J. A. Phillips, S. E. Thorsett, & S. R. Kulkarni (San Francisco, CA: ASP), 267–73

Collier Cameron, A., Guenther, E., Smalley, B., et al. 2010, MNRAS, 407, 507

Döllinger, M. P., Hatzes, A. P., Pasquini, L., et al. 2007, A&A, 472, 649

Galland, F., Lagrange, A. M., Udry, S., et al. 2005, A&A, 443, 337

Glebocki, R., Gnacinski, P., & Stawikowski, A. 2000, AcA, 50, 509

Han, I., Lee, B. C., Kim, K. M., et al. 2010, A&A, 509, A24

Hartmann, M., & Hatzes, A. P. 2015, A&A, 582, A84

Hatzes, A. P., & Cochran, W. D. 1992, in European Southern Observatory Conf. and Workshop Proc., Vol. 40, ed. M.-H. Ulrich, 275

Hatzes, A. P., Cochran, W. D., Endl, M., et al. 2006, A&A, 457, 335

Hatzes, A. P., & Zechmeister, M. 2007, ApJ, 670, L37

Hatzes, A. P., Zechmeister, M., Matthews, J., et al. 2012, A&A, 543, A98

Hatzes, A.P. 2016, in Astrophysics and Space Science Library, Vol. 428, Methods of Detecting Exoplanets, ed. V. Bozza, L. Mancini, & A. Sozzetti (Berlin: Springer), 3

Hrudková, M., Hatzes, A., Karjalainen, R., et al. 2017, MNRAS, 464, 1018

Johnson, J. A., Fischer, D. A., Marcy, G. W., et al. 2007, ApJ, 665, 785

Kraft, R. P. 1967, ApJ, 150, 551

Lehmann, H., Guenther, E. W., Sebastian, D., Hartmann, M., & Mkrtichian, D. E. 2015, A&A, 578, L4

Mahadevan, S., Ramsey, L. W., Terrien, R., et al. 2014, Proc. SPIE., 9147, 91471G

Niedzielski, A., Goździewski, K., Wolszczan, A., et al. 2009, ApJ, 693, 276

Queloz, D., Mayor, M., Sivan, J. P., et al. 1998, in ASP Conf. Ser. 134, Brown Dwarfs and Extrasolar Planets, ed. R. Rebolo, E. L. Martin, M. R. Zapatero Osorio, et al. (San Francisco, CA: ASP), 324

Quirrenbach, A., Reffert, S., Trifonov, T., Bergmann, C., & Schwab, C. 2015, AAS/ESS 3, 502.01

Quirrenbach, A., Amado, P. J., Ribas, I., et al. 2018, Proc. SPIE, 10702, 107020W

Reffert, S., Quirrenbach, A., Mitchell, D. S., et al. 2006, ApJ, 652, 661

Reiners, A., Zechmeister, M., Caballero, J. A., et al. 2018, A&A, 612, A49

Sato, B., Izumiura, H., Toyota, E., et al. 2008, PASJ, 60, 539

Setiawan, J., Pasquini, L., da Silva, L., von der Lühe, O., & Hatzes, A. 2003, A&A, 397, 1151

Wittenmyer, R. A., Endl, M., Wang, L., et al. 2011, ApJ, 743, 184

Chapter 4

Simultaneous Wavelength Calibration

You are an eager exoplanet hunter, and you want to find exoplanets using the radial velocity (RV) method. Taking everything you have learned from the previous chapter, you design a spectrograph with enough resolution, wavelength coverage, etc. to achieve an RV precision of, say $10\,\mathrm{m\ s^{-1}}$. You prepare a target list of suitable stars, i.e., bright, late-type, and slowly rotating stars. You start making measurements, and after a time, you realize that the scatter of your measurements, even for stable stars, is far worse than your estimated uncertainty. What went wrong? Most likely, instrumental shifts have introduced an unwanted and large source of errors.

A CCD detector only records the intensity of light as a function of pixel location. To measure a Doppler shift, you need to know the intensity of light as a function of wavelength. Thus, you have to put a wavelength scale on your spectrum using a suitable calibration source. A good wavelength calibrator should have a high density of spectral features with measured wavelengths well spread across the spectral range of your Doppler measurements. The more features you have, the better the mapping between pixel location and wavelength will be.

But that is not the whole story. We have seen in Chapter 3 that a Doppler shift is a tiny displacement on your detector. It does not take much of a mechanical shift to mimic a shift of the spectral line. The problem is that you most likely observed the calibration source at a different time than when you made your stellar observation. Furthermore, the light from the hollow cathode lamp always goes through a different optical path than your starlight, and this might introduce a systematic error. If you are getting large uncertainties in your Doppler measurements, much higher than is predicted by photon statistics, then the likely cause is instrumental shifts.

If you want to minimize the effects of instrumental shifts on your Doppler measurement, it is essential that you observe your wavelength calibration at the same time as your stellar observation. Unless your spectrograph is extremely stable, there is a good chance that something has moved—optical elements, the detector,

doi:10.1088/2514-3433/ab46a3ch4

etc.—between the time you observed your star and the time you observed your calibrator.

In this chapter, we examine the various methods of simultaneous wavelength calibration, both historic and modern, that have been employed to minimize the effects of instrumental shifts on the RV measurement.

4.1 Instrumental Shifts

Instrumental shifts occur because the traditional way of performing wavelength calibration is to observe a calibration source either before or after your science observation. At optical wavelengths, this is typically a thorium–argon (Th–Ar) hollow cathode lamp (see below). In the data reduction process, you identify emission lines with wavelengths that have been measured in the laboratory. A fit to the pixel versus wavelengths of these lines using a high-order polynomial provides the mapping between pixel and wavelength space. This function is then applied to the stellar spectrum.

The problem is that the stellar spectrum is taken at a different time to that of the calibration source. Mechanical shifts of the spectrograph and detector or temperature and pressure changes in the spectrograph room can occur between the time you take your calibration and that of the stellar observations. Table 3.1 shows that for an R = 100,000 spectrograph, a Doppler shift of 1 m s^{-1} will cause a physical shift at 10^{-5} cm at the detector, or about one-fifth of the wavelength of the incoming light. It does not take much of an instrumental shift of the spectrograph or detector to obliterate this signal. These instrumental shifts can dominate the measurement error due to simple photon statistics.

Let's look into the instrumental shifts for modern spectrographs using measurements taken with the Tautenburg Coudé Echelle spectrograph (TCES). The TCES is an example of a modern, well-designed echelle spectrograph with good stability for standard spectroscopic work. The spectrograph has a resolving power of R = 67,000. Cross-dispersion is accomplished using three grisms so as to access the full wavelength region between 3000 Å to 10,000 Å. The spectrograph has only one part that is moved on a regular basis and that is the translation slide for the grisms. The focus of the instrument is also stable. As of this writing, the spectrograph has not had to be focused since the installation of the new echelle grating and CCD detector in 2014.

The spectrograph is a "classic" instrument in the sense that it resides in a room one floor below the telescope. The room is not thermally stabilized, but because it is built "in the ground," temperature variations throughout the year never exceed a few tenths of degrees Celsius. The floor of the spectrograph room is decoupled from the telescope dome so as to minimize mechanical vibrations. In short, the TCES is as good a stable spectrograph you can have using a traditional design for a high-resolution spectrograph.

It is of interest to calculate the expected "best case" RV precision of the TCES. The best way to do this is by taking what we learned in the previous chapter and scaling to the performance of a known stable spectrograph. For this example, we will take HARPS, which has a best RV precision of ≈1 m s^{-1} for a signal-to-noise ratio, S/N = 200. For our measurements, we will be using flat lamp exposures with S/N ≈ 400. This will give us a factor of 2 lower error over HARPS due to the higher S/N. The resolving power of HARPS is R = 110,000. Taking our "nominal" relationship $\sigma = R^{-1.2}$, the TCES should have approximately twice the error just from the decreased resolution. The usable wavelength of HARPS is ~2000 Å. For our experiment, we will be using single spectral orders, which have a wavelength coverage of ≈80 Å. We thus have a factor of 5 increased measurement error due to the wavelength coverage. So, for the measurements we describe below, we expect a $\sigma \approx 5$ m s^{-1} for the TCES.

In 2018 the Tautenburg observing school measured the instrumental shifts of a white-light source taken through a glass cell filled with molecular iodine (a technique we will discuss below). As we shall shortly see, molecular iodine has absorption features that roughly span the wavelength region 5000–6000 Å. For these observations, the CCD detector covered 50 spectral orders in the range 4520–7550 Å. Pixel shifts of the molecular iodine absorption lines were calculated using spectral lines from spectral orders m = 112 (5080 Å) to m = 129 (5990 Å) or covering roughly the center one-third of the detector. These pixel shifts were then converted to an equivalent Doppler shift using the mean dispersion of the stellar order.

The left panels of Figure 4.1 show the instrumental shifts for three representative spectral orders. The shifts are relatively constant for the first four hours, after which they show an overall increase of about 40 m s^{-1} hr^{-1}. Over the course of 9 hr, if you were measuring the location of a spectral line on your detector, it would have shifted

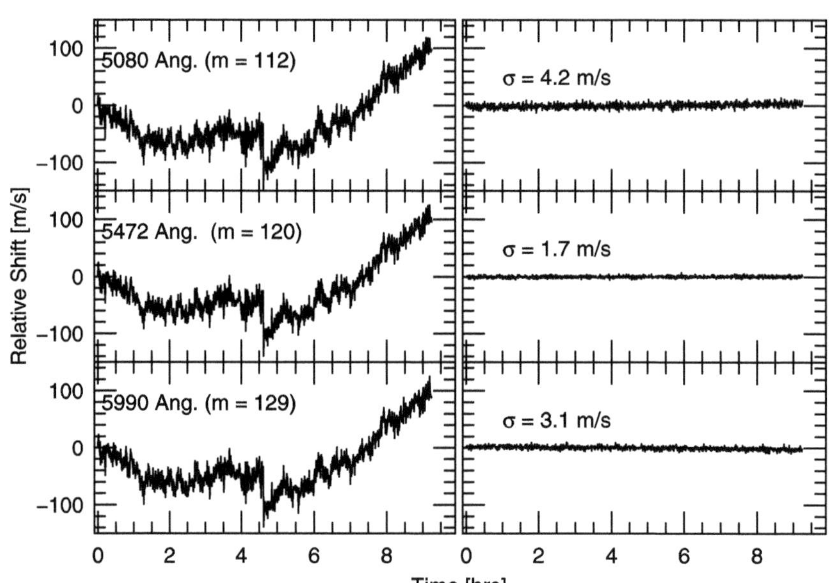

Figure 4.1. (Left) Instrumental shifts of the Tautenburg Coudé Echelle spectrograph. Shown are three wavelength regions (5080 Å, 5472 Å, and 5990 Å) corresponding to relative spectral orders m = 112, 120, and 129, respectively. The residual velocity shifts after subtracting the average of the pixel shifts of all orders 112 through 129 and then converting this to a velocity. The rms scatter is below 5 m s^{-1} in all orders.

by the equivalent of 200 m s^{-1} just due to the mechanical and thermal shifts of your instrument. This is a factor of about 50 larger than our expectations. These long-term shifts are most likely due to slow temperature and pressure changes in the spectrograph room during the course of the observations.

Figure 4.2 shows that these instrumental shifts can occur even on rather short timescales. The shifts can be as large as $\Delta V \approx 40$ m s^{-1} over a few minutes—disastrous if you want to detect an exoplanet with a velocity amplitude of 10 m s^{-1}. These short-term shifts are most likely due to vibrations and mechanical shifts of optical components and the detector.

Remarkably, the shifts in each spectral order appear to track each other extremely well. The right panels of Figure 4.1 show the residual instrumental shifts after subtracting the mean pixel shift calculated from 20 spectral orders. The shifts of all orders show no trend, and the rms scatter of $\sigma = 2-4$ m s^{-1}, i.e., much closer to our expectations.

If you want to make precise RV measurements with a spectrograph, it would be instructive to perform time series measurements like those shown in Figure 4.1. These will establish the timescales and magnitude of instrumental shifts and will point to ways to improve the RV stability (e.g., better thermal and mechanical stability).

This experiment demonstrates that the largest instrumental shifts are "global" ones at the detector. One can use pixel shifts measured for one spectral order to correct for those in adjacent orders where you cannot measure instrumental shifts and thus achieve a substantial improvement in the RV precision. This is important

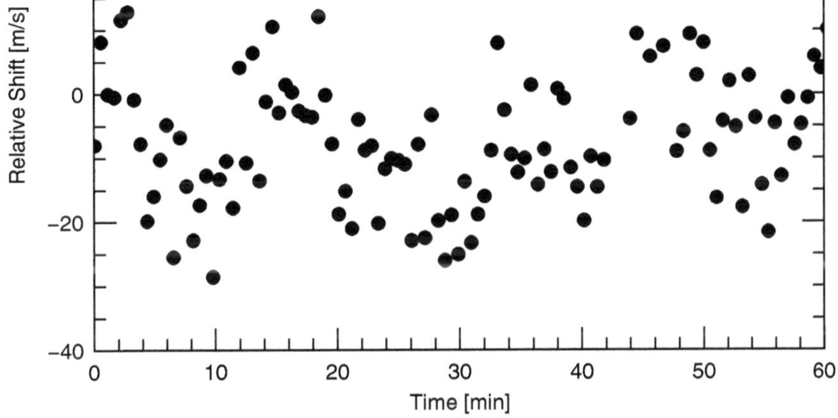

Figure 4.2. Short-term instrumental shifts of the TCES. These can be as high as ±40 m s^{-1} on timescales of minutes.

for such techniques as the telluric method (see below), where the wavelength calibration features only appear in one spectral order.

There are a number of "moving parts" in an astronomical observatory. The telescope moves, and the dome rotates during the course of the observations. There could be support staff moving about the facility closing doors, moving machinery, etc. Some locations are seismically active, and earthquakes that are not felt but can still result in large instrumental shifts. It is of interest to see possible origins of large instrumental shifts.

As an experiment during the time series of iodine cell measurements, the students of the Tautenburg observing school moved the telescope dome (65 tons), the telescope (25 tons), and for good measure, jumped up and down outside the spectrograph room. The resulting shifts are shown in Figure 4.3. As expected, moving the dome results large shifts of up to 100 $m\,s^{-1}$. This is a large, heavy structure that creates considerable vibrations when it moves. You should make sure that the dome does not rotate during the exposure.

The rapid motion of the telescope seems to have minimal impact on the measured shifts. This is also expected as telescopes are well-balanced structures, and their motion is smooth and accompanied by little vibration. One barely notices when a telescope moves, but not so for the dome. Fortunately, mischievous students jumping up and down outside the spectrograph room caused no greater shifts of the instrument than the times they were better behaved.

If you design a spectrograph specifically for RV stability, you can minimize the magnitude of these instrumental shifts considerably. This is the strategy behind

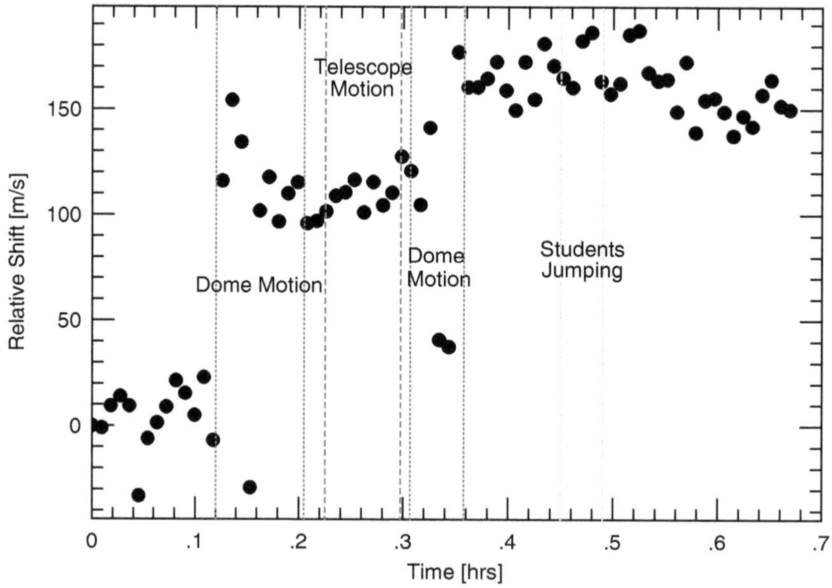

Figure 4.3. Instrumental shift caused by dome (between dotted red lines) and telescope movement (between blue dashed lines), and students jumping up and down outside the spectrograph room (between the green dashed–dotted line).

Figure 4.4. The vacuum tank that encloses the HARPS spectrograph. The vessel is open so that one can see the optical bench and the echelle grating. Photo credit: European Southern Observatory.

the design of the High Accuracy Radial velocity Planetary Searcher (HARPS) spectrograph (Mayor et al. 2003) on the 3.6 m telescope of the European Southern Observatory at La Silla, Chile. The entire spectrograph sits in an evacuated tank that is temperature and pressure stabilized. Care has been taken to minimize mechanical vibrations. Unlike the TCES, which is a classic slit-fed spectrograph, HARPS is fed with two optical fibers, one for the stellar observation and another for the calibration source. Figure 4.4 shows the vacuum tank and echelle grating for HARPS.

Figure 4.5 shows the instrumental shifts of HARPS measured over a comparable time span to the TCES measurements. For HARPS, the shifts were calculated using emission lines from a thorium–argon lamp. The shifts shown for two fibers A and B (a science fiber and a calibration fiber when performing real observations) have a peak-to-peak variation of about 1 m s^{-1}. If one takes the difference (i.e., one is tracking the instrumental shifts), one gets an rms scatter of only 10 cm s^{-1}. This stability comes with a price as the HARPS spectrograph costs a factor of 20 or more than TCES.

4.2 Hollow Cathode Lamps

Hollow cathode lamps (HCLs) are frequently employed as a wavelength reference. An HCL consists of a glass tube containing a cathode, an anode, and a buffer gas (Figure 4.6). A large voltage is applied across the anode and cathode, which ionizes the buffer gas. These ions are then accelerated to the cathode, where they dislodge atoms from the cathode element. Both the buffer gas and the sputtering cathode atoms will be excited by collisions and will emit photons from electrons decaying to lower energy states. The HCL will thus produce emission lines from both the buffer gas and the sputtering cathode atoms.

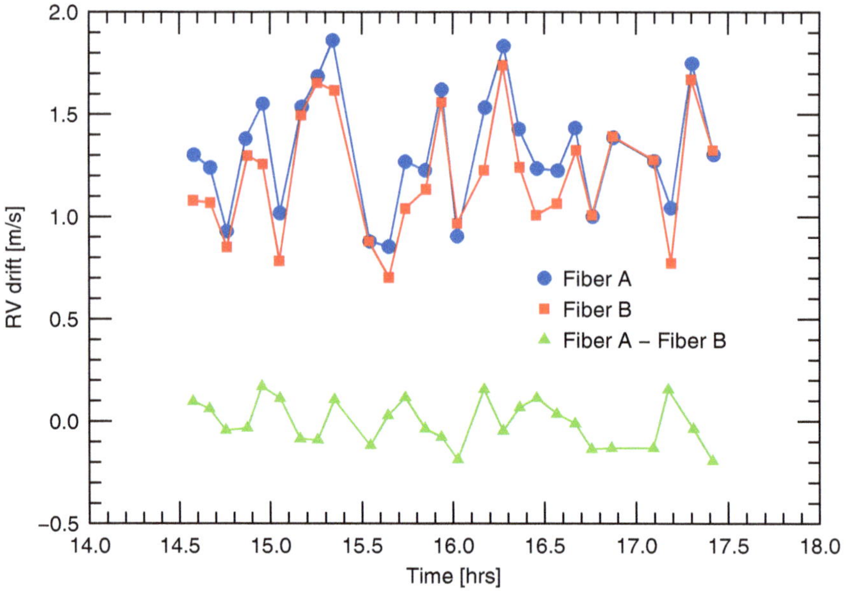

Figure 4.5. Instrumental shifts of the HARPS spectrograph. Shown are the drifts in the stellar fiber (A) and the simultaneous Th–Ar reference (fiber B). The lower curve is the difference, which has a scatter of 10 cm s^{-1} (from Mayor et al. 2003).

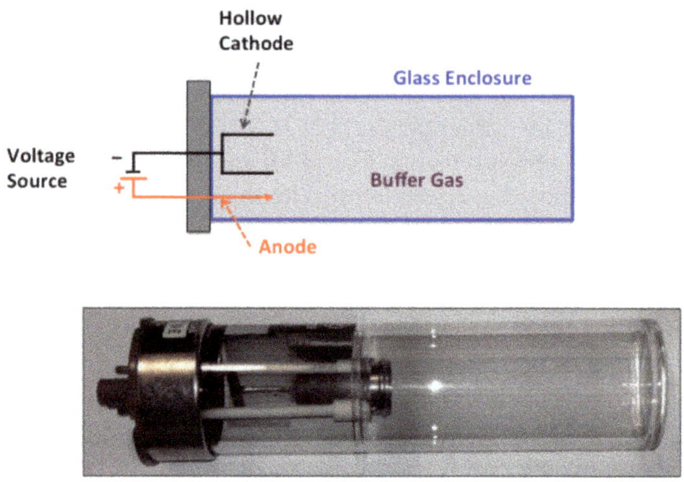

Figure 4.6. (Top) A schematic of a hollow cathode lamp. (Bottom) Photo of an actual hollow cathode lamp.

Cathodes that have been employed for HCLs include iron, thorium, aluminum, sodium, and uranium to name a few. The buffer gas is almost always a noble gas with argon and neon the most commonly used. The choice of cathode depends on the wavelength range and density of emission lines one needs.

4.2.1 Th–Ar

The "traditional" HCL for astronomical purposes is Th–Ar, and these have been used for many decades as a wavelength calibration source. Thorium makes an excellent cathode as it has a single isotope and a dense spectrum of narrow spectral lines covering the visible wavelengths (Figure 4.7).[1]

To provide simultaneous calibration with Th–Ar and in order to minimize the instrumental shifts, you have to record your Th–Ar spectrum at the same time as your stellar one. The development of optical fibers allowed astronomers to do this. You simply use one fiber optic to feed light from the star into the spectrograph, and a second to feed light from a calibration lamp. The calibration spectrum is thus recorded on the CCD detector adjacent to the stellar spectrum, so any instrumental shifts will affect both equally. One of the first instruments to do this was the ELODIE spectrograph (Baranne et al. 1996), the spectrograph that discovered the planet to 51 Peg. Currently, this technique is best exemplified by the HARPS spectrograph (Mayor et al. 2003; Pepe et al. 2000).

Figure 4.8 shows a stellar spectrum recorded using the simultaneous Th–Ar technique. This spectrum was recorded using the HARPS spectrograph of ESO's 3.6 m telescope at La Silla. The continuous bands represent the stellar spectrum. In between these one can see the emission spectrum from the Th–Ar fiber. If you look carefully you will see that the images of the thorium emission lines have a circular shape because they are an image of the fiber.

The Pros and Cons of Th–Ar

The use of traditional Th–Ar HCLs has a number of advantages when used as a wavelength calibration.

Advantages of Th–Ar:

1. *Ease of use.*

   Th–Ar hollow cathode lamps are standard calibration sources that have been employed by astronomers for the past several decades. Over the years,

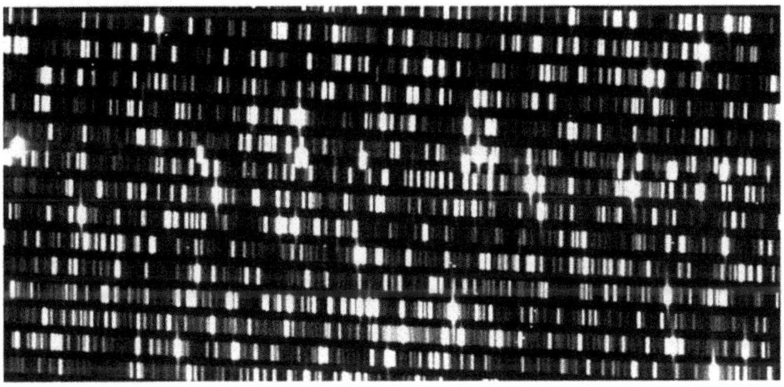

Figure 4.7. An echelle spectrum of a Th–Ar hollow cathode lamp.

[1] Astronomers typically use Th–Ar in the wavelength range ≈3000–8000Å.

Figure 4.8. A spectrum recorded with the HARPS spectrograph. The solid bands are from the star fiber. The emission-line spectrum of Th–Ar above the stellar one comes from the calibration fiber. Note the contamination of the stellar spectrum from strong Th lines in the lower left and upper center.

considerable effort has been made to identify the wavelengths of emission features in these lamps. Computation of the RVs is also straightforward. Once you have a spectrum of intensity versus wavelength, there are a number of standard data reduction packages for calculating the RV. The details of this will be left to the next chapter.

2. *Large wavelength coverage for calibration.*

   In order to have a good wavelength calibration, one needs a high density of emission lines with known wavelengths. Thorium has useful emission lines that cover fairly well the optical region 4000–7000 Å. Figure 4.9 shows the number density of thorium emission lines in the wavelength range 3000–10,000 Å. It has roughly six to eight emission lines per 10 Å out to about 6500 Å. After approximately 6500 Å, the number density of thorium emission lines drops by a factor of 2–3. Therefore, Th–Ar hollow cathode lamps are not well suited for wavelength calibration in the near-infrared—there are just too few emission lines.

3. *No loss of light or contamination of stellar spectra.*

   Other methods, to be discussed shortly, require putting a gas absorption cell into the optical path in order to provide the wavelength calibration. This not only results in a loss of light, but it also contaminates the stellar spectrum. One can therefore use the stellar spectra for other analyses such as abundance or line-shape studies.

Disadvantages of Th–Ar:

Despite its long use as a wavelength calibrator, Th–Ar does have some important drawbacks when performing precise stellar RV measurements.

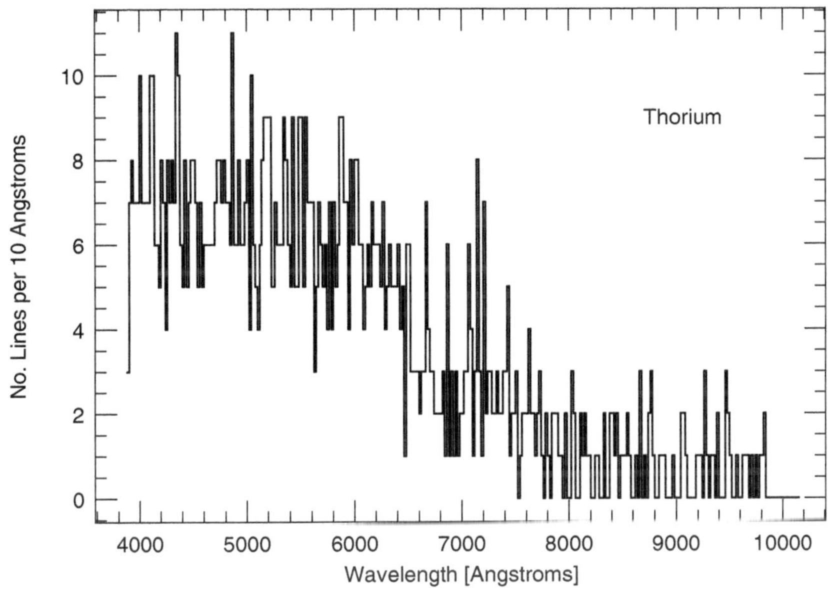

Figure 4.9. The number density of thorium emission lines with identified wavelengths per 10 Å as a function of wavelength.

1. *Active devices and aging.*

   Th–Ar hollow cathode lamps are what we call "active" devices when it comes to wavelength calibration. These lamps are devices where you have to apply a high voltage to the lamp to get emission lines. Slight changes in the voltage may result in changes in the emission spectrum of the lamp. The danger is that over time, your wavelength calibration will be slowly changing.

   Because they are active devices, Th–Ar lamps will age and change their spectrum they produce. Figure 4.10 shows two exposures taken in 1995 and 2005 of the same Th–Ar hollow cathode lamp used at McDonald Observatory. After seven years of use, the emission spectrum has noticeably changed. If one is interested in precise RV measurements at the submeter per second, then even slight changes in the relative intensity of thorium lines may introduce significant systematic errors.

   Due to aging, Th–Ar hollow cathode lamps are not recommended if one is interested in long-term RV stability over several decades.

2. *Failure of devices.*

   HCLs do not live forever; at some point they will fail. In fact, the spectrum in the lower panel of Figure 4.10 was one of the last spectra taken with this lamp before failing. Broken lamps need to be replaced, and one cannot guarantee that the new lamp will have the same spectrum. This almost certainly will introduce instrumental offsets of the new data compared to those taken with the previous lamp. As far as precise RV measurements are concerned, it will be like making measurements with a

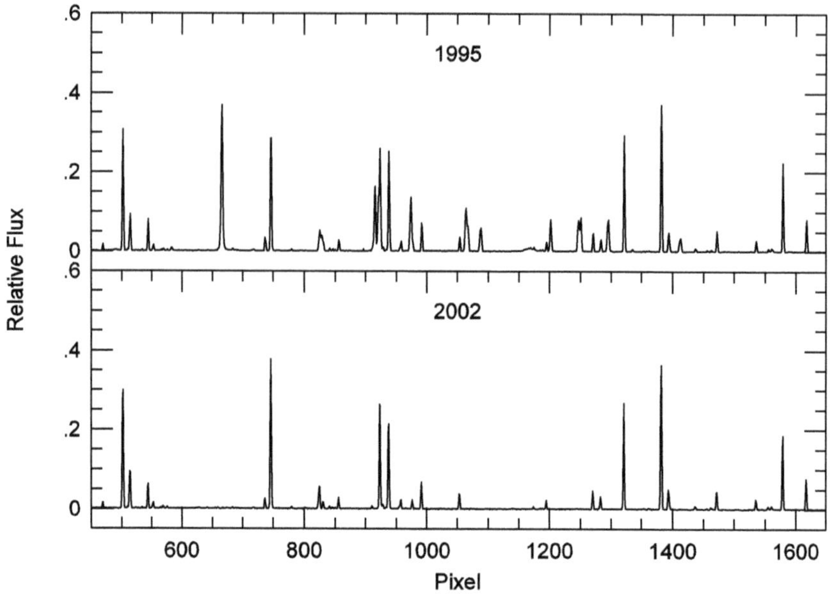

Figure 4.10. A Th–Ar lamp used at McDonald Observatory. (Top) An exposure from 1995. (Bottom) The same lamp as it appeared in 2002.

completely new instrument, one where you have to compute relative offsets between the old and new data.

One can attempt to anticipate a failure of a lamp by taking calibrations with both the old and the "future" lamp so that one can match up RV data over a long time span, but some level of systematic errors may always be present. In short, if you want to perform precise RV measurements over a long time span, say decades, Th–Ar calibration may not be the best option.

3. *Contamination and cross-talk.*

   If you are using simultaneous Th–Ar, strong lines may spill light into the star fiber and contaminate its spectrum. This can be seen in Figure 4.8. This contamination is not easy to model out.

   A larger problem is the emission lines from the buffer gas, typically argon or neon (see Figure 4.11). These lines are generally present beyond 6700 Å, and they are very strong lines which are saturated for most exposures that require good signal in the thorium lines. The saturated lines spill over several spectral orders and are a source of contamination.

4. *Wavelength calibration not in situ.*

   You will notice in Figure 4.8 that the Th–Ar fiber, even though it is taken at the same time as the stellar spectrum, is not recorded at the same place on the detector as the stellar spectrum. Thus, the wavelength calibration is not done in situ to the stellar spectrum, but rather adjacent to it. One has to have faith that the same wavelength calibration applies to the slightly different region of the detector.

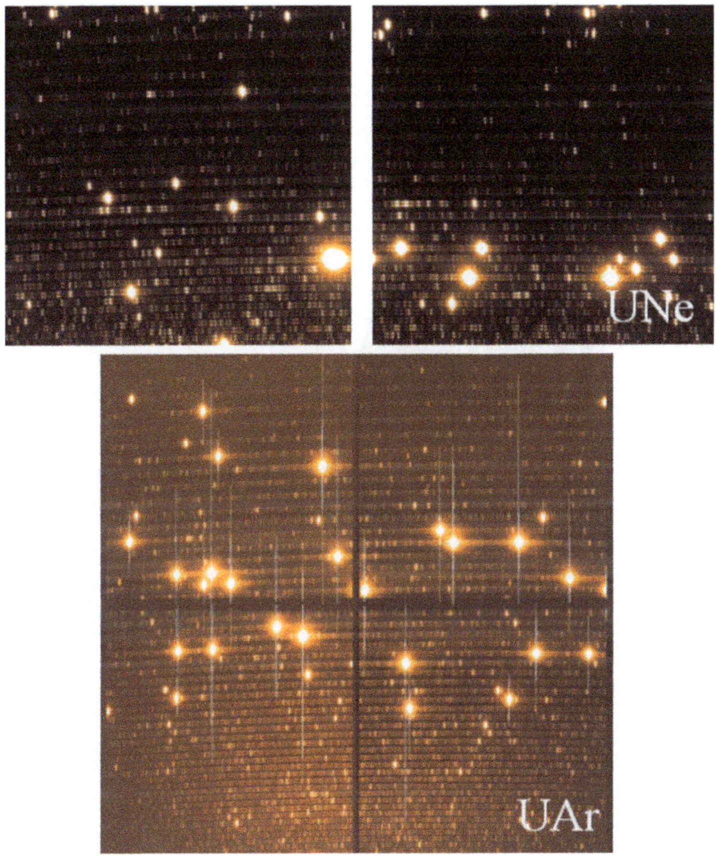

Figure 4.11. (Top) A spectrum of a uranium–neon HCL taken through the visual channel (VIS) of the CARMENES spectrograph and recorded on two detectors. (Bottom) A spectrum of uranium–argon recorded on the mosaic (four detectors) of the near-infrared (NIR) channel of CARMENES. The strong, saturated lines are due to the buffer gas in the HCL (neon and argon). (Image courtesy of Mathias Zechmeister and the CARMENES consortium.)

5. *No modeling of the instrumental profile.*

   In Chapter 6, we will discuss at length the instrumental profile and how it can affect the RV precision of your measurement. For now, suffice it to say that in order to improve on your precision, it is important to model any changes in the instrumental profile. This cannot be done with Th–Ar calibration as it requires features that are unresolved by the spectrograph that you use. Thorium emission lines are intrinsically broader than the resolution of the spectrograph.
6. *Availability of Th–Ar lamps.*

   As of this writing, the future of Th–Ar hollow cathode lamps as a calibration source is uncertain. Manufacturers are having difficulties in finding the thorium wires needed for fabrication and have thus ceased producing Th–Ar lamps. Other companies have taken on manufacturing

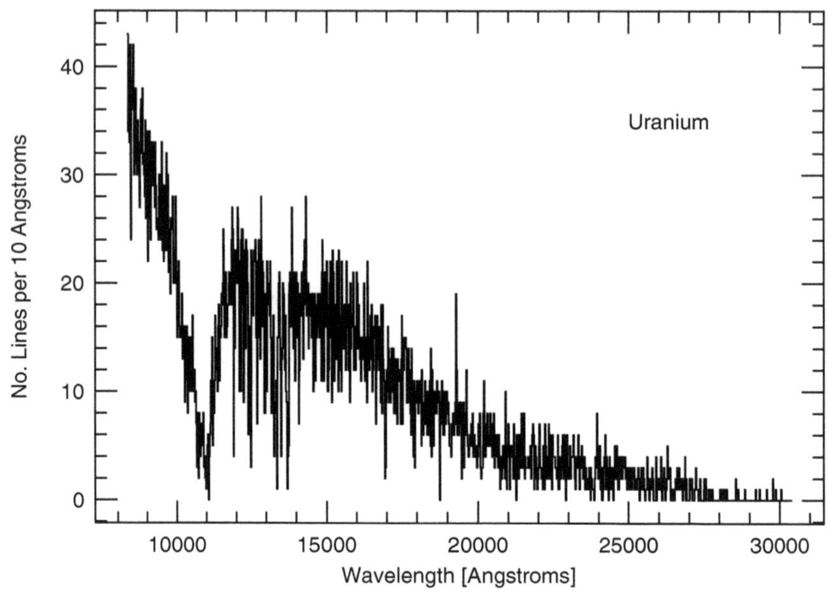

Figure 4.12. The number density of uranium emission lines with identified wavelengths per 10 Å as a function of wavelength.

Th–Ar lamps, but these have not had the quality for precise RV work. Complicating matters. the European Union has regulations on the handling of thorium and have classified it as as a "nuclear" material. Astronomers may thus be forced to choose other means for "standard" wavelength calibration.

4.2.2 HCL in the Infrared

As we have seen, thorium has too sparse a line density beyond about 8000 Å to be an effective wavelength calibrator at infrared (IR) and near-infrared (NIR) wavelengths (>1 μm). An alternative is to use uranium lamps, which have a much higher line density out to about 2 μm. Figure 4.11 shows spectra of a hollow uranium HCL (Ne and Ar) used by the CARMENES spectrograph. Note the strong, saturated lines due to neon and argon. Figure 4.12 shows the number density of uranium emission lines from 8000–30,000 Å using the line list from Redman et al. (2011).

4.3 The Telluric Method

Griffin & Griffin (1973) were one of the first to realize that instrumental shifts, and in particular the fact that the calibration source went through a different optical path than the stellar light, was the limiting factor for the Doppler precision. They proposed the simple solution of using telluric and O_2 features at 6300 Å (Figure 4.13). This method, but not always using the same telluric features, is used to this day. It provides an easy way to improve the RV precision by providing a

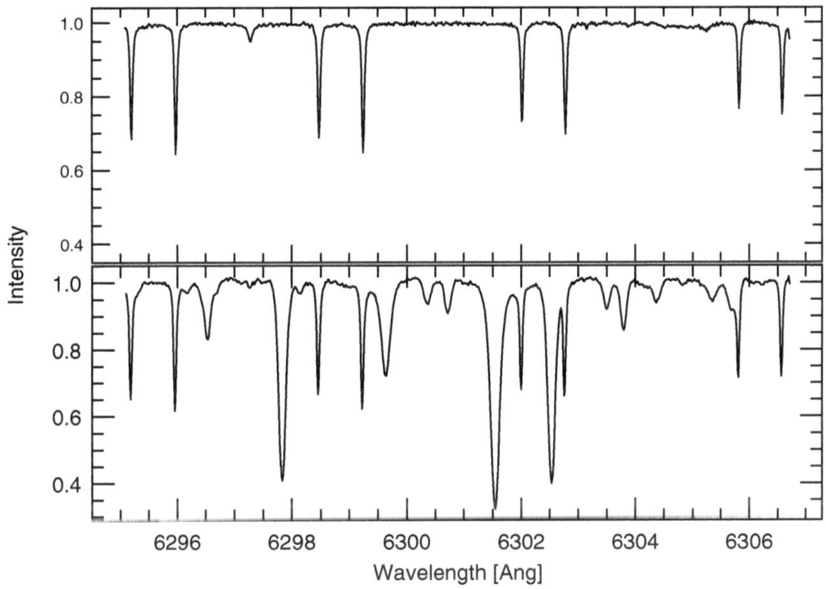

Figure 4.13. (Top) The oxygen bands at 6300 Å obtained by observing a hot A-type star. The star has no spectral features in this range. (Bottom) A spectrum of the Sun-like star μ Her. Doppler shifts of the stellar lines are measured with respect to the telluric features.

rough correction to the instrumental shifts because these will affect the telluric and spectral lines in the same way.

Griffin & Griffin (1973) suggested that an RV precision of 15–20 m s^{-1} was possible with this method, and this is largely the case. Cochran et al. (1991) used this method to confirm the giant planet candidate around HD 114762 discovered by Latham et al. (1989). Figure 4.14 shows the RV data and orbital solution for HD 114762b using the telluric method. Table 4.1 compares the orbital solutions (see Chapter 8 for the definition of the parameters) from the telluric and more "traditional" RV measurements. The telluric method yields a factor of 10 improvement in the RV precision, which means that you can get more accurate orbital parameters with 1/10th of the measurements. Only the orbital period determined from the traditional measurements has a smaller error, but that is due primarily to the longer time span for these measurements.

This method can also be extended to other wavelengths, for example, the oxygen *A* bands (Figure 4.15) at 6860–6930 Å or the *B* band (see Guenther & Wuchterl 2003).

Although the method is simple and inexpensive to implement, it has the large disadvantage in that it covers a rather limited wavelength range. There is also the problem that one cannot control Earth's atmosphere. Pressure and temperature changes, as well as winds in Earth's atmosphere ultimately limit the measurement precision. It is probably difficult to achieve an RV precision better than about 20 m s^{-1} with this method. However, if one does not have other options, it is still a convenient method if one wants to get a substantial improvement in the RV measurement error over traditional calibration methods.

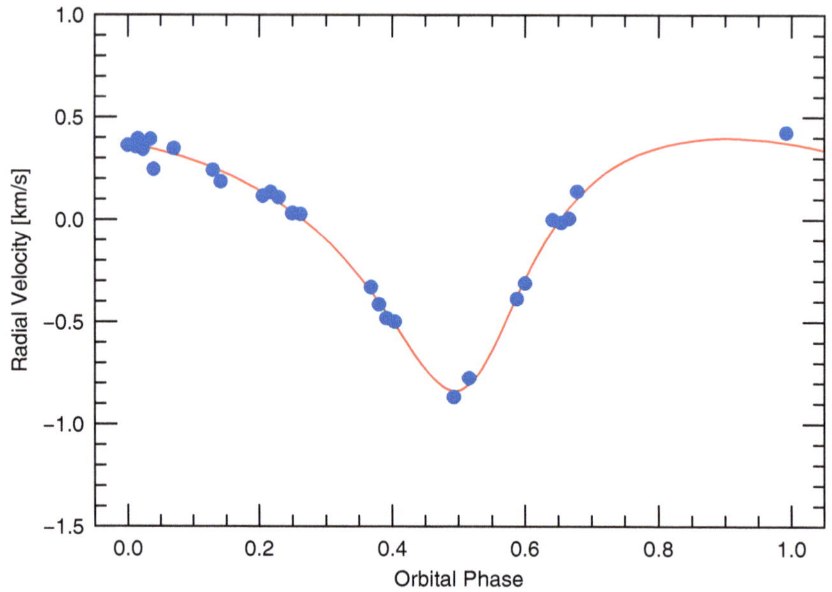

Figure 4.14. The orbital radial velocity variations (red line) of HD 114762 caused by its planetary companion. The blue points are measurements using the telluric method (Cochran et al. 1991).

Table 4.1. Orbital Solutions for HD 114762b

Parameter	Traditional	Telluric
N	280	28
σ (m/s)	412.7	34.4
Period	84.05 ± 0.08	83.91 ± 0.13
K (m/s)	565.3 ± 35.6	617.0 ± 16.5
Eccentricity	0.253 ± 0.064	0.380 ± 0.015
ω	234.6 ± 14.9	199.8 ± 4.4
T_0	2,447,380.5 ± 14.9	2,447,371.6 ± 4.4

4.4 Gas Absorption Cells

An improvement on the telluric method could be achieved if you could eliminate the systematic errors caused by the variations in Earth's atmosphere by controlling the absorbing gas. This is the principle behind the gas absorption cell, which in a sense is the "laboratory" extension of the telluric method. The idea is simple, one takes a chemical gas that produces absorption lines not found in either the stellar spectrum or Earth's atmosphere and fills it in a glass cell. This is then permanently sealed and operated at a fixed temperature and pressure. When used, it is placed in the optical path of the telescope, generally before the entrance slit or fiber to the spectrograph. As the starlight passes through the cell, the absorbing gas will impose a set of

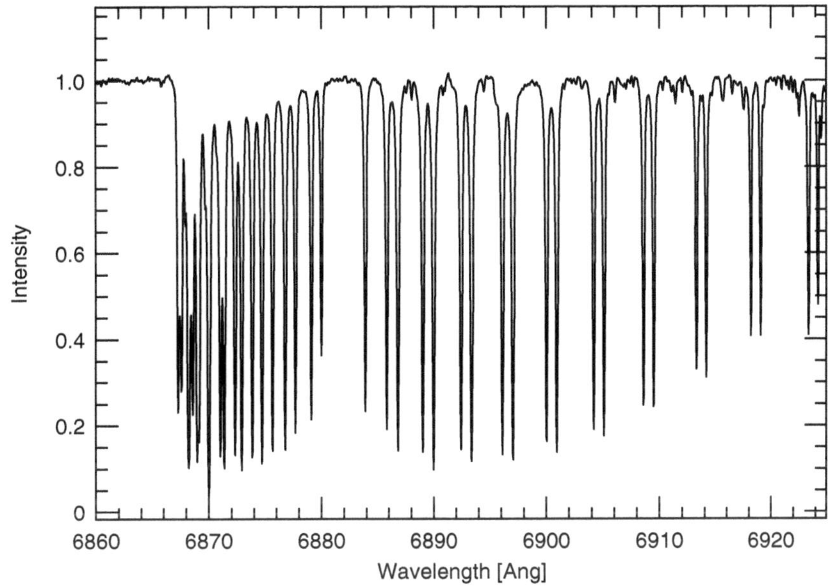

Figure 4.15. The oxygen A band near 6900 Å.

absorption features on top of the stellar spectrum. As in the telluric method, Doppler shifts of the stellar lines are then measured with respect to the gas cell features.

4.4.1 The Hydrogen Fluoride Cell

In their pioneering work, Campbell & Walker (1979) first used a gas cell for planet detection with precise RV measurements. Their choice of absorbing gas was hydrogen fluoride (HF), which has the 3–0 band R branch at 8670–8770 Å (Figure 4.16). These sharp sets of HF absorption lines were used to provide the velocity metric. Figure 4.17 shows an image of the HF cell before the entrance slit to the spectrograph. With this method, Campbell & Walker (1979) achieved an RV precision of 13 m s^{-1} in 1979. Keep in mind that this was the same RV precision achieved in the discovery of 51 Peg 15 years later (Mayor & Queloz 1995). Although the survey is considered not to have found any exoplanets, this is not strictly the case. It uncovered the RV variations due to the giant companion to the primary of the binary star γ Cep A (Campbell et al. 1988). Regrettably, Walker et al. (1992) later attributed the RV variations to rotational modulation from the star. Hatzes et al. (2003) ultimately demonstrated that the presence of a planetary companion was the cause of the RV variations about the binary orbit. The HF program also found the first hints of the planet around ε Eri (Campbell et al. 1988), which also proved to be due to a companion (Hatzes et al. 2000).

Although the HF method was able to achieve good results and was capable of detecting exoplanets, it suffered from several drawbacks:

1. The absorption features of HF only covered about 100 Å, so relatively few spectral lines could be used for the Doppler measurement.

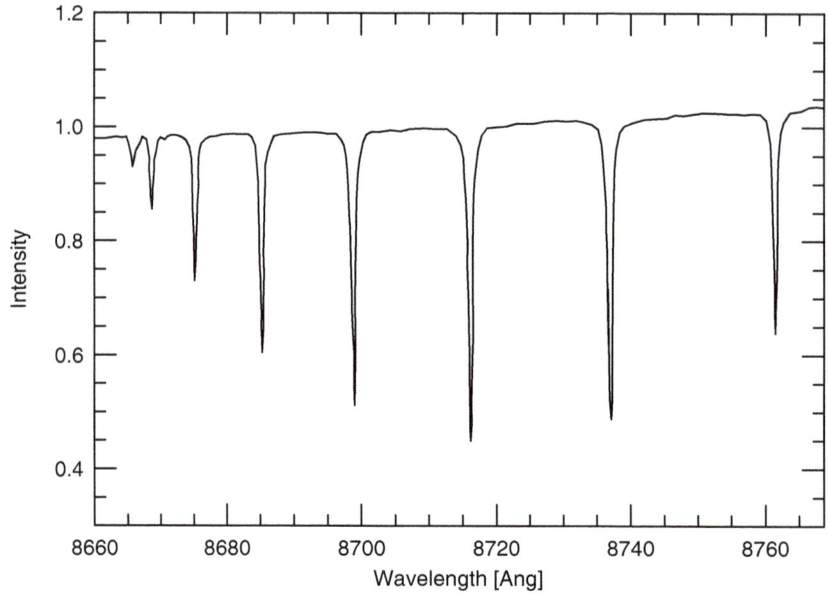

Figure 4.16. Absorption spectrum of the hydrogen fluoride 3–0 band R branch.

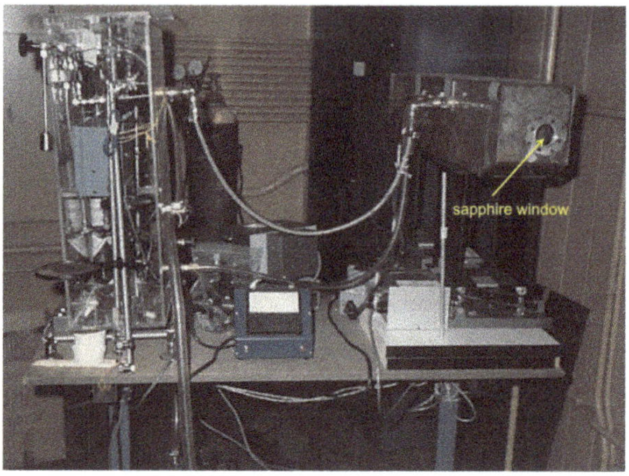

Figure 4.17. The HF cell in the lab. To the right is the cell, which is 1 m in length. The sapphire entrance window is marked. To the left are the equipment for temperature control and monitoring the pressure of the device. (Photo courtesy of Gordon Walker.)

2. HF is sensitive to pressure shifts.
3. To produce suitable HF absorption lines, a large path length ($\approx$1 m) for the cell was required. This could present problems if your spectrograph has space restrictions.
4. The cell has to be filled for each observing run because HF is highly corrosive and would damage the cell. Because HF dissolves glass, special sapphire windows had to be employed.

5. HF is a toxic chemical and represents a health hazard. In order to reduce the risk of injury to personnel when the Walker team conducted observations, during every exposure, personnel were evacuated from the telescope dome except for the astronomer chosen to wait for the exposure to end (G. A. H. Walker, 2003, private communication).

In Chapter 1, we saw examples of RV data taken with the HF cell.

4.4.2 The Iodine Absorption Cell

A safer alternative to the HF is to use molecular iodine (I_2) as the absorbing gas. The use of iodine gas cells as a wavelength reference for precise RV measurement was first proposed by Beckers (1977) for solar observations. The application to stellar work was pioneered by Marcy & Butler (1992). It has become a popular and cost-effective way to have simultaneous wavelength calibration on high-resolution spectrographs that were not necessarily built for precise stellar RV work.

Molecular iodine has a rich spectrum of narrow absorption lines roughly in the wavelength range 4000–6000 Å, which is a factor of 10 larger than that provided by an HF cell. Figure 4.18 shows a spectrum of iodine taken by observing a continuum source through an iodine cell. Note the rich forest of molecular iodine lines.

Advantages of the Iodine Cell:

1. *Compact and ease of use.*

   The device is very compact (Figure 4.19). A typical path length for an I_2 cell is about 10 cm, but cells have been constructed with lengths of 4–15 cm. This is in contrast to the HF cell, which requires a path length of 1 m to produce sufficiently strong HF lines. The main criterion for the length is that the cell should easily fit in front of the entrance slit of most spectrographs.

   Iodine cells are easy to construct (see Chapter 6) and are operated at modest temperatures of 50°C to 80°C. It is a simple matter to slide a cell in and out of the light path. No additional optics (a fiber feed for instance) are needed. Once an iodine cell has been constructed, virtually no maintenance is required to keep in operating. In short, iodine absorption cells are "turnkey devices" and can be put in use immediately.

Figure 4.18. A spectrum of molecular iodine obtained by observing a white-light source through an iodine absorption cell.

Figure 4.19. (Left) An iodine absorption cell. (Right) An iodine absorption cell shown in front of the dome of ESO's 3.6 m telescope. The housing is for insulation and temperature control. Note the purplish color of the gas. The housing of the cell is for temperature stabilization.

2. *Insensitive to pressure and temperature changes.*

   Unlike HF, molecular iodine is relatively insensitive to temperature and pressure changes. Changing the operating temperature of an iodine cell by several degrees produces no noticeable difference in the resulting spectrum. Maintaining a temperature stability to a few tenths of a degree is generally sufficient to achieve an RV precision of 3–5 m s^{-1}, if not better.

3. *Long-term stability.*

   Once filled, an iodine is permanently sealed and can be used for an indefinite period of time. Unlike a hollow cathode lamp, it is rare for a cell to fail, although this has happened in one case (Fischer et al. 2014). When they do, it is often by accident (breakage, running at extremely high temperatures, etc). The iodine cell at McDonald Observatory was installed in 1990, and as of this writing, it is still used for precise stellar RV measurements.

   The spectrum produced by this cell can remain unchanged for decades. Figure 4.20 shows a segment of the spectrum of the iodine cell used for the McDonald Observatory Planet Search Program. It compares an observation taken in 1998 to that of one taken in 2018. There are no changes in the spectrum.

4. *A safe device.*

   Unlike HF, molecular iodine is a relatively benign gas and poses no real health hazard when used. An iodine cell was built for the HARPS spectrograph and it broke due to a flaw in its construction.[2] Iodine contaminated the optics (much to the anger of the builders!), but no staff suffered health consequences. The optics were cleaned, and the spectrograph is still in use as of this writing.

5. *Modeling of the instrumental profile.*

   One of the major advantages in the use of an iodine absorption cell is that it is one of the few wavelength calibrators that can also monitor change in the instrumental profile. This ability will be discussed in greater detail in Chapter 6.

[2] The cause of the failure was having a too narrow feedthrough tube into the cell, which proved fragile. The author has the distinction of being the first and last user of the HARPS iodine cell.

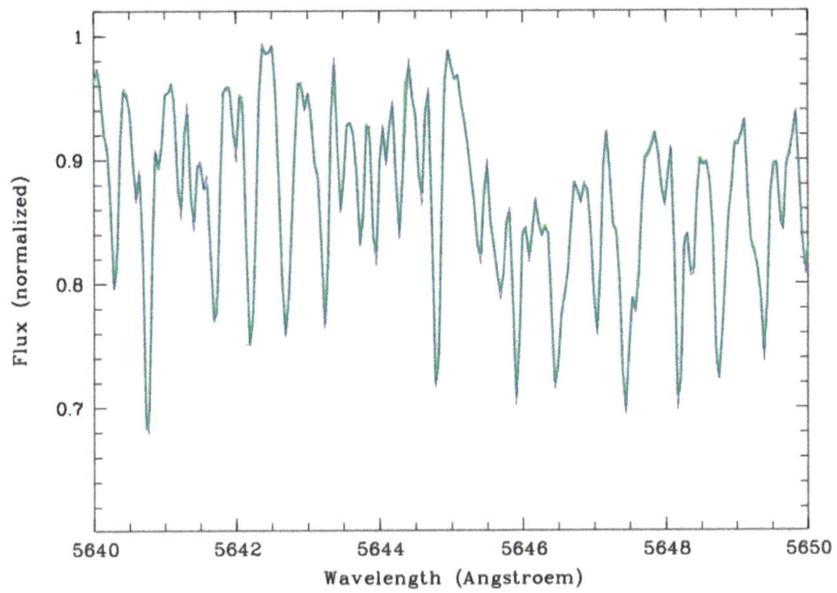

Figure 4.20. The green line shows a wavelength segment of molecular iodine from the McDonald Observatory cell taken in 1998. The blue line shows the same wavelength spectrum from an observation taken in 2018. The blue line is not visible because the two spectra are identical. (Figure courtesy of Michael Endl.)

Disadvantages of the Iodine Cell:

Although the iodine cell is a simple and elegant solution for turning any existing high-resolution spectrograph into an "RV machine," there are some drawbacks:

1. *Restricted wavelength range in the visible.*

   Although I_2 offers a generous wavelength range for precise RV measurements, it is difficult to extend the wavelength coverage beyond the nominal 5000–6000 Å range. The molecular band head kicks in at 5000 Å, so it is impossible to have iodine lines below this wavelength—you are limited by quantum physics.

   For wavelengths beyond 6000 Å, the iodine lines become too weak to be useful. It is possible to extend the absorption features farther into the red by heating the cell to higher temperatures (several hundred degrees Celsius). However, this causes additional problems such as a large heat source in the optical path, difficulties in maintaining temperature stability, etc. Another problem is that the efficacy of the cell when used at longer wavelengths will be diminished due to the prevalence of telluric features (see Chapter 12). To date, no one has employed a high temperature iodine cell.

2. *Contamination of the stellar spectrum.*

   Iodine absorption lines will contaminate the stellar spectrum. This makes it more difficult to perform additional analyses on the spectrum such as abundance studies, the measurement of stellar parameters, and in particular an examination of the spectral line shapes. As we shall see in Chapter 10, line-shape information is an important means for confirming the nature of RV signals. To perform these spectral studies will require observations without the cell, and this will add increased overhead to the observational program. The iodine spectrum is stable, and in principle, it should be possible to remove the contribution of the iodine lines; however, this removal may not be perfect.

3. *Lower throughput.*

   As the name "absorption cell" implies, the cell will absorb starlight as it passes though the cell, and this results in a lower S/N for a given exposure time. Your spectrograph is thus less efficient.

   There are two principal sources of light loss. First, you have four optical interfaces due to two windows (air to glass and glass to iodine gas). Antireflection coatings may minimize the reflective losses on the outer sides of the windows, but it is not wise to have these coatings on the inside where they can react with the iodine. Second, the iodine has features that absorb light. Depending on how much iodine is in the cell, the total losses when going through a cell can be 20%–50%. For the TCES, the light loss in passing through the cell is about 30%.

4.4.3 Absorption Cells at Infrared Wavelengths

Absorption cells working at IR and NIR wavelengths (>1 μm) have not seen as extensive use as their optical counterparts. This is largely due to the fact that there were few high resolution IR and NIR spectrographs operating up until now, but also in finding a suitable absorbing gas. Currently, a number of NIR/IR spectrographs are being used for high-precision RV measurements, so absorption cells in use in the IR may become more common.

One of the first high-resolution IR spectrographs, ESO's CRyogenic high-resolution InfraRed Echelle Spectrograph (CRIRES; Moorwood et al. 2003) used N_2O and CO gas absorption cells for wavelength calibration. Care had to be taken because the Earth's atmosphere also has these gases. Bean et al. (2010) constructed an ammonia (NH_3) gas cell for CRIRES and demonstrated a precision of 5 $m\,s^{-1}$.

The *JHKL* IR bands span a wavelength region of over 2 μm (Table 4.2), and it is difficult to find a single chemical element that spans such a broad region. This is demonstrated in Figure 4.21, which shows the wavelength range of absorption lines from a variety of molecular features (figure reproduced from Guelachvili & Rao 1986).

The criteria for finding a suitable gas that works in the IR is the same as finding cells for the optical regions:

- It should be gaseous at operating temperature.
- It should be relatively safe to use (i.e., non-toxic).
- It should require a short path length (high absorption cross sections).
- It should operate at modest temperatures.
- It should be chemically stable and not react with other chemicals in the cell.
- It should have no naturally occurring species from the Earth's atmosphere.

Table 4.2. Photometric Bands in the Infrared

Band	Wavelength	Bandwidth
J	1.220 μm	0.213 μm
H	1.630 μm	0.307 μm
K	2.190 μm	0.390 μm
L	3.450 μm	0.472 μm

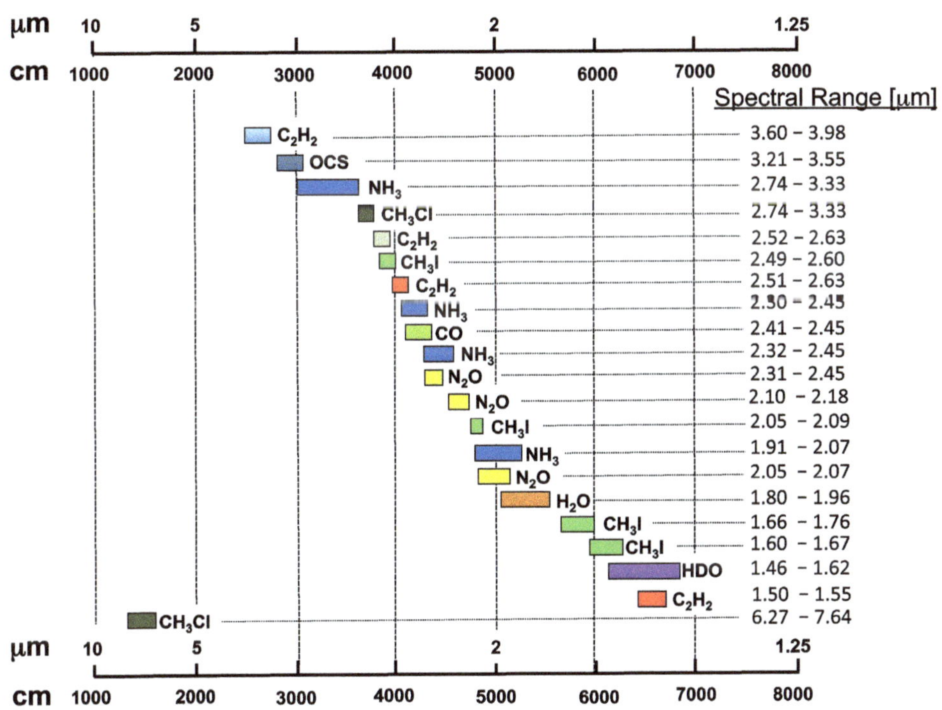

Figure 4.21. Various molecular gases and the wavelength range for which there are absorption features.

The problem is that one substance usually cannot fill all of these requirements. Our friend iodine does have molecular species (e.g., CH_3I) that operate in some infrared regions (Figure 4.21). Indeed, the spectrum of these show the same rich density of lines that one sees for I_2 in the visible region. The problem is that these species have very small absorption cross sections. One would need a cell of untenable lengths (~ many meters) or would have to be operated at several hundreds of degrees (U. Seeman, 2019, private communication). The latter is generally not good for an infrared spectrograph!

Because no single species has proven to be the "perfect" gas for an IR absorption cell, one has to use a combination of gases. Valdivielso et al. (2010) developed a cell composed of a mixture of acetylene, nitrous oxide, ammonia, chloromethanes, and hydrocarbons. The absorption lines covered most of the H (1.63 μm) and K (2.19 μm) bands. Absorption lines in the J band (1.22 μm) were too weak to be used for wavelength calibration. The authors presented no RV measurements using the cell.

Figure 4.22 shows the spectra of the component gases that are used for the absorption cell of CRIRES$^+$ for the K band. Figure 4.23 shows a region of the cell where NH_3 provides the absorption features.

4.5 Laser Frequency Combs

In the RV business, one is always in search of the perfect wavelength calibrator. This is especially true if you want to achieve a Doppler precision of 1–10 cm s^{-1}. This requires a dense spacing of features of known wavelengths such that one can determine the wavelength scale on short intervals and with exquisite precision. As we have seen, Th–Ar does not provide a dense or uniform spacing of emission lines, and these become sparser toward redder wavelength regions (>6000 Å). Molecular iodine provides a rich spectrum of narrow absorption lines, but only in the 5000–6000 Å wavelength interval. A significant step toward improved wavelength calibration can be made with laser frequency combs (LFCs; Udem et al. 2002; Murphy et al. 2007).

An LFC consists of thousands of laser peaks that are equally spaced in frequency over a broad bandwidth of several terahertz (THz). It is based on the properties of the femtosecond (fs) mode-locked lasers (Figure 4.24). In the time domain, these produce pulses with a repetition rate, T, and a pulse duration (τ, which is of order fs. In the frequency domain (after performing a Fourier transform), this produces a frequency comb with a repetition frequency T^{-1}. The comb has a spectral width given by τ^{-1}, about several hundred terahertz. The mode spacing is constant in frequency, and the pulse repetition rate is synchronized to an atomic clock.[3] It has several advantages over traditional calibration methods:

1. The absolute wavelength of each peak is known a priori without the need for laboratory measurements.
2. It has long-term stability and reproducibility.
3. It has a very high precision, limited only by the reference signal.
4. In principle, it can be developed to work over a wide range of wavelengths.

The problems for Doppler measurements are that an LFC produces too many peaks for wavelength calibration. A typical LFC has peak spacings of ≈250 MHz, which at 6000 Å translates to a spacing in wavelength of about 0.003 Å. For a typical high-resolution spectrograph with $R = 100{,}000$, this is about 10 comb peaks per CCD pixel! These are so dense that with high-resolution spectrographs used in astronomy, the comb peaks blend together and are inseparable. One needs a Fabry–Pérot (F–P) filter cavity (see below) to reduce the number of comb features. The final result, however, is a rich spectrum of equally spaced emission peaks. Figure 4.25

[3] Ted Hänsch and John Hall received the 2005 Nobel Prize in Physics for their pioneering work in LFCs.

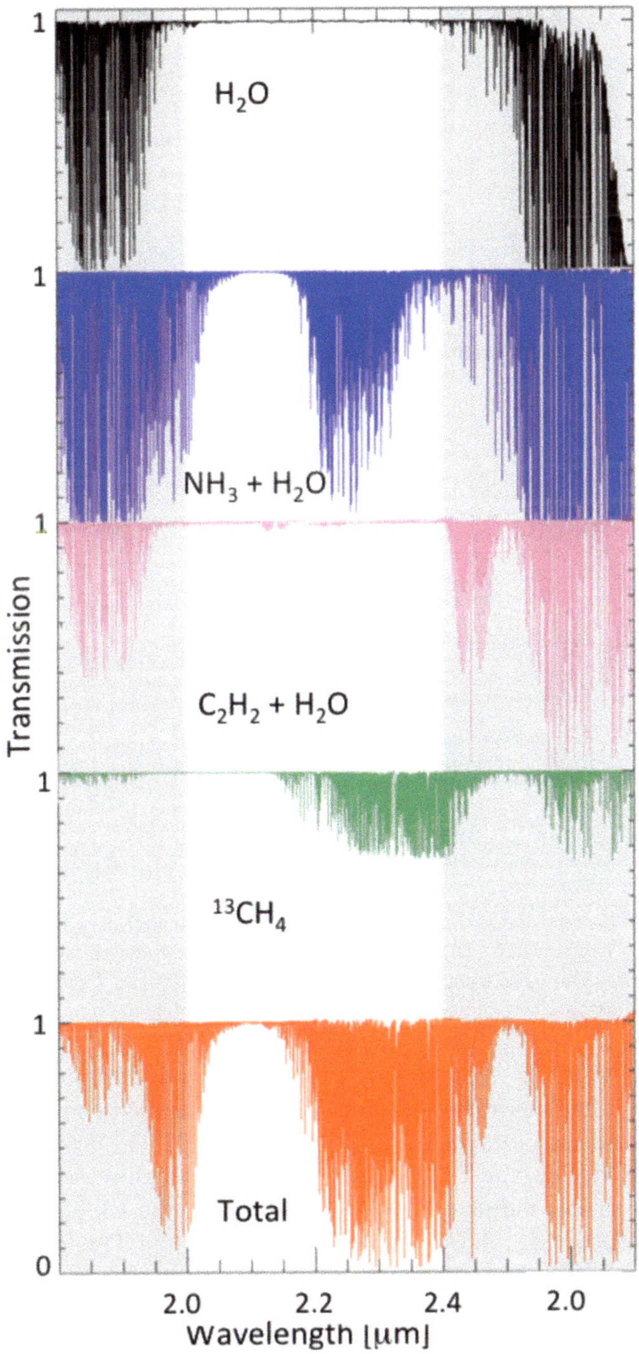

Figure 4.22. Spectrum from the *K*-band absorption cell used in the CRIRES+ IR spectrograph. The absorption lines from the individual gas constituents: (from top to bottom) H_2O, NH_3 + H_2O, C_2H_2 + H_2O, and $^{13}CH_4$. (Bottom) The total absorption from all components. (Courtesy of U. Seeman.)

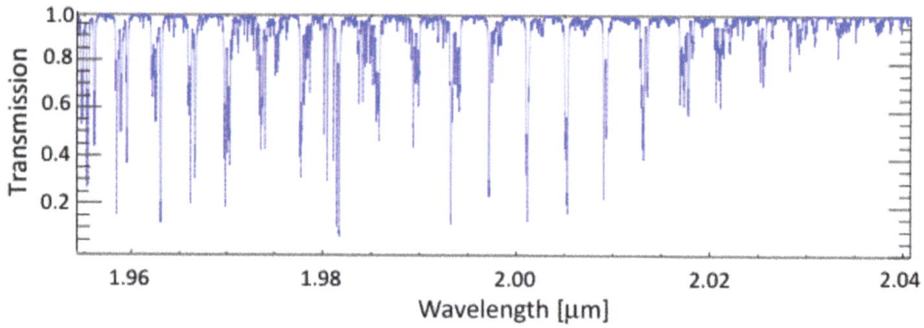

Figure 4.23. NH_3 absorption lines from the CRIRES$^+$ *K*-band cell (courtesy of U. Seeman).

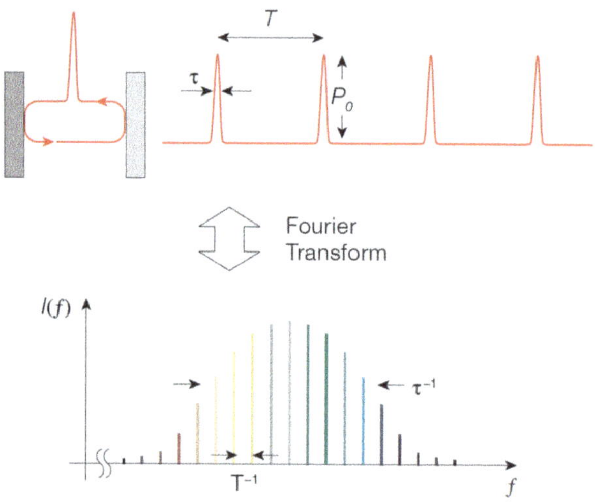

Figure 4.24. (Top) In the time domain, a mode-locked laser has a characteristic repetition rate, *T*, and a pulse duration, (τ (or order femtoseconds). (Bottom) In the frequency domain the pulses of the mode-locked laser appears as a frequency comb. The comb has a repetition frequency T^{-1} and a spectral width of τ^{-1}. (From Araujo-Hauck et al. 2007.)

compares the LFC to Th–Ar and iodine. The spacing is comparable to the spacing of the absorption lines of iodine, but more uniform in shape and spacing.

There are two disadvantages to the use of the LFC. The first is complexity. The main part of the device, the LFC plus F–P filter (see below), is considerably more complex than a simple iodine cell. Figure 4.26 shows the LFC as it is implemented on the HARPS spectrograph (Araujo-Hauck et al. 2007). The frequency comb with a spacing of 250 MHz passes through the F–P filter, which filters out most of the combs to produce a final spacing of 18 GHz (spacing ≈0.2 Å). The spectrum then passes through an amplifier followed by a fiber coupling as well as a fiber scrambler, before entering the HARPS spectrograph. Unlike the iodine cell, which is relatively maintenance free, the operation of the LFC for astronomical purposes probably requires a technical staff to maintain its operation. It is currently not a "turnkey" device.

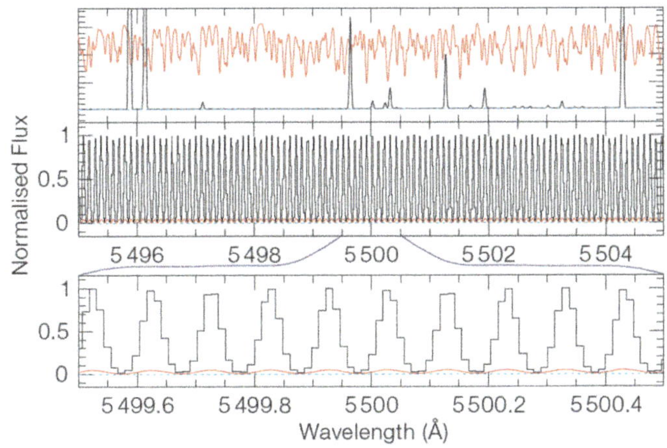

Figure 4.25. A comparison of the spectum of iodine (top in red) and Th–Ar (top in black) at the 5500 Å spectral region. The middle panel shows the same region of the LFC. The bottom panel is a zoom of a 1 Å. window of the comb spectrum. (From Araujo-Hauck et al. 2007.)

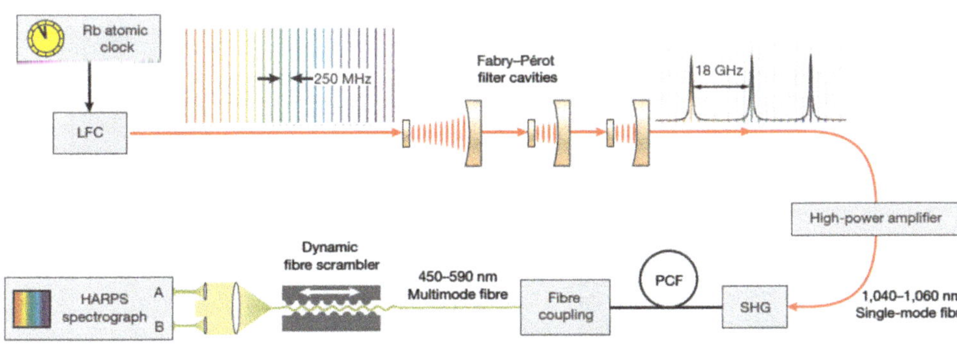

Figure 4.26. A schematic of the LFC system for the HARPS spectrograph. An Yb fiber laser serves as the comb generator and the Fabry–Pérot cavities are used to increase the line spacing (from Lo Curto et al. 2012b).

The second disadvantage is the costs of an LFC. As of this writing, an LFC costs of order several hundred thousand Euros, and this represents a significant cost to the design of a spectrograph. By comparison, an iodine absorption cell and associated heater can be implemented for less than 1000 Euros. Hopefully in the near future the cost and ease of use of an LFC will be more reasonable.

Figure 4.27 is a spectrum of a star taken with the LFC installed at the HARPS spectrograph. One can see that compared to the Th–Ar spectrum (Figure 4.8), the laser combs provide a much denser set of calibration lines. First use of the LFC at HARPS indicates that an RV precision of ~cm s^{-1} is possible (Lo Curto et al. 2012a).

4.6 Fabry–Pérot Etalons

An F–P etalon consists of two plane-parallel, highly reflective surfaces separated by a distance, *d*. A portion of the beam gets transmitted, while another portion is

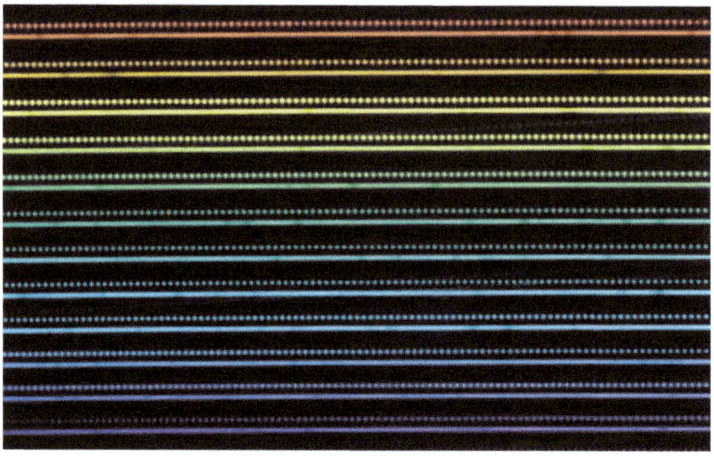

Figure 4.27. The laser frequency comb at the HARPS spectrograph. The lower rows are the orders of the stellar spectrum. Just above these are a series of emission peaks produced by the laser frequency comb. Note the higher density of calibration features compared to Th–Ar (see Figure 4.8). Figure courtesy of ESO.

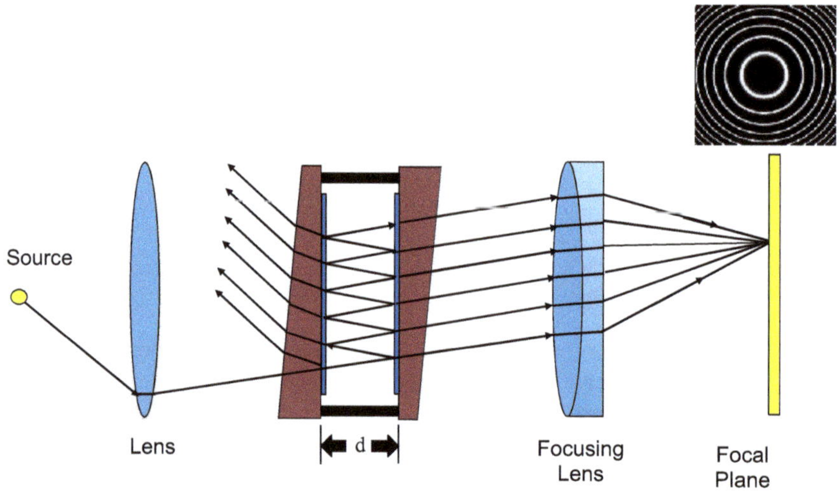

Figure 4.28. Schematic of a Fabry–Pérot etalon. Shown here is an imaging Fabry–Pérot which produces a series of circular interference rings.

reflected between the two plates before emerging (Figure 4.28). The transmission maxima fulfill the interference condition:

$$m\lambda_m = 2d. \tag{4.1}$$

The effective cavity width is $dn \cos \theta$, where d is the separation of the two mirrors, n is the index of refraction, and θ is the incident angle.

The top panel of Figure 4.29 shows the F–P transmission maxima compared to emission lines from a Th–Ar lamp. The F–P clearly provides a much denser spacing

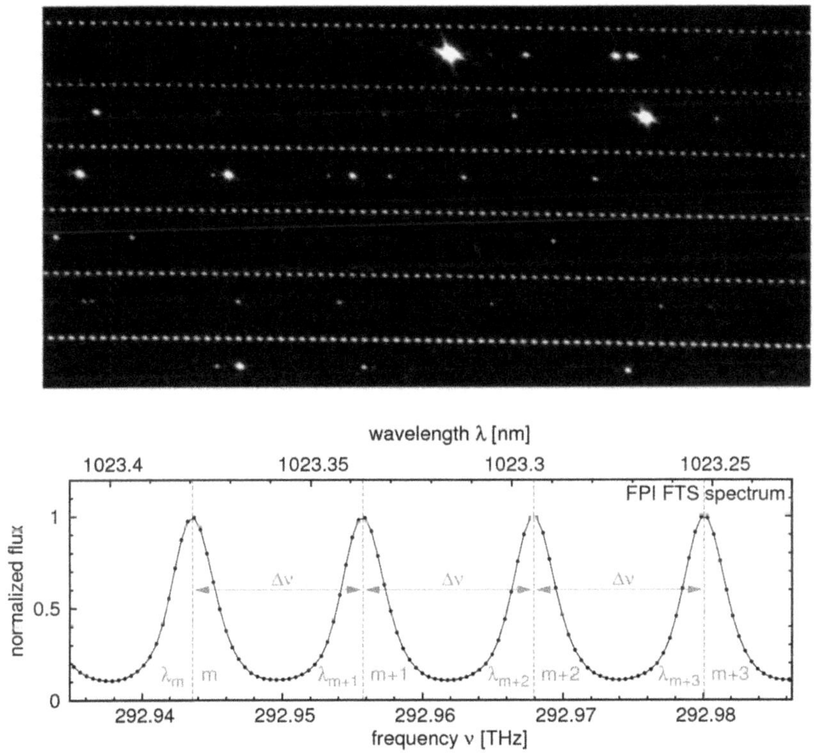

Figure 4.29. (Top) Image spectra of an Fabry–Pérot etalon illuminated by a white-light source and a Th–Ar lamp which appears below each F–P spectra (from Schwab et al. 2015). (Bottom) High-resolution spectrum of the Fabry–Pérot interferometer taken with a Fourier Transform Spectrometer showing the shapes of the individual peaks (from Bauer et al. 2015, reproduced with permission © ESO).

of emission lines for wavelength calibration and is comparable to the LFC. The lower panel shows a high-resolution (R = 500,000) spectrum of an F–P interferometer taken with a Fourier transform Sspectrometer. The transmission peaks are equally spaced in frequency, ν.

One of the first uses of a Fabry–Pérot interferometer (FPI) for RV measurements was by McMillan et al. (1993), who used a F–P in transmission to detect Doppler shifts from changes in the flux of light on the slopes of stellar absorption lines. McMillan et al. (1994) surveyed 20 solar-type stars for exoplanets and achieved an RV precision of $\approx$20 m s^{-1}, typical for precise RV measurements from that time.

In practice, the use of an F–P for wavelength calibration can be challenging. One needs to know the exact wavelengths of the interference maxima, and this depends on the effective cavity width. For a wavelength calibration to achieve a precision of 1 m s^{-1}, this requires an accuracy of $\delta/d \approx 3 \times 10^{-9}$ (Bauer et al. 2015). However, the mirror separation is only known to an accuracy of 1 μm, so one must calibrate the width d. Complicating matters is the fact that the mirror surfaces are coated, which means that photons of different energies penetrate to different depths—the effective d thus varies with wavelength. Frequency drifts can occur for a number of reasons, so one needs to cross-check the calibration of the FPI against traditional methods. Because of these shortcomings, F–P etalons have up until now mostly been used to monitor nightly spectrograph drifts or in combination with HCLs to improve the wavelength calibration (Bauer et al. 2015). It has proven to be an effective way of correcting nightly instrumental drifts for the CARMENES spectrograph (Reiners et al. 2018).

In order to improve the wavelength, Reiners et al. (2014) and Schwab et al. (2015) proposed actively controlling the FPI cavity externally using the ^{8}7Rb D_2 atomic line. The drift of the cavity is monitored relative to an external cavity diode laser that is compared to the atomic frequency from the atomic transition. It is proposed that a Doppler precision of ≈3 cm s^{-1} may be possible using this laser-lock concept.

4.7 The RV Precision of Modern Spectrographs

Fischer et al. (2016) presented a nice review of the various programs using precise RV measurements to search for exoplanets. Their table is largely reproduced in Table 4.3. We have supplemented the programs presented in Fischer et al. (2016) with results from the UVES M-dwarf program (Endl et al. 2006) and the visual channel (VIS) of the CARMENES spectrograph (data kindly provided by A. Reiners). These programs used either the simultaneous Th–Ar or the iodine absorption cell. The histogram of the RV precisions for many of these programs is shown in Figure 4.30. The value σ_{200} in the table is the best possible precision reported by the various groups for an S/N = 200. Also listed is the median value of the histograms.

Table 4.3. Spectrographs for Precision RVs

Spectrograph	Slit Fiber	R ($\lambda/\delta\lambda$)	Calibrator	σ_{200} (m s^{-1})	Median (m s^{-1})	Predicted (m s^{-1})
CARMENES	f	94,600	HCL/FP	1.3	3.5	2.0
CHIRON	f	90,000	Iodine	1.0	2.5	4.0
Hamilton	s	50,000	Iodine	3.0	7.5	8.0
HARPS	f	115,000	Th–Ar	0.8	2.5	2.0
HARPS-N	f	115,000	Th–Ar	0.8	3.5	2.0
HIRES	s	55,000	Iodine	1.5	5.0	7.0
HRS	s	60,000	Iodine	3.0	7.0	6.5
LEVY	s	110,000	Iodine	1.5	3.5	3.0
PARAS	f	67,000	Th–Ar	1.0	2.5	4.0
PFS	s	76,000	Iodine	1.2	3.0	5.0
SOPHIE	f	75,000	Th–Ar	1.1	3.0	3.5
SONG	s	90,000	Iodine	2.0	5.0	4.0
UCLES	s	45,000	Iodine	3.0	5.5	9.0
TCES	s	67,000	Iodine	1.9	9.0	6.0
Tull	s	60,000	Iodine	5.0	6.5	6.5

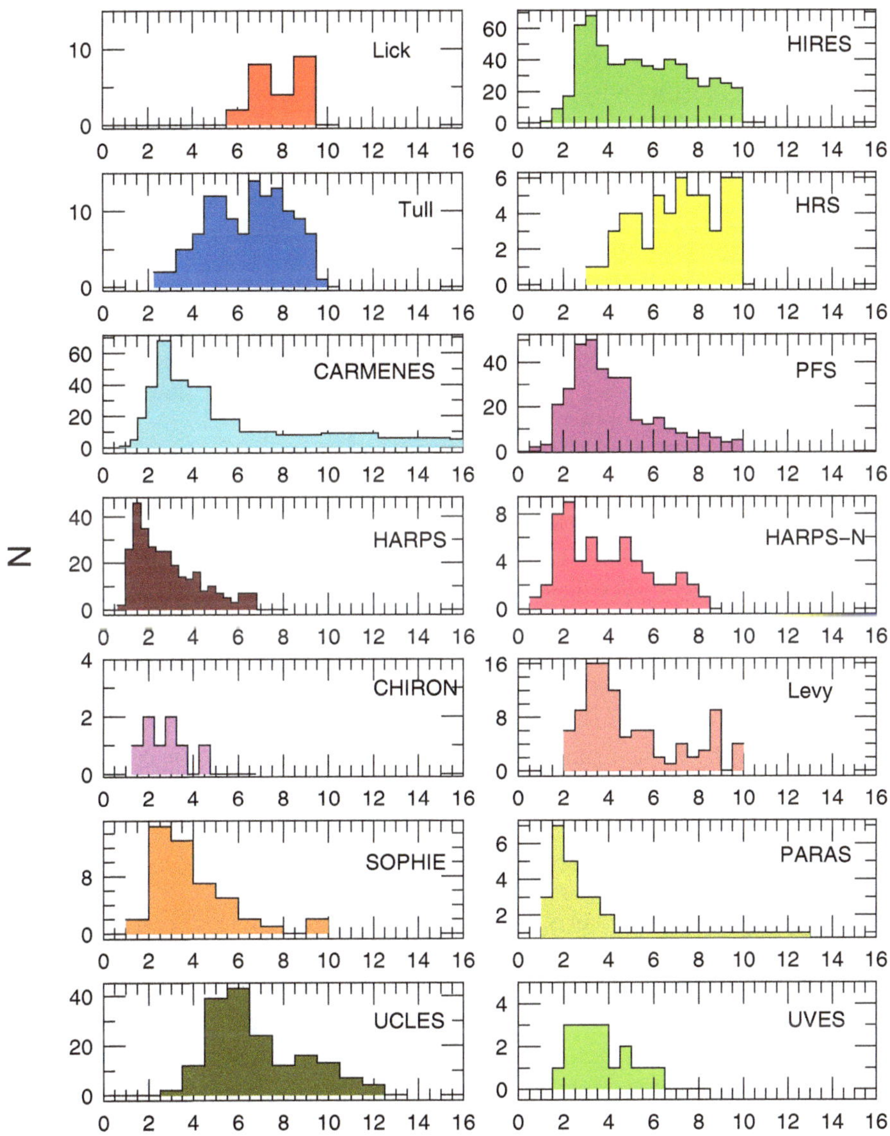

Figure 4.30. The histogram of RV uncertainties from various RV programs taken from Fischer et al. (2016). Also included are values from CARMENES and UVES.

There are two takeaway messages from Table 4.3 and Figure 4.30. First, it is very difficult to get an RV precision better than 2 m s^{-1} on a typical star. Even for HARPS, the spectrograph that arguably achieves the best possible precision has a median σ of 2.5 m s^{-1} and a peak in the histogram at 1.5 m s^{-1}. Although σ_{200} indicates that instruments are able to achieve an exquisite RV precision, this is only for a few stars. The "working precision," i.e., the RV uncertainty one achieves on a typical program star observed under typical conditions, is significantly worse, by about a factor of 2.

There are two reasons why the working precision is worse than the "best case" precision. First, although it is relatively easy to achieve exquisite precision in a single night, it is much more difficult to achieve this from night to night, month to month, and year to year. Changes in the instrument, particularly in the instrumental profile which will be discussed in more detail in the next chapter, introduce systematic errors which are hard to correct.

Second, and most importantly, is stellar variability—the stars are just not well behaved when it comes to RV stability. As we shall see in Chapters 9 and 10, such phenomena as stellar activity, convection, pulsations, etc. introduce a significant contribution to the measured RV. It is difficult to find stars that are intrinsically stable to a level of better than 0.5–1 m s^{-1}.

Another important message from Table 4.3 and Figure 4.30 is that the iodine absorption cell method does remarkably well. For stabilized spectrographs using simultaneous Th–Ar the average quoted "best error" is 1 m s^{-1} compared to an average error of 2.4 m s^{-1} for the iodine method. The median error on a sample of stars is 2.9 m s^{-1} for Th–Ar and 5.5 m s^{-1} for iodine, still only about a factor of 2 worse than for Th–Ar calibration.[4] It is important to note that many of the spectrographs employing the iodine method were not designed for ultra-stability like HARPS, which is stabilized in terms of temperature, pressure and mechanical vibrations. Furthermore, many of these are instruments for general use, which often have moving gratings and usually no fiber links. The absorption cell method is an inexpensive way to improve the RV measurement precision of all high-resolution spectrographs.

We realize when comparing the performance of different spectrographs with different designs that they have different targets and achieve different S/Ns for a typical program star. Furthermore, the spectrographs have different resolving powers, wavelength coverage, and different levels of stability (thermally and mechanically). Ideally we would like to compare the performance of each spectrograph observing the same star and at the same S/N level, but that is impossible. However, we can make a comparison to give us an overview of the typical RV precision of modern spectrographs observing the typical type of stars for exoplanet searches.

We can now use the performance of the various spectrographs to derive the proportionality constant in Equation (3.7). To do so we must make some assumptions. First, let us assume that the RV uncertainty scales with resolving power as $R^{-1.2} (\alpha = 1.2)$. We tried using the σ_{200} values to derive a value of α, but these showed too much scatter. A value of $\alpha = 1.2$ represents a good compromise of the various values reported in the literature. All of the spectrographs have a

[4] The wavelength coverage for the iodine method is about a factor of 2 less than that for the Th–Ar technique. Correcting for this the iodine method would have a median error of 3.9 m s^{-1} if it had the same wavelength coverage.

resolving power within a factor of 2 within each other, and it turns out that the proportionality constant can absorb much of differences due to varying α.

Second, we do not know the types of stars from the different RV programs, but let us assume that they are typical ones for planet search programs, i.e., slowly rotating solar-type stars. Third, the S/N for a typical exposure is not known, but let us assume it is not too high and not too low, say S/N = 75.

Finally, we use σ_{median} as the measurement error. Under these assumptions, the average value of the proportionality constant from all programs is $C = 8.3 \times 10^9$. This results in an RV precision of, including all parameters,

$$\sigma\ [\mathrm{m\ s^{-1}}] = (8.2 \times 10^9)\Delta\lambda^{-1/2}(S/N)^{-1}R^{-1.2}f(V)g(T), \tag{4.2}$$

where R is the resolving power, S/N the signal-to-noise ratio, T the effective temperature of the star, and $\Delta\lambda$ the wavelength coverage in angstroms.

The functions f and g are

$$f(V) = 0.62 + (0.21 \log R - 0.86)V + (0.00260 \log R - 0.0103)V^2, \tag{4.3}$$

where V is the projected rotational velocity of the star in km s^{-1}, and

$$g(T) = 0.16e^{1.79(T/5000)}, \tag{4.4}$$

where T is the effective temperature of the star. For effective temperatures below about 5000 K, one can simply assume $g = 1$.

Equation (4.2) does reasonably well in predicting the RV working precision of modern, state-of-the-art spectrographs ("Predicted" in Table 4.3). This expression can be used to estimate the typical RV uncertainty of a spectrograph probably to within a factor of 2. If your spectrograph can achieve a precision as good as or better than the estimated value then it is performing about as well as most modern spectrographs using simultaneous wavelength calibration.

Currently, the CARMENES instrument is the only high-resolution spectrograph capable of making RV measurements simultaneously at visual (VIS) and near-infrared (NIR) wavelengths, so it is of interest to compare the performance in the two wavelength bands. The VIS channel has a resolving power of $R = 94{,}600$ in the wavelength range 520–960 nm whereas the NIR covers the 960–1710 nm region at $R = 80{,}400$. Both channels are calibrated using HCLs. In addition, F–P etalons are used to monitor simultaneously the spectrograph shifts during the night as well as to interpolate the wavelength solution. The CARMENES program is targeting M dwarfs.

Figure 4.31 shows the internal errors of RV measurements from both the VIS and NIR arms. These errors are determined during the calculation of the RVs and represent the best RV precision one can achieve due to photon statistics and no contribution due to long-term systematic errors or intrinsic stellar variability.

The internal precision of the VIS channel has 1.2 m s^{-1}, in line with the performance of other state-of-the-art RV instruments. The NIR channel, on the other hand, has a median error of 5.9 m s^{-1}, or about a factor of 5 worse than the VIS channel. There are two reasons for this.

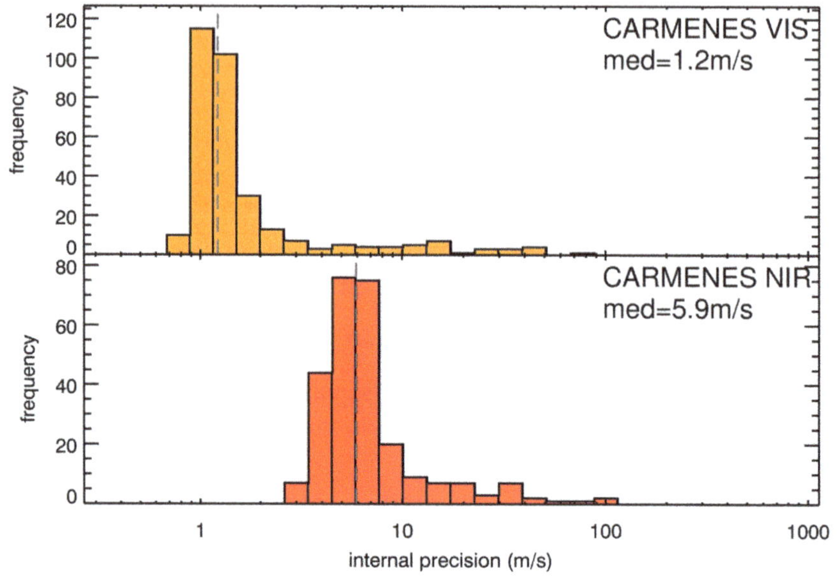

Figure 4.31. The internal RV errors of the CARMENES VIS channel (top) and NIR channel (bottom). (Figure courtesy of Ansgar Reiners.)

First, there is simply a higher content of spectral features in the VIS, which outweigh the decrease in flux at optical wavelengths. Furthermore, the contamination of telluric features becomes more of a problem at NIR wavelengths (see Chapter 12). In fact, the predicted uncertainty for RV measurements made with the NIR channel is ≈6 m s^{-1} (A. Reiners, 2018, private communication), so the NIR channel of CARMENES is performing as expected.

Second, CCD detectors tend to have better performance and stability. This is due to the decades of development of such devices. The NIR channel of CARMENES does not use CCDs, but rather HAWAII-2RG infrared detectors. The level of performance in terms of stability and noise characteristics of infrared devices is still not at the level of their optical counterparts.

Of course, what is of interest to an observer is not the internal precision of the instrument but rather the "working precision," i.e., the measurement error that you get on real stars. This is the external error and is shown for the VIS and NIR channels in Figure 4.32. The median external error for the VIS channel is 3.5 m s^{-1}, or a factor of 3 worse than the internal error. This represents the contribution of stellar activity, which adds an RV component to the measured values. M dwarfs are generally active stars, and the RV from intrinsic stellar variability dominates the scatter of the measured RV values. The subject of activity-related RV signals will be dealt with at length in the ensuing chapters.

The median external error of the NIR is 8.9 m s^{-1}, or only about 50% higher than the external errors. At face value, one would think that just a slight increase in scatter is what you would expect due to different contrast of temperatures between the magnetic and nonmagnetic features when observing at longer wavelengths.

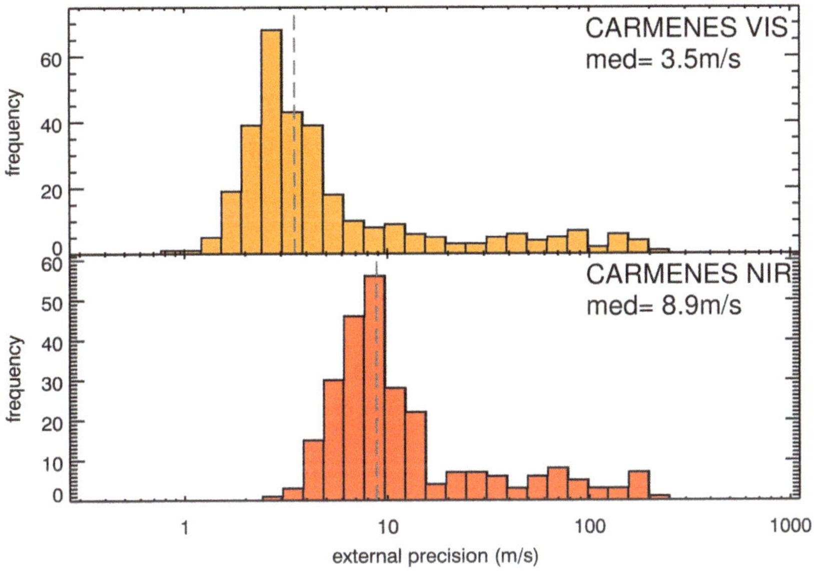

Figure 4.32. The external RV errors of the CARMENES VIS channel (top) and NIR channel (bottom). (Figure courtesy of Ansgar Reiners.)

However, this does not seem to be the case. We can remove the contribution of the internal errors (in quadrature) in order to isolate the contribution of intrinsic variability, σ_{var}. This results in σ_{var} = 3.3 m s^{-1} and σ_{var} = 6.6 m s^{-1} for the VIS and NIR arms, respectively. So it seems that at least for M dwarfs, the stars show higher intrinsic variability in the NIR. The CARMENES program is still in its early stages, so these results may change. However, this is consistent with Reiners et al. (2010), who argued that the NIR offers only marginal advantages to a spectrograph working at visual wavelengths.

One final remark—one may think that because one can achieve an RV precision of ≈1 m s^{-1} (the reality is more like ≈2 m s^{-1}) with a visual spectrograph, one can simply build a spectrograph for the NIR that has the same performance. This is simply not the case. The NIR and IR have a completely different set of problems: different spectral content, more telluric line contamination, different detectors with different stability and noise characteristics, different wavelength calibration. Decades of work have been invested in getting optical RV measurements down to 1 m s^{-1}. A similar amount of time has not been invested in improving the performance of RVs at infrared wavelengths. Currently, the best one can hope to achieve is an RV measurement error of ≈5 m s^{-1} with an infrared spectrograph.

References

Araujo-Hauck, C., Pasquini, L., Manescau, A., et al. 2007, Msngr, 129, 24
Baranne, A., Queloz, D., Mayor, M., et al. 1996, A&AS, 119, 373
Bauer, F. F., Zechmeister, M., & Reiners, A. 2015, A&A, 581, A117
Bean, J. L., Seifahrt, A., Hartman, H., et al. 2010, ApJ, 713, 410

Beckers, J. M. 1977, ApJ, 213, 900
Campbell, B., & Walker, G. A. H. 1979, PASP, 91, 540
Campbell, B., Walker, G. A. H., & Yang, S. 1988, ApJ, 331, 902
Cochran, W. D., Hatzes, A. P., & Hancock, T. J. 1991, ApJ, 380, L35
Endl, M., Cochran, W. D., Kürster, M., et al. 2006, ApJ, 649, 436
Fischer, D. A., Marcy, G. W., & Spronck, J. F. P. 2014, ApJS, 210, 5
Fischer, D. A., Anglada-Escude, G., Arriagada, P., et al. 2016, PASP, 128, 066001
Griffin, R., & Griffin, R. 1973, MNRAS, 162, 255
Guelachvili, G., & Rao, N. 1986, Handbook of Infrared Standards (Amsterdam: Elsevier)
Guenther, E. W., & Wuchterl, G. 2003, A&A, 401, 677
Hatzes, A. P., Cochran, W. D., McArthur, B., et al. 2000, ApJ, 544, L145
Hatzes, A. P., Cochran, W. D., Endl, M., et al. 2003, ApJ, 599, 1383
Latham, D. W., Mazeh, T., Stefanik, R. P., Mayor, M., & Burki, G. 1989, Natur, 339, 38
Lo Curto, G., Manescau, A., Avila, G., et al. 2012a, Proc. SPIE, 8446, 84461W
Lo Curto, G., Pasquini, L., Manescau, A., et al. 2012b, Msngr, 149, 2
Marcy, G. W., & Butler, R. P. 1992, PASP, 104, 270
Mayor, M., & Queloz, D. 1995, Natur, 378, 355
Mayor, M., Pepe, F., Queloz, D., et al. 2003, Msngr, 114, 20
McMillan, R. S., Moore, T. L., Perry, M. L., & Smith, P. H. 1993, ApJ, 403, 801
McMillan, R. S., Moore, T. L., Perry, M. L., & Smith, P. H. 1994, Ap&SS, 212, 271
Moorwood, A. F. M., Biereichel, P., Brynnel, J., et al. 2003, Proc. SPIE, 4841, 1592
Murphy, M. T., Udem, T., Holzwarth, R., et al. 2007, MNRAS, 380, 839
Pepe, F., Mayor, M., Delabre, B., et al. 2000, Proc. SPIE, 4008, 582
Redman, S. L., Lawler, J. E., Nave, G., Ramsey, L. W., & Mahadevan, S. 2011, ApJS, 195, 24
Reiners, A., Banyal, R. K., & Ulbrich, R. G. 2014, A&A, 569, A77
Reiners, A., Bean, J. L., Huber, K. F., et al. 2010, ApJ, 710, 432
Reiners, A., Zechmeister, M., Caballero, J. A., et al. 2018, A&A, 612, A49
Schwab, C., Stürmer, J., Gurevich, Y. V., et al. 2015, PASP, 127, 880
Udem, T., Holzwarth, R., & Hänsch, T. W. 2002, Natur, 416, 233
Valdivielso, L., Esparza, P., Martín, E. L., Maukonen, D., & Peale, R. E. 2010, ApJ, 715, 1366
Walker, G. A. H., Bohlender, D. A., Walker, A. R., et al. 1992, ApJ, 396, L91

AAS | IOP Astronomy

The Doppler Method for the Detection of Exoplanets

A P Hatzes

Chapter 5

Calculating the Doppler Shifts: The Cross-correlation Method

The next two chapters will be devoted to the details of how radial velocities (RVs) are calculated from spectra. The standard technique for the calculation of RVs is the cross-correlation method (Tonry & Davis 1979) and is the one employed, in some form or another, with the simultaneous Th–Ar method. The cross-correlation method will be the subject of this chapter. This technique is standard, and many data reduction packages, like the Image Reduction and Analysis Facility (IRAF) have routines to calculate RVs using cross-correlation.

The other method is to model the observed spectrum plus calibration spectrum using χ^2 fitting. This is the method generally employed for the absorption cell method, in particular the iodine cell. This will be the subject of the next chapter.

5.1 Mathematical Formalism

If $s(x)$ is your stellar spectrum as a function of pixels and $t(x)$ is a template spectrum, then the cross-correlation function (CCF) is defined as

$$\mathrm{CCF}(\Delta x) = s(\Delta x) \otimes t(\Delta x) = \int_{-\infty}^{+\infty} s(x)t(x + \Delta x)dx. \tag{5.1}$$

Because we are dealing with discretely sampled spectra with CCFs that are calculated numerically, we use the discrete form of the CCF,

$$\mathrm{CCF}(\Delta x) = \sum_{x=1}^{N} s(x)t(x + \Delta x)dx. \tag{5.2}$$

Δx is called the lag of the CCF.

The CCF is a measure of the similarity of two signals and is thus most sensitive to Δx when s and t are identical. The CCF will be maximum for a Δx that matches

doi:10.1088/2514-3433/ab46a3ch5

both functions. For this reason, the CCF is often called a matching or detection filter.

Unfortunately, the Doppler shift is nonlinear because at different wavelengths, the shift in pixels will be different. This can be remedied by rebinning the linear wavelength scale onto a logarithmic one, thus transforming the Doppler formula to

$$\Delta \ln \lambda = \ln \lambda - \ln \lambda_0 = \ln\left(1 + \frac{v}{c}\right). \tag{5.3}$$

The lag of the CCF is then $\Delta x = \Delta \ln \lambda = \ln(1 + v/c)$, which is a constant for a given velocity, v.

The CCF can be very effective at detecting the Doppler shift of stellar lines even in the presence of substantial noise. Figure 5.1 shows a single spectral line at various levels of signal-to-noise ratio (S/N; left panels) and the corresponding CCF using a synthetic line profile with no noise as the template (right panels). For S/N = 100, there is almost no need to calculate the CCF. You can probably get a comparable result by fitting the core of the original spectral line profile. However, this procedure would not be effective for the S/N = 10 profiles, yet the CCF function is still well defined. Remarkably, the CCF is able to detect the RV shift of the spectral line even for S/N = 5, where you are hard-pressed even to identify the correct spectral feature in the data.

We have shown the example of only using a single line to compute the CCF. The real power of the method comes from using a large wavelength region in computing

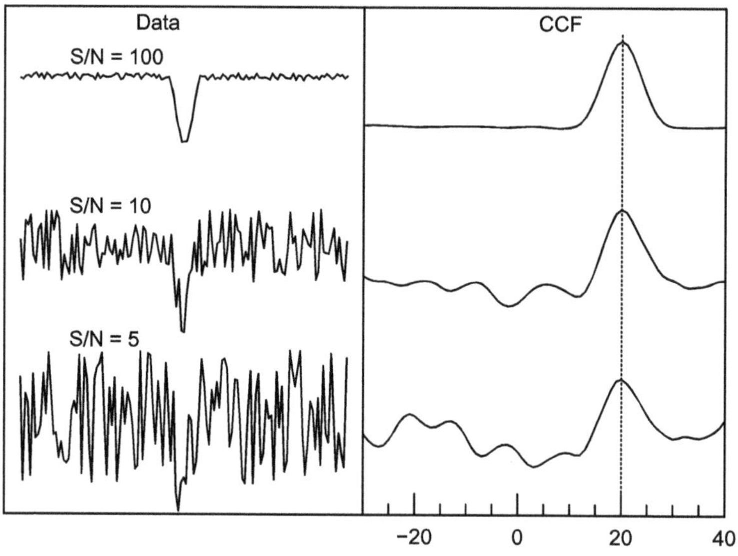

Figure 5.1. Example of the use of the CCF to detect a Doppler shift of a spectral line. (Left) Synthetic spectra generated at three levels of signal-to-noise ratios (S/N = 100, 10, and 5) and with a Doppler shift of +20 km s^{-1}. (Right) The CCF of the noisy spectra cross-correlated with a noise-free synthetic spectral line. Even at low S/N levels, the correct Doppler shift is recovered (vertical dashed line).

the CCF. In this case, one increases the number of spectral lines used for determining the Doppler shift. This produces a clean and well-defined CCF.

We note that when one uses a synthetic spectrum of the star, the CCF represents the mean shape of a typical spectral line profile. This is an effective way to increase the S/N of a line profile for studies involving the spectral line shape. This will be important when we discuss spectral line bisectors as a means of confirming exoplanet discoveries.

A detailed description of the mathematical formalism of the CCF as applied to RV measurements can be found in Murdoch & Hearnshaw (1991).

5.2 Choice of Template

The key to a good RV measurement with the CCF method is having a low-noise template that is a close match to the spectrum of your target star. The closer the match, the stronger and well-defined the CCF will be. Clearly, you do not want to measure the Doppler shift of a hot early-type using the template of a cool solar-type star, which has different spectral features.

Figure 5.2 shows the effects of using mismatched stellar and template spectra for the CCF calculation. The black line in the left panel is the CCF of a "stellar" spectrum consisting of a single stellar line with rotational broadening at a level of 20 km s^{-1}. A Doppler shift of +30 km s^{-1} was also applied. The synthetic spectrum has S/N = 50. To produce the CCF, a template consisting of a single line with little rotation ($v \sin i = 3$ km s^{-1}) is used. If one can also match the width of the stellar lines in the template, one gets an even narrower CCF. This is the case for the CCF shown by the red line where both template and target star have the same spectrum with the same rotational broadening. Note that the CCF peak of the "perfect match" is higher and narrower and has a higher contrast. In fact, the width of the

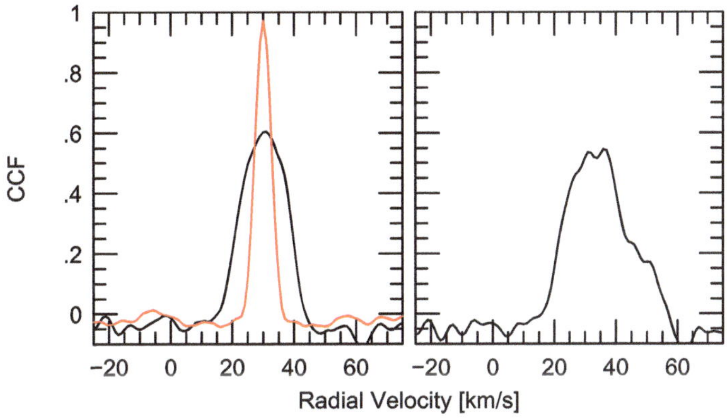

Figure 5.2. Example of the resulting CCF depending on the template one uses. (Left) A CCF (black line) of a rapidly rotating star ($v \sin i = 20$ km s^{-1}) using a template of a slowly rotating ($v \sin i = 3$) star. The S/N of the "star" spectrum is 50. The red line shows the CCF of the slowly rotating star (S/N = 50) using a stellar line with no noise and with the same rotational velocity. (Right) CCF of a mismatched star. In this case, the template spectrum has two narrow lines of different depths with slow rotation. The "stellar" spectrum is a single line with rapid rotation.

CCF as shown in the black line can be used as a measure of the projected RV (Queloz et al. 1998), so long as this is calibrated against stars with well-known rotational velocities (Melo et al. 2001).

The right panel of Figure 5.2 shows the results of a much larger mismatch between stellar and template spectra. In this case, the template spectrum consisted of two narrow spectral lines ($v \sin i = 3$ km s^{-1}) separated by 0.38 Å and with relative depths differing by a factor of 2. The "stellar" spectrum consisted of the single, rotationally broadened profile used for the left panel. This simulation mimics trying to use a template of a late-type star to cross-correlate with an early, rapidly rotating star with fewer spectral lines. In this case, the CCF is more distorted, resulting in a degraded RV precision. The larger the mismatch between stellar and template spectrum, the poorer the RV measurement precision.

5.2.1 Standard Stars

A common practice is to use spectra of so-called standard stars as the template for the CCF. These are stars which are believed to be RV constant, at least to the measurement errors, and for which an absolute RV has been measured. You can thus put your relative RV on an absolute scale. Table 5.1 list standard stars often employed for RV measurements (Udry et al. 1999). You simply take a high-S/N spectrum of one of these standard stars and use it for the cross-correlation of your target star.

As a historical aside, the International Astronomical Union (IAU) once designated a list of K giant stars which could serve as RV standards, i.e., the IAU deemed that these stars were RV constant. The RV survey of Campbell & Walker used such standards to assess the performance of the HF method. It turned out that many of these were RV variables with amplitudes of 30–300 m s^{-1} (Walker et al. 1989). The variations ultimately were shown to be due to stellar oscillations, rotational modulation, and planetary companions. With sufficient measurement precision, no star is constant.

5.2.2 Synthetic Masks

The problem with using a standard star spectrum for the cross-correlation method is that it will introduce additional noise, which may influence the shape of the CCF. For better RV precision, it is common practice to use digital templates. These are masks that have zero values at all wavelengths except for those covering stellar lines. There are two advantages to using a digital mask:

1. You have a completely noise-free template. All errors due to photon uncertainties come only from your stellar spectrum.
2. You have a choice of which stellar lines to use for the RV measurement. If you have some stellar lines which have more intrinsic RV noise, say from stellar activity, these can simply be masked off.

Alternatively, one can use a synthetic stellar spectrum generated using stellar model atmospheres.

Table 5.1. Radial Velocity Standard Stars

HD	V (mag)	SpType	RV (km s^{-1})	R.A. (hr)	Dec. (deg)
3765	7.36	K2V	−63.30	00:40:49.270	+40:11:13.82
10780	5.63	K0V	2.70	01:47:44.835	+63:51:09.00
32923	5.60	G4V	20.50	05:07:27.006	+18:38:42.19
38230	7.36	K0V	−29.25	05:46:01.886	+37:17:04.73
42807	6.45	G2V	6.00	06:13:12.503	+10:37:37.72
50692	5.74	G0V	−15.05	06:55:18.668	+25:22:32.51
62613	6.56	G8V	−7.85	07:56:17.230	+80:15:55.95
65583	7.00	K0V	14.80	08:00:32.129	+29:12:44.48
73667	7.58	K2V	−12.10	08:39:50.792	+11:31:21.62
79210	7.62	K7V	10.65	09:14:22.793	+52:41:11.85
82106	7.20	K3V	29.75	09:29:54.824	+05:39:18.48
82885	5.34	G8V	14.40	09:35:39.502	+35:48:36.48
90343	7.29	K0	9.55	10:35:11.265	+84:23:57.56
109358	4.25	G0V	6.25	12:33:44.545	+41:21:26.93
125184	6.47	G5	−12.40	14:18:00.727	-07:32:32.60
128165	7.23	K3V	11.25	14:33:28.867	+52:54:31.64
131977	5.72	K4V	26.85	14:57:28.000	−21:24:55.71
139323	7.60	K3V	−67.20	15:35:56.566	+39:49:52.02
140538	5.86	G2.5V	19.00	15:44:01.820	+02:30:54.62
144579	6.67	G8V	−59.45	16:04:56.793	+39:09:23.43
145742	7.55	K0	−21.85	16:00:36.674	+80:37:40.07
151541	7.56	K1V	9.40	16:42:38.577	+68:06:07.81
154345	6.77	G8V	−46.95	17:02:36.404	+47:04:54.77
158633	6.43	K0V	−38.60	17:25:00.099	+67:18:24.14
159222	6.56	G1V	−51.60	17:32:00.993	+34:16:16.13
164922	6.99	G9V	20.15	18:02:30.862	+26:18:46.81
168009	6.30	G1V	−64.65	18:15:32.464	+45:12:33.54
182488	6.37	K0V	−21.55	19:23:34.013	+33:13:19.08
182572	5.16	G7IV	−100.35	19:24:58.200	+11:56:39.90
190007	7.46	K5V	−30.40	20:02:47.045	+03:19:34.28
190404	7.27	K1V	−2.60	20:03:52.128	+23:20:26.47
193664	5.93	G3V	−4.50	20:17:31.328	+66:51:13.27
196850	6.75	G1V	−21.05	20:38:40.190	+38:38:06.33
197076	6.44	G5V	−35.40	20:40:45.141	+19:56:07.93
210667	7.24	G9V	−19.50	22:11:11.913	+36:15:22.79
221354	6.74	K0V	−25.20	23:31:22.209	+59:09:55.86

5.2.3 Self-templates

Because we are primarily interested in relative Doppler shifts, in principle the best possible template to match the star is to use a spectrum of the star itself. There are two drawbacks to doing this. First, you do not get an absolute measure of the RV of

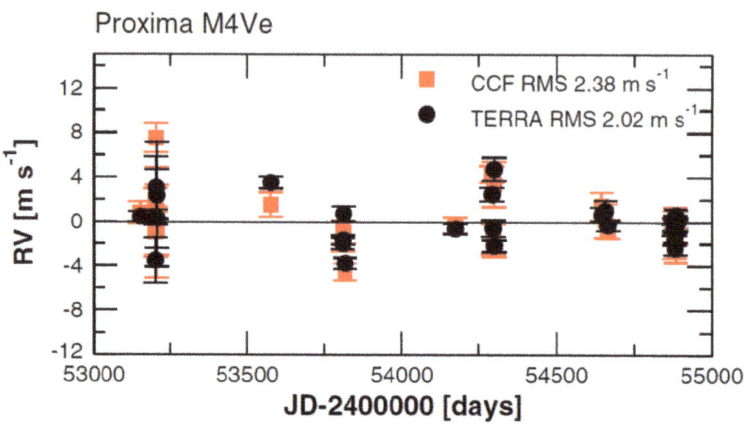

Figure 5.3. Radial velocities of Proxima Centauri calculated using HARPS–TERRA (black dots) compared to those obtained with the standard CCF method (red squares). (Reproduced from Anglada-Escudé & Butler 2012. The American Astronomical Society. All rights reserved.)

the star. This is generally not a concern as we are searching for companions and are only interested in the relative changes of the stellar RV. A more serious issue is that it is difficult to get a very high signal-to-noise spectrum of the star to be used as the template. This should have a much higher S/N than the stellar spectrum for which you want to calculate the RV.

As an improvement to the standard HARPS reduction pipeline which uses the standard CCF and a digital mask, Anglada-Escudé & Butler (2012) proposed using the so-called HARPS–TERRA method. Basically, you produce a master, high signal-to-noise stellar template by co-adding all of the stellar observations. Because these are generally taken over a long time span, you must correct for any Doppler shifts of the star as well as put all spectra on the same wavelength scale. It is thus an interactive process. You first go through all the observations, calculating the Doppler shifts from a single template spectrum of the star, typically the one with the highest S/N ratio. The Doppler-shift-corrected spectra are then all co-added and you run through the calculation a second time using the master, coadded, high-S/N master spectrum as the template. The same strategy is also behind the SERVAL RV data reduction software for CARMENES (Zechmeister et al. 2018).

One can use the CCF method on this master template, but an improvement in the RV performance can be achieved by performing a least-squares fitting of the stellar observation with the template spectrum. Both HARPS–TERRA and SERVAL find the best fit, in a χ^2 sense, to the observed spectrum using the high-S/N template. Figure 5.3 compares the RVs obtained on Proxima Centauri using TERRA–HARPS and the standard CCF method. A slight improvement of $\sigma = 2.02$ m s^{-1} from $\sigma = 2.38$ m s^{-1} is realized using TERRA–HARPS. The improvement is slight, but in the RV business, one must fight for every fraction of m s^{-1} improvement in the precision.

5.2.4 Mismatched Template and Stellar Spectra

When performing a CCF using a template that is not a perfect match to the stellar spectrum, you may end up with a slightly different or even discrepant result, depending on the type of star for the cross-correlation. The magnetic Ap stars offer us an extreme case that is useful for instructive purposes.

The magnetic Ap stars have spectral types between B0 and F5, and constitute 15% of the main-sequence stars in that interval. The distinguishing characteristics of these stars are the anomalous abundances of Cr or Si, as well as Eu, and other rare earth elements. They also possess large dipole fields that are usually inclined to the rotation axis of the star (oblique rotator). The magnetic field organizes these elements into abundance spots, and each element can have a different spot distribution (e.g., Hatzes 1991). Thus, the results may depend on the template one uses to compute the CCF.

Hartmann (2019) surveyed a sample of Ap stars for exoplanets using RV measurements taken with the HARPS spectrograph. He calculated the RVs using both the standard CCD method and HARPS–TERRA. Two program stars were HR 1217 (=HD 24712) and HD 126515.

HR 1217 is an A9 SrEuCr star (Renson & Manfroid 2009) with a known rotational period of 12.45 days determined through photometry (Kurtz et al. 2005) and magnetic field measurements (Bagnulo et al. 1995, Bychkov et al. 2005).

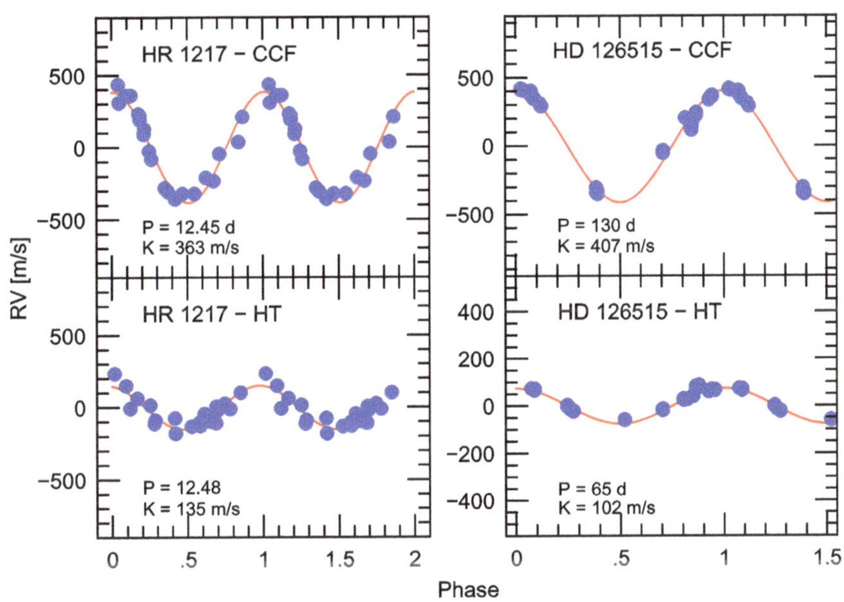

Figure 5.4. Comparison of RVs for two Ap stars, HR 1217 and HD 188041, calculated with the normal CCF and HARPS–TERRA. (Upper left) For HR 1217, the CCF finds a period of 12.45 days and a K amplitude of 363 m s^{-1}. (Lower left) HARPS–TERRA finds the same period, but $K = 135$ m s^{-1}. (Upper right) For the Ap star HD 126515, CCF finds a period of 130 days and $K = 407$ m s^{-1}. (Lower right) HARPS–TERRA finds half the period for this star, and one-fourth the amplitude ($K = 102$ m s^{-1}). Data courtesy of Michael Hartmann.

Figure 5.4 compares the results of RV measurements from standard CCF (HARPS pipeline) and the HARPS–TERRA (HT). The CCF method using the template of a late-type star finds the rotation period of HR 1217 and a velocity K amplitude[1] = 363 m s^{-1}. HARPS–TERRA also finds the correct rotational period, but a lower amplitude of K = 135 m s^{-1}.

The discrepancy occurs due to the differences in the template spectra employed and the spectral lines used in the RV calculation. It could be that iron on HR 1217 is more inhomogeneously distributed, possibly concentrated in one large spot. Because the HARPS pipeline uses the mask appropriate for a late-type star, which is dominated by iron lines, it measures a larger amplitude. HARPS–TERRA uses the actual spectrum of the star, and these are weighted more toward lines of Cr, Sr, and Eu—the dominant spectral features in HR 1217. Some elements might be located in one spot, but at different longitudes, others might be located in multiple spots or rings (e.g. Hatzes 1991a, 1991b). This can result in a diminished amplitude when looking at the integrated RVs of all spectral lines.

HD 126515 is an A2 CrSrEu (Renson & Manfroid 2009) with a rotational period of 129.95 days (Mathys & Hubrig 1997). The CCF finds the correct rotational period and K = 407 m s^{-1} (top-right panel of Figure 5.4). However, HARPS–TERRA finds half the rotational period (65 days) and K = 102 m s^{-1} (lower right panel of Figure 5.4). In this case, the spectral features picked out by the CCF mask come from elements concentrated in a single spot, which produces RV variations at the rotation period. HARPS–TERRA is more sensitive to spectral features from elements most likely distributed in two spots on opposite sides of the star. It detects half the rotation at one-fourth the K amplitude.

So, for magnetic Ap stars, choosing the appropriate mask for the RV calculation can be a powerful tool for probing the distribution of abundance features on these stars.

5.3 CCF Detection of Spectroscopic Binaries

One advantage of the CCF is that it can also detect additional components of spectroscopic binaries (the so-called SB2). If your template has only stellar lines from a single star, additional components will appear as a second peak in the CCF.

Figure 5.5 shows the CCF of an SB2. In the first observation, the CCF appears at a single peak, but there is a hint of an asymmetry toward negative velocities due to a secondary. In the second observation, the two components are clearly resolved and one sees a double CCF. By fitting the peaks individually, one gets the RV measurements of both components.

This works well if both components of the SB2 have similar spectral types to the mask. If there is a large mismatch between the spectral types—for instance an A-type star with a G-type star companion, then the method would be less sensitive. In this case, it is better to first perform the cross-correlation using a template for the hotter star, and then a second time using the template of the cooler companion. The separation of the peaks will give you the RV shift between the two stars.

[1] Traditionally, the RV amplitude is called the "K-amplitude" (see Chapter 8).

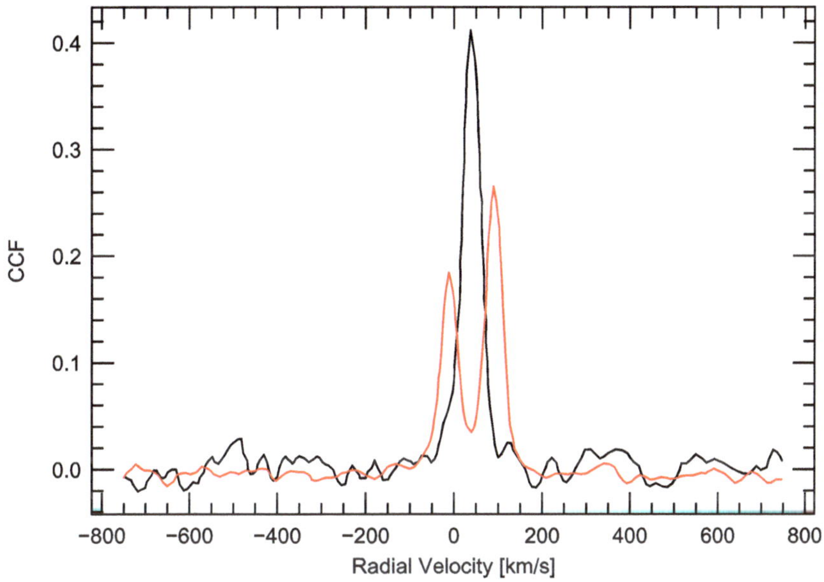

Figure 5.5. CCF of an SB2 spectroscopic binary observed at two epochs. The CCF shown by the black line shows a hint of asymmetry, possibly indicating a second component. The CCF shown by the red line shows the two components clearly separated. (Figure courtesy of Petr Kabath.)

If the two peaks, however, are too close, and they have a large height ratio, then the respective CCF peaks may not be easily resolved. It is better to use a "two-dimensional" version of the CCF. The TwO Dimensional CORrelation or TODCOR algorithm does exactly this (Zucker & Mazeh 1994). TODCOR has proven to be an efficient tool for measuring RVs of two components of a binary system (Mazeh et al. 1995; Torres et al. 1995; Zucker et al. 2003, 2004). An application of TODCOR has also been extended to multi-order spectra (Zucker et al. 2003, 2004).

TODCOR assumes that the observed spectrum is composed of two known spectra, but with unknown shifts. It calculates the cross-correlation against a combination of all possible radial velocity shifts. The peak in the two-dimensional CCF reveals the velocity of the respective components.

Zucker & Mazeh (1994) demonstrated the efficacy of the method using a simulated spectrum consisting of a spectrum of a double-lined spectroscopic binary with A- and G-type components. Each star has a rotational velocity of 40 km s^{-1}, and the G-type star had a Doppler shift, V_G, of 20 km s^{-1} with respect to the primary, which has an absolute Doppler shift, V_A, of zero. The intensity ratio of the stars was 0.25, and the S/N level of the synthetic spectrum is 20.

The left panels of Figure 5.6 show the results of the one-dimensional cross-correlation, first using the A-star template and then the G-star template. The CCF peak of the primary (A star) is easily detected, but that of the G-type star is indistinguishable from the primary peak.

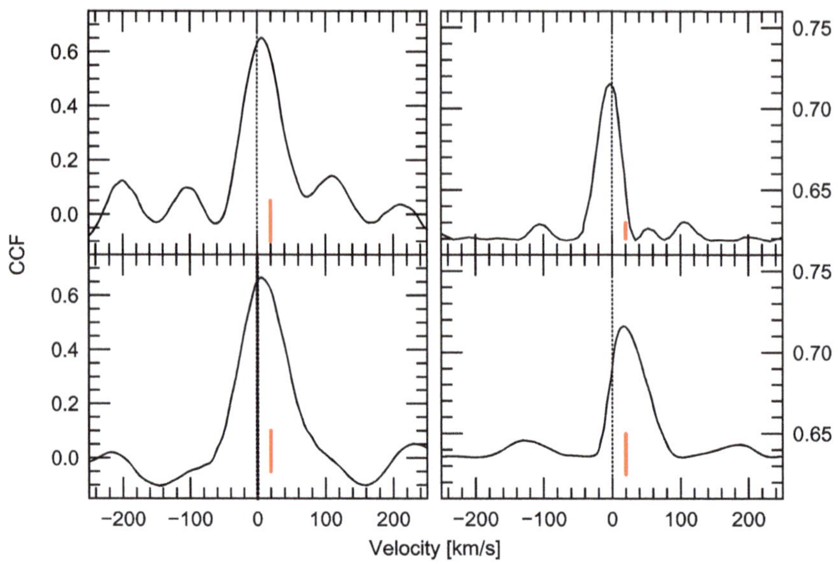

Figure 5.6. Cross-correlations of a synthetic composite spectra consisting of an A-type plus a G-type star. The dashed vertical line indicates the velocity of the A star and the red vertical line marks the velocity of the G star. A cross-correlation with an A-star template easily finds the A-star component (top left). A cross-correlation with a G-star template fails to find this component (bottom left). The right panels are cross sections of the two-dimensional correlation function. It easily finds the A-star component (top) and the G-star component (bottom). (From Zucker & Mazeh 1994.)

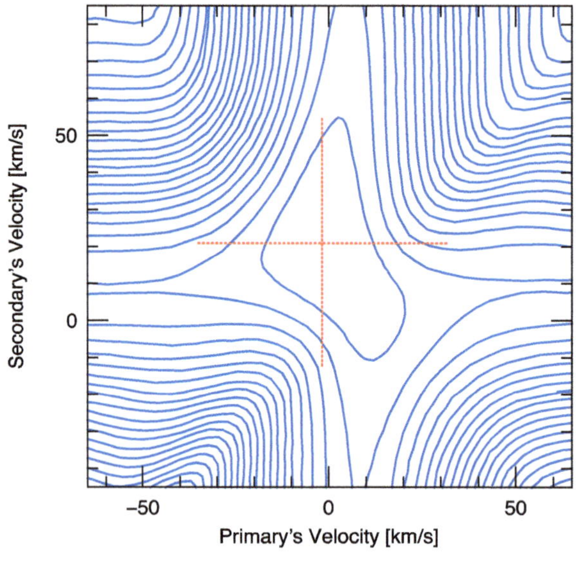

Figure 5.7. Contour plot of the two-dimensional cross-correlation function of the simulated spectrum of A and G stars. The dashed cross is centered on the maximum. (From Zucker & Mazeh 1994, The American Astronomical Society. All rights reserved.)

The contour of the two-dimensional CCF (Figure 5.7) shows a maximum $V_A = 0.0 \pm 0.7$ km s^{-1} and $V_G = 20.5 \pm 0.7$ km s^{-1}. No specification was given for the intensity ratio, but the algorithm found a value of 0.27 ± 0.04, in excellent agreement with the true value.

The right panels of Figure 5.6 show cross sections of the two-dimensional CCF. Each CCF is a one-dimensional cut along the dashed line in Figure 5.7, freezing the spectrum. In this case, the peak of the secondary component is easily identified. This method has shown that it is able to detect secondary components that are a thousand times fainter than the primary.

TODCOR can be a powerful tool for eliminating false-positive detections of transiting candidate planets found by photometric surveys. The most common types of false positives come from eclipsing binaries (EBs). where the secondary component cannot be seen in the spectrum. There are three cases for these: (1) the EB is a faint companion in orbit around the primary star, (2) an EB that is in orbit around the primary as part of a hierarchical system, and (3) a background EB that is not physically bound to the primary.

The first case is easily identified with a few RV measurements as this will show a large reflex motion (~km s^{-1}). The last two cases can be somewhat insidious as the primary star will show little reflex motion in the case of a long-period hierarchical system, or no Doppler motion for a background EB. TODCOR may reveal the presence of the faint companion.

Tal-Or et al. (2011) used a version of TODCOR (called TOMOR in the paper) to show that a transit candidate found by the *CoRoT* space mission (Baglin et al. 2006) was actually a hierarchical triple system. The star, LRa_E2_0121 (hereafter C0121), was discovered as a transiting candidate by *CoRoT*. The transit light showed a transit depth of 0.3% occurring every 36.3 days (Tal-Or et al. 2011). This was consistent with a Neptune-size companion. RV measurements of C0121 showed this to be a member of a binary system in a long period (top-right left panel of Figure 5.8). Still, this did not rule out a planetary nature for the transit curve.

An application of TODCOR (right panel Figure 5.8) easily detected the secondary companion, C0121B. This could then be used to measure the RV variations of the companion to C0121B, which revealed a stellar companion with an amplitude of 62 km s^{-1}. So here was a case where the transit-like event was due to an EB in a long-period orbit and with eclipse light curve diluted by the light of the primary star, C0121A.

5.4 Fahlman–Glaspey Shift Detection

An alternative to the classic cross-correlation method is the shift detection method of Fahlman & Glaspey (1973). In calculating Doppler shifts, we can assume that two observations of the same star, $a(\lambda)$ and $b(\lambda)$, represent the same spectrum, $S(\lambda)$, but that $b(\lambda)$ is shifted by a small amount $\Delta\lambda_0$:

$$a(\lambda) = S(\lambda);\ b(\lambda) = S(\lambda + \Delta\lambda_0). \tag{5.4}$$

If we apply a small shift, $\Delta\lambda_s$, to $b(\lambda)$ and construct the difference spectrum,

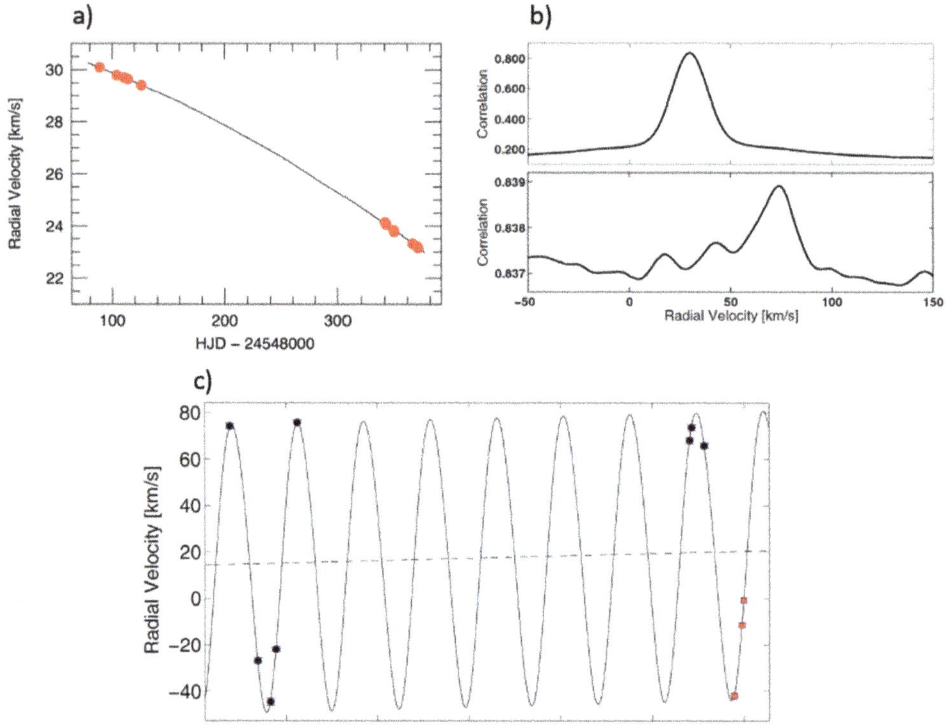

Figure 5.8. Panel a: RVs of the primary in C0121 (red circles). The black line is a second-order polynomial fit. Panel b: TODMOR plot of spectrum of C0121. The upper and lower panels represent cuts of the secondary and primary in the two-dimensional cross-correlation function. Panel c: RVs of C0121Bb. The solid line is the Keplerian orbit. From Tal-Or et al. (2011), reproduced with permission © ESO.

$$d(\lambda, \Delta\lambda_0; \Delta\lambda_s) = a(\lambda) - b(\lambda - \Delta\lambda_s), \tag{5.5}$$

which to first order becomes

$$d(\lambda, \Delta\lambda_0; \Delta\lambda_s) = \Delta\lambda_s \frac{dS(\lambda)}{d\lambda} - \Delta\lambda_0 \frac{dS(\lambda)}{d\lambda} = (\Delta\lambda_s - \Delta\lambda_0)\frac{dS(\lambda)}{d\lambda}. \tag{5.6}$$

Now, consider the quadratic function

$$y(\Delta_s) = \sum_\lambda d^2(\lambda, \Delta\lambda_0, \Delta\lambda_s), \tag{5.7}$$

where the summation extends over the line profile or spectral region.

To lowest order, this expression becomes

$$y(\Delta\lambda_s) = (\Delta\lambda_s^2 - 2\Delta\lambda_s\Delta\lambda_0 + \Delta\lambda_0^2)\sum_\lambda \left[\frac{dS(\lambda)}{d\lambda}\right]^2. \tag{5.8}$$

In other words, $y(\Delta\lambda_s)$ is a parabola. So, if $a(\lambda)$ is our observation for which we want to measure the Doppler shift, we merely take $S(\lambda)$ (your reference spectrum),

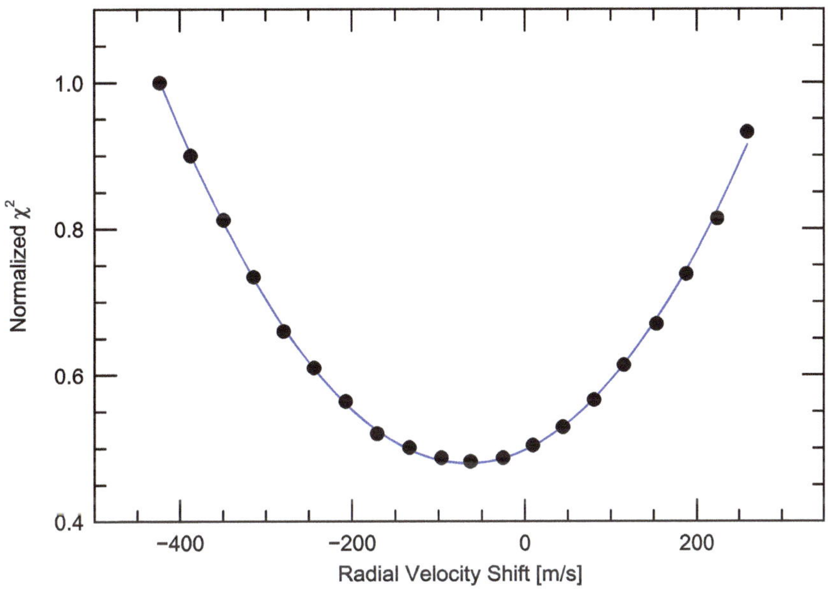

Figure 5.9. Doppler shift detection using the Fahlman–Glaspey algorithm. For small shifts, the relative change varies according to a second-order polynomial. A fit to these data yields the minimum value of the parabola.

apply a trial shift, $\Delta\lambda_s$, to produce $b(\lambda)$, take the difference squared, and sum over the entire spectral region of interest, and plot this as a function of your trial shift. The result should ideally be a parabola with a minimum at $\Delta\lambda_s = \Delta\lambda_0$ (Figure 5.9). The parabolic minimum defines the shift which minimizes the differences between $a(\lambda)$ and $b(\lambda)$ in a least-squares sense.

This method has the advantage over the standard cross-correlation method in that it produces a well-known function (parabola) that can be fit. The CCF may have an approximate Gaussian shape, but this may not always be the case.

Another advantage of the Fahlman–Glaspey method is that it can find minimum values for parameters other than a Doppler shift. For example, if you did not know the exact dispersion in a region of your observed spectra, you can simply vary the dispersion and compare that to a template spectra of known dispersion, and fit a parabola near the minimum χ^2.

However, this method only works if the shift, $\Delta\lambda_0$, is small. This means for large Doppler shifts, you need to find a coarse shift with another method (e.g., cross-correlation) and then refine $\Delta\lambda$. The Fahlman–Glaspey cross-correlation is the basis for the RV reduction program of the McDonald Observatory Planet Search Program (Cochran & Hatzes 1994).

References

Anglada-Escudé, G., & Butler, R. P. 2012, ApJS, 200, 15

Baglin, A., Auvergne, M., Boisnard, L., et al. 2006, COSPAR Meeting, 36th COSPAR Scientific Assembly, 3749

Bagnulo, S., Landi Degl'Innocenti, E., Landolfi, M., & Leroy, J. L. 1995, A&A, 295, 459

Bychkov, V. D., Bychkova, L. V., & Madej, J. 2005, A&A, 430, 1143
Cochran, W. D., & Hatzes, A. P. 1994, Ap&SS, 212, 281
Fahlmann, G. G., & Glaspey, J. W. 1973, in Astronomical Observations with Television Type Sensors, ed. W. Glaspey (Vancouver: Institute of Astronomy and Space Science)
Hartmann, M. 2019, PhD thesis, Friedrich Schiller Univ.
Hatzes, A. P. 1991a, MNRAS, 248, 487
Hatzes, A.P. 1991b, MNRAS, 253, 89
Kurtz, D. W., Cameron, C., Cunha, M. S., et al. 2005, MNRAS, 358, 651
Mathys, G., & Hubrig, S. 1997, A&AS, 124, 475
Mazeh, T., Zucker, S., Goldberg, D., et al. 1995, ApJ, 449, 909
Melo, C. H. F., Pasquini, L., & De Medeiros, J. R. 2001, A&A, 375, 851
Murdoch, K., & Hearnshaw, J. B. 1991, Ap&SS, 186, 169
Queloz, D., Allain, S., Mermilliod, J. C., Bouvier, J., & Mayor, M. 1998, A&A, 335, 183
Renson, P., & Manfroid, J. 2009, A&A, 498, 961
Tal-Or, L., Santerne, A., Mazeh, T., et al. 2011, A&A, 534, A67
Tonry, J., & Davis, M. 1979, AJ, 84, 1511
Torres, G., Stefanik, R. P., Latham, D. W., & Mazeh, T. 1995, ApJ, 452, 870
Udry, S., Mayor, M., & Queloz, D. 1999, in ASP Conf. Ser. 185, IAU Colloq. 170: Precise Stellar Radial Velocities, ed. J. B. Hearnshaw, & C. D. Scarfe (San Francisco, CA: ASP), 367
Walker, G. A. H., Yang, S., Campbell, B., & Irwin, A. W. 1989, ApJ, 343, L21
Zechmeister, M., Reiners, A., Amado, P. J., et al. 2018, A&A, 609, A12
Zucker, S., & Mazeh, T. 1994, ApJ, 420, 806
Zucker, S., Mazeh, T., Santos, N. C., Udry, S., & Mayor, M. 2003, A&A, 404, 775
Zucker, S., Mazeh, T., Santos, N. C., Udry, S., & Mayor, M. 2004, A&A, 426, 695

AAS | IOP Astronomy

The Doppler Method for the Detection of Exoplanets

A P Hatzes

Chapter 6

The Iodine Cell Method

In calculating the Doppler shifts with the iodine cell method, one can also use the cross-correlation method. You simply cross-correlate your observations of a star observed through the iodine cell with an observation of the star observed without the iodine cell (template). You then cross-correlate your observation with a spectrum of pure iodine (fiducial) typically done by observing a white-light source through the cell. This gives you the instrumental shifts, which can be applied to the Doppler shifts of your observations. This will largely correct the large mechanical shifts like those seen in Figure 4.1.

The real power of the iodine cell method, however, comes from using information about the instrumental response of the spectrograph, or the so-called instrumental profile. Depending on the stability of the spectrograph, these can be rather large and are not taken into account when using simple cross-correlation.

Before we delve into the details of the iodine method, we first should give some important background knowledge on the instrumental profile and how it can influence the radial velocity (RV) measurement.

6.1 The Instrumental Profile

The instrumental profile[1] (IP) represents the instrumental response of your spectrograph. Imagine that you have a monochromatic beam that has an infinitesimally small width in wavelength, i.e., a δ function. If the spectrograph were a perfect instrument, after passing through all of its optics, the δ function will not have changed (top panel of Figure 6.1). What you actually see is a blurred function of finite width (lower panel of Figure 6.1). This is typically a Gaussian-like profile, and for a properly designed spectrograph, its FWHM should span at least two detector

[1] Some people like to use the term "line profile" to signify that it affects the shape of the spectral lines. This is analogous to the "point-spread function" used in imaging. I will use instrumental profile as it results from the response of your instrument.

doi:10.1088/2514-3433/ab46a3ch6

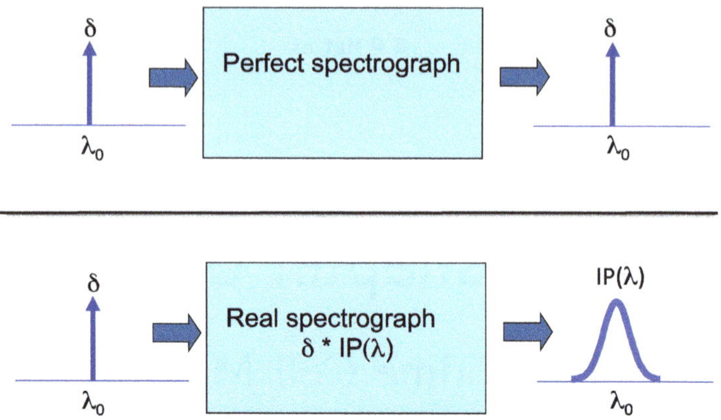

Figure 6.1. Schematic of the IP function. (Top) A perfect spectrograph with no instrumental response: a monochromatic δ function produces a δ function after passing through the spectrograph. (Bottom) For a real spectrograph with an instrumental function, $IP(\lambda)$, a δ function is recorded as $I(\lambda)$, i.e., the convolution of the δ function and $IP(\lambda)$.

pixels. This IP is convolved with the spectrum of the incoming light. (In Chapter 7, we will discuss the process of convolution in greater detail.) For an incoming δ function, the observed spectrum is the IP itself. For stars, all features in the stellar spectrum are convolved with the IP; if it has an asymmetric shape, this asymmetry will be introduced in all of the spectral lines.

For the detection of exoplanets, we are measuring relative Doppler shifts. This means we really do not care about the absolute shape of the IP—it can even be asymmetric. All we care about is that this asymmetry with the same shape is present in every observation we make. The problem for precise RV work is if the shape of the IP changes.

This is demonstrated in an extreme case in Figure 6.2. The left panel shows an IP that is an asymmetric Gaussian with a centroid that is located +0.17 pixels from the centroid of a symmetric Gaussian (dashed line). This is the equivalent of a velocity shift of +250 m s^{-1} for a spectrograph with resolving power R = 100,000. Because the stellar spectrum is convolved with this IP, this asymmetric shape will be imposed on all spectral lines, and all will have a relative Doppler shift of +250 m s^{-1} with respect to stellar spectrum had it been observed with a spectrograph having a symmetric IP.

For the RV detection of planets, this shift is irrelevant because it will introduce a zero-point offset of +250 m s^{-1} to every observation. However, if the IP were to change into the blue asymmetric profile (right panel of Figure 6.2), then the centroid of each spectral line would shift by −0.17 pixels (−250 m s^{-1}) with respect to the symmetric IP, or a total of +500 m s^{-1} from the first observation. This velocity change is not from the star, but from changes in the shape of the IP and would be present even if no other instrumental (e.g., mechanical) shifts were present.

It is important to emphasize that unless you have a spectrograph that is thermally and mechanically stabilized and resides in a vacuum tank held at low pressure, the IP will be different for every stellar observation that you make due to variations

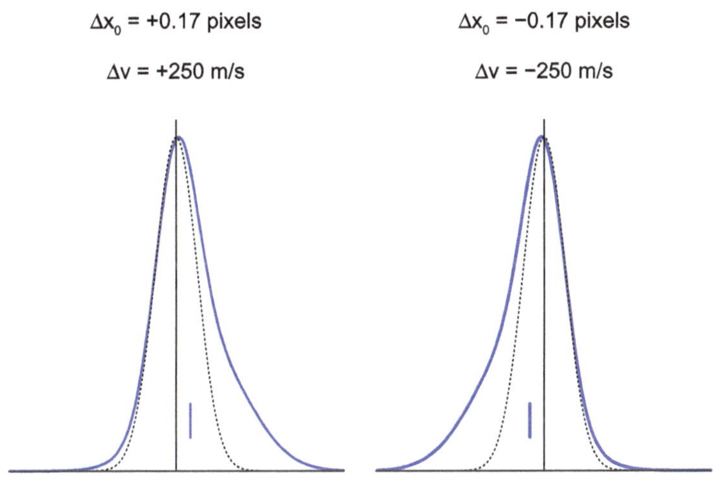

Figure 6.2. (Left) The solid line is an asymmetric instrumental profile (IP). The dashed line is symmetrical IP profile. The vertical line represents the centroid of the asymmetric IP. It is shifted by +0.17 pixels or +250 m s^{-1} for an $R = 100{,}000$ spectrograph. (Right) An IP profile that is asymmetric toward the blue side. Such change in the IP would introduce a total instrumental shift of $\Delta v = 500$ m s^{-1} in the RV measurement.

(e.g., temperature) in the spectrograph, instrumental shifts, etc., and this will result in an instrumental Doppler shift.

Figure 6.3 shows two 45 minute windows of the instrumental shifts of the Tautenburg Coudé Echelle spectrograph (TCES; shown in Figure 4.1). Again, pixel shifts have been converted to a Doppler shift in velocity. If you want to observe a faint star with a 45 minute exposure, the spectral lines will be moving by this amount on your detector, and it will affect the shape of the spectral lines (measured in the IP). The instrumental shifts have an rms of $\sigma = 13$ m s^{-1} and are random about a zero-velocity shift. This will result in slightly more broadened line profiles, but these would remain more or less symmetrical—a best-case scenario in terms of affecting the RV precision.

This is not the case if, during your exposure, the spectrograph had the instrumental shifts shown in the lower panel. First of all, the rms of these shifts is a factor of 2 higher at $\sigma = 26$ m s^{-1}. More importantly, the instrument would have taken a +100 m s^{-1} jump in a very short time and would then have had relatively small shifts for about 15 minutes before undergoing a linear change of +40 m s^{-1} over the remaining exposure. This type of motion of the spectral lines on the detector would surely introduce asymmetries into the spectral line profiles. Whether such could be actually measured depends on their magnitude, but if they are large enough to introduce a detectable Doppler shift, one should see accompanying changes in the IP.

Because the influence of instrumental shifts on the IP, most high-precision, state-of-the-art spectrographs, like HARPS, are operated in a vacuum and extraordinary measures are taken to ensure the thermal, pressure, and mechanical stability of the instrument. This result is a stable IP profile (hopefully), but this comes at

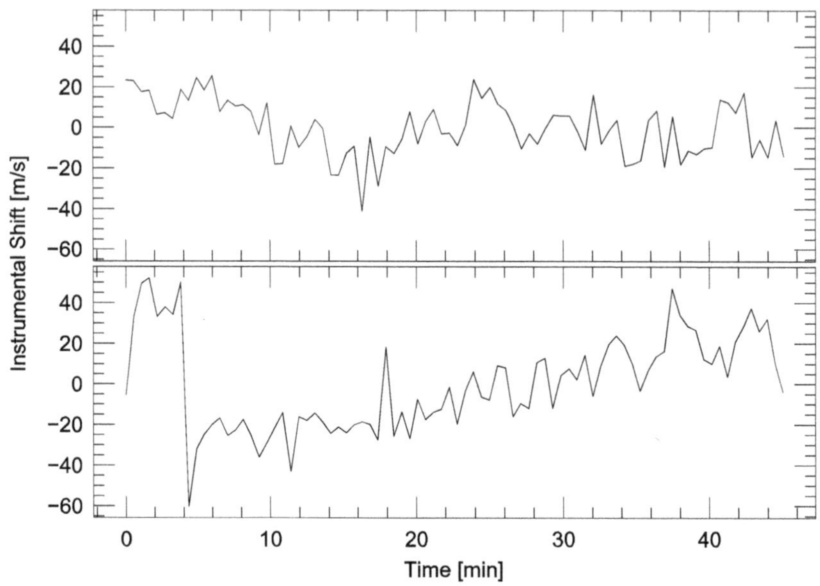

Figure 6.3. Instrumental shifts over two 45 minute time spans measured with the Tautenburg Coudé Echelle Spectrograph. (Top) The shifts have an rms scatter of 13 m s^{-1} about a mean value of zero. These shifts would blur the IP, but it would still remain more or less symmetrical. (Bottom) These shifts show a sharp change of $\approx$100 m s^{-1}, then stay relatively flat for approximately 10 minutes, followed by a trend. Observations of a star taken during this time span would result in a more asymmetric IP.

considerably higher costs for the development and construction of the spectrograph. What can be done if you do not have access to such a stable spectrograph? For instance, it is an existing instrument that was not really designed for precise RV measurements. Your only recourse is to correct for the changes in the IP.

A tremendous advantage of the iodine method over the simultaneous Th–Ar method is that one can use information in the iodine lines to model the IP. This can be done because iodine lines are unresolved even at a resolving power of $R = 100{,}000$ (see Figure 2.10). Thus, they carry information about the IP of the spectrograph. This is not the case for thorium emission lines from a hollow cathode lamp that have an intrinsic width comparable to, if not greater than, the width of the IP.

Figure 6.4 shows the improvement of RV measurements when treating the changes in the IP. It shows RV measurements of 51 Peg taken with the TCES and an iodine absorption cell. Without the IP modeling (simple cross-correlation), one sees the orbital motion, but the scatter about the orbital solution is $\sigma = 20$ m s^{-1}. This is most likely due to the instrumental shifts during the exposure. Including IP modeling, this scatter is reduced to 3.4 m s^{-1}. The simple "cross-correlation" method is inadequate for correcting the RV shifts due to this changing IP function.

6.2 Modeling the IP with the Iodine Cell Method

How do you measure your IP in practice? What you would like to do is feed your spectrograph with a monochromatic δ function of light and record the output. This

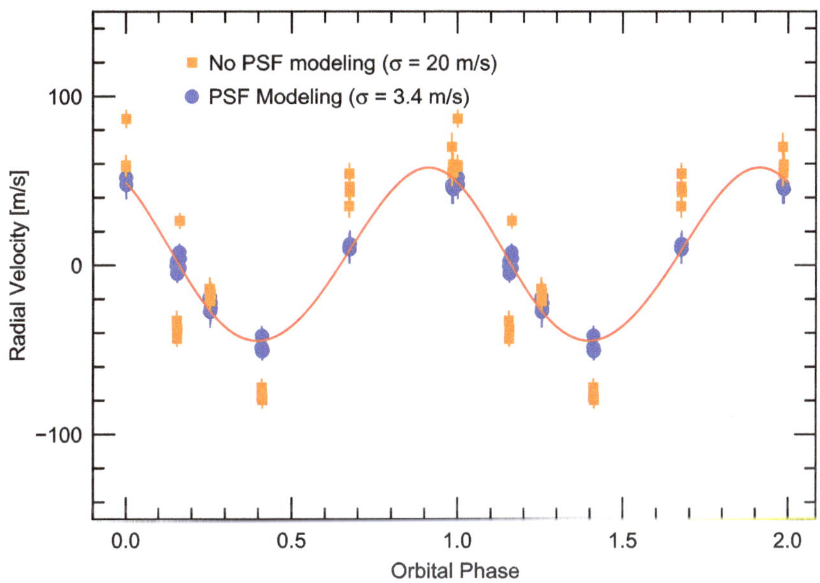

Figure 6.4. RV measurements of 51 Peg using an iodine absorption cell but without modeling of the IP (squares). Points are repeated for the second orbital cycle. The rms scatter about the orbital solution (curve) is 20 m s^{-1}. The same, but including IP modeling (dots). The rms scatter has been reduced to 3.4 m s^{-1}.

of course would have to be done over the wavelength range of the spectrograph because the IP is probably a function of wavelength. This is simply impossible.

Valenti et al. (1995) proposed a clever method to model the IP in practice. Recall the FTS in Chapter 2, which is a type of spectrometer that can achieve a very high resolving power of R = 500,000–1,000,000, or about a factor of 10 higher than the resolving power of virtually all spectrographs used for high-precision RV work. Valenti et al. proposed taking a very high resolution, high signal-to-noise ratio spectrum of iodine (an observation of a white-light source taken through the cell) with an FTS (see Figure 2.10). You then rebin this FTS spectrum to the same dispersion as your RV spectrograph typically taken with R = 60,000–100,000. This binned FTS spectrum is thus a good representation of the monochromatic δ function response of your spectrograph. You then find a model for the IP that when convolved with this super high-resolution iodine spectrum will produce the observed iodine spectrum.

Figure 6.5 shows a section of the iodine spectrum, shown as dots, taken with the coudé spectrograph of the 2.7 m telescope at McDonald Observatory. The dashed red line is the FTS iodine spectrum of the iodine cell binned to the same dispersion as the data. Clearly, the actual observations have a much lower spectral resolution. The solid line shows the binned FTS spectrum after it was convolved with a suitable model IP function.

It is important to have a good mathematical representation of the IP. Most data reduction pipelines for iodine cell follow the prescription first described by Valenti

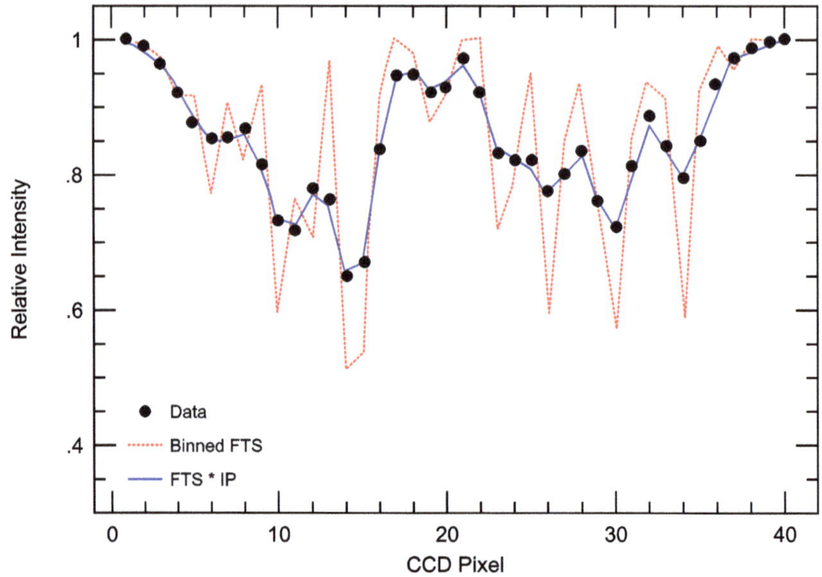

Figure 6.5. (Dots) Measurements of the spectrum of I_2 over a short wavelength region. (Dashed line) Spectrum of I_2 taken at high resolution using an FTS and binned to the dispersion of the data. (Solid line) The FTS I_2 spectrum convolved with the model IP.

et al. (1996), who proposed modeling the IP as a sum of several Gaussian functions. Gaussian profiles are chosen because the IP, to first order, is a Gaussian profile, and the addition of satellite Gaussian components makes it easy to introduce asymmetries. One has a central Gaussian profile and 2–10 satellite Gaussians. The number of satellite Gaussians depends on the complexity of the IP profile and how rapidly it changes with time. The central Gaussian located at the origin in IP space has a height of unity. Each satellite Gaussian has its own position, width, and height.

One also typically oversamples the IP by a factor of 5, the so-called "IP space," so that it has a finer grid than the "detector space." This means that you have to rebin your IP-space Gaussian by a factor of 5 in order to replicate the observed profile.

The IP parameterization process for on echelle order of the TCES on the Tautenburg 2 m Alfred Jensch Telescope is illustrated in Figure 6.6. The IP in this case is modeled by a central Gaussian (red line) plus four satellite Gaussians. The black line (tall Gaussian-like profile) represents the sum of the Gaussians.

You will note that the IPs in the top panel of the figure have a slight dip in the center caused by the contribution of the two close-in Gaussians. One characteristic of the TCES is that Th–Ar lines appear to be a bit flat topped. This is easily understood when one recalls that a spectrograph produces an image of the slit on the detector. A slit is box shaped, so it is reasonable to expect that the IP is a bit flat topped (actually a convolution of a Gaussian with a box function). The only way to reproduce a flat-topped Gaussian is by having two close-in Gaussians about the central one.

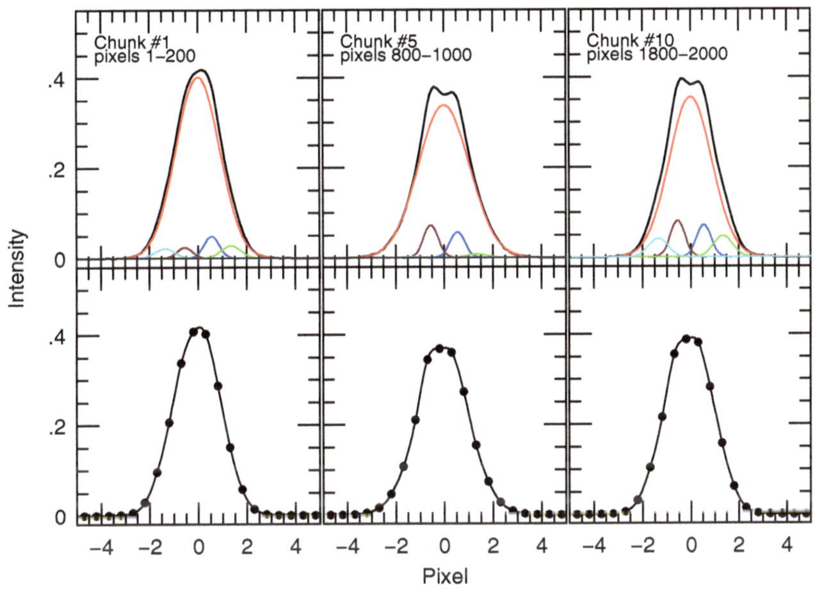

Figure 6.6. A model of the IP using multi-Gaussian profiles. The smaller Gaussians represent satellite Gaussians that are combined with a larger central Gaussian shown in red. The black line represents the final IP (tallest Gaussian) that is a sum of the central plus satellite Gaussians. The model IP is shown for the blue side (left), central (center), and red side (right) of the spectral order. The IP was calculated using 200 chunk pixels of an I_2 spectrum. Note that one pixel on the CCD ("detector" space) corresponds to five pixels in "IP space."

Note that the dip in the IP appears only in the oversampled version. When one rebins to detector space and samples it like the real data (lower panel of Figure 6.6), the IP appears a bit flat topped. So, in this case, the IP modeling process is doing a decent job of reproducing the expected IP shape. It is important to remember that large changes in the center of the IP will have relatively little effect on the Doppler measurement. Most IP shifts result from an asymmetric profile, and these are governed largely by the satellite Gaussians.

6.3 Influence of Changes in the IP

The IP can change across the spectral format and even across a single spectral order and not properly treating these changes can degrade your measurement precision. This is best seen in the modeling of the IP across spectral orders for the Hamilton Echelle spectrograph (Vogt 1987). Built in the mid-1980s, it was one of the first large-wavelength echelle spectrographs that challenged fabrication techniques at the time. Early RV measurements with the spectrograph could rarely achieve an RV precision better than about 30 m s^{-1} in spite of the best efforts to model the IP (G. Marcy, 1994, private communication). The problem was traced to a wrongly configured optical component which required a strong aspherical shape. Measurement techniques at the time could not determine the shape of the element with sufficient accuracy. Once better measurement techniques were available, it was found that this element had the wrong shape (S. Vogt, 1994, private

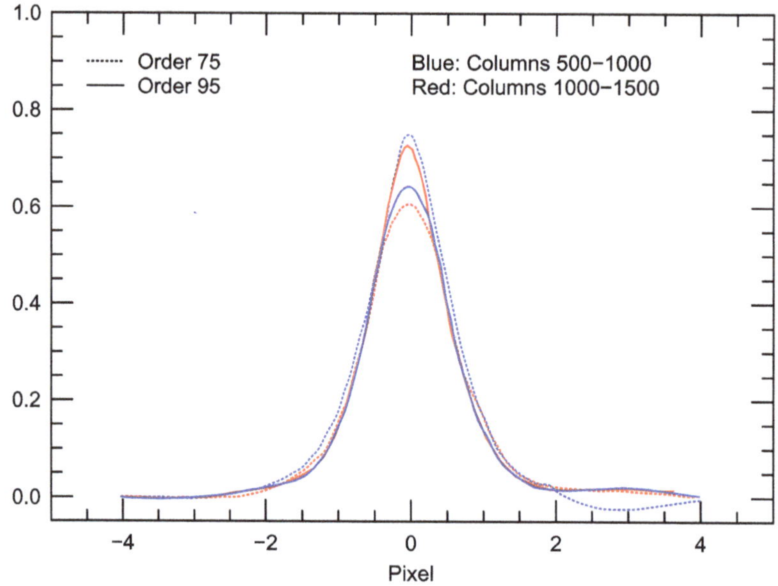

Figure 6.7. The instrumental profiles from the Hamilton Echelle spectrograph analyzed over 500 CCD columns. The dashed lines are from spectral order 75 and the solid lines for spectral order 95. Red is for CCD columns 500–1000 and blue for columns 1000–1500 (from Valenti et al. 1995).

communication). Correcting the component to the proper shape resulted in an improved RV precision by an order of magnitude. This only emphasizes that the foundation for any precise RV spectrograph is a good spectral design that works to specifications. Modeling of the IP improves the RV measurement precision, but it can only go so far—it is not a perfect modeling process. Figure 6.7 shows the IP modeled for various orders and locations of the Hamilton spectrograph. One can see that there can be large changes not only from order to order, but also along a spectral order (column).

The IP not only can show spatial variations, but also temporal ones, and these can mimic the variations of a planetary signal in the RV if these are not treated properly. The left panel of Figure 6.8 shows the changes of the IP for the Coudé Echelle spectrograph (CES) that was used at the La Silla Observatory of the European Southern Observatory up until the mid-1990s. These were derived using the iodine method (Endl et al. 2000). One can see large changes in the IP over the course of five years. The central panel shows the RV measurements calculated without incorporating changes in the IP. There are clear, long-term, sine-like variations. Indeed, a periodogram analysis (see Chapter 7) reveals a peak at a period of ≈2000 days. This signal has a false-alarm probability that it is caused by noise of a mere 0.03%—a highly significant signal. An orbital solution (curve) yields a companion mass of 2.2 M_{Jup} orbiting with a period of 2020 days, a Jupiter analog!

Even though these variations are real, they are not due to a planetary companion, but to instrumental shifts caused by the changing IP (left panel of Figure 6.8). The right panel Figure 6.8 shows the RV variations after Endl et al. (2000) included IP

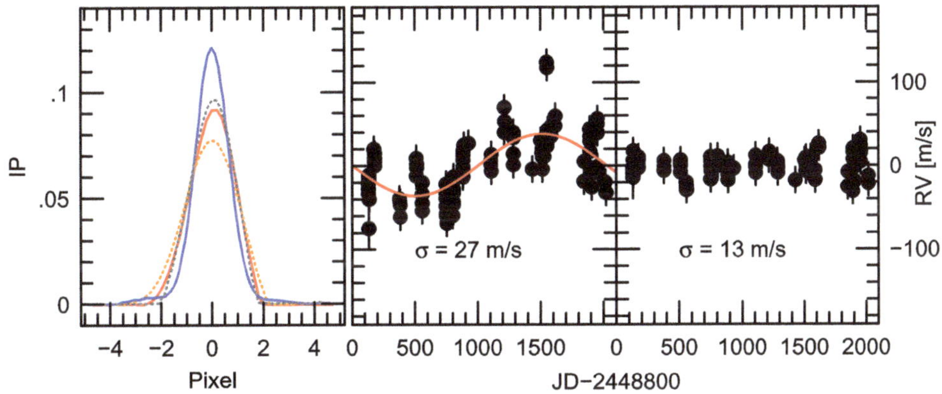

Figure 6.8. (Left) Radial velocity measurements of τ Cet taken with the former CES at La Silla, Chile. No IP modeling was done in calculating the Doppler shifts. (Right) The calculated Doppler shifts using the same data but with IP modeling (from Endl et al. 2000).

modeling in calculating the RVs. The sine-like variations have disappeared, and the rms scatter has been reduced from 27 m s^{-1} to 13 m s^{-1}.

6.4 Ingredients for the Iodine Cell Method

There are three main ingredients (besides an iodine cell) needed to produce stellar RVs with the method:

1. A high-resolution spectrum of iodine taken with the cell. This is your *fiducial*.
2. A high-resolution spectrum of your star taken without the iodine cell. This is your *template*.
3. A spectrum of your star taken through the cell and for which you want to calculate a Doppler shift. This is your *data*.

Figure 6.9 shows a segment of the iodine spectrum along the stellar spectrum and the spectrum produced by observing the star through the iodine cell.

6.4.1 The Fiducial

After you have constructed the iodine cell, the next step is to scan this (i.e., observe a white-light source through the cell) using an FTS. This should be done at the highest resolving power possible, at least $R = 500{,}000$, although $R = 1{,}000{,}000$ is better. The cell should be scanned at its operating temperature, but it is wise to do this over a range bracketing the nominal operating temperature. In this way, you can assess changes in the iodine spectrum with temperature variations.

If you lack an FTS spectrum of your iodine cell, you can still perform the following analysis by simply using an iodine spectrum taken with your RV spectrograph as your fiducial. This generally degrades the RV precision by only a few m s^{-1}.

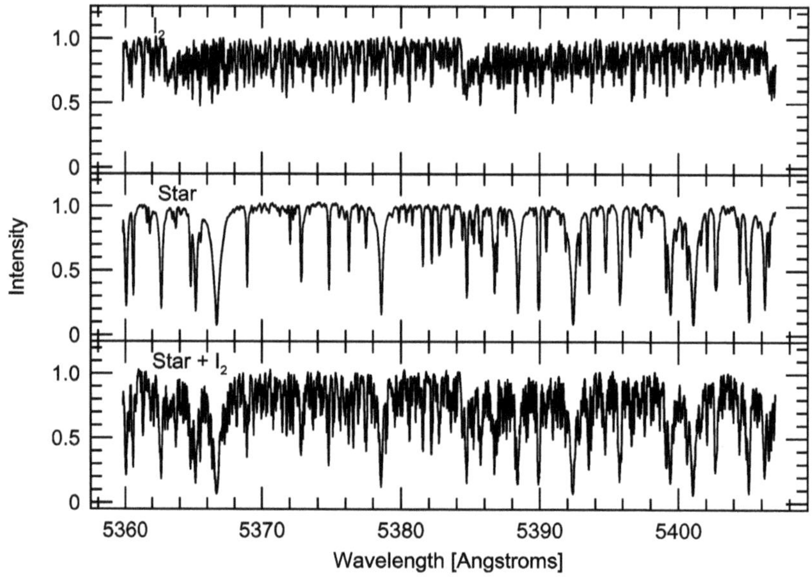

Figure 6.9. (Top) A segment of the spectrum of I_2. (Middle) The same spectral region of a star. (Bottom) The same star as observed through the iodine cell.

6.4.2 The Template

The simplest template that one can use is a spectrum of your program star taken without the iodine cell. This should have as a high signal-to-noise ratio as possible. However, if one wants to achieve the highest precision RV possible, then one should use a spectrum of the star with the IP removed.

In calculating the RVs, we will combine the fiducial I_2 spectrum and convolve the results with the IP. If you use a template spectrum taken with your spectrograph, then this has already been convolved with the IP, so in the data reduction you want to avoid convolving the stellar spectrum twice. We thus need a deconvolved stellar spectrum with the IP removed. This deconvolved spectrum must also be over-sampled, just like our IP function.

Deconvolution is a tricky procedure and entire books can be devoted to the subject (e.g., Jansson 1997). A detailed discussion is certainly beyond the scope of this book. The danger of deconvolution is that it can introduce artifacts and noise into the deconvolved spectrum.

Many deconvolution techniques are done in the Fourier domain. The Fourier transform will be covered in greater detail in the next chapter, but briefly, the complicated process of deconvolution of two functions in the spatial domain reduces simply to the product of their individual Fourier transforms in the Fourier domain. Let $s_{\rm int}$ be the intrinsic stellar spectrum before passing through your spectrograph and its resulting Fourier transform $S_{\rm int}$, and $s_{\rm obs}$ the observed spectrum with transform $S_{\rm obs}$. If i is the instrumental profile with the associated transform I, then what we observe is simply

$$s_{obs} = s_{int} * i, \tag{6.1}$$

which in the Fourier domain reduces to the product of the individual transforms:

$$S_{obs} = S_{int} \cdot I. \tag{6.2}$$

So, the deconvolution is in principle simple; you take the individual transforms of the stellar spectrum and divide this by the transform of the IP and then take the inverse transform.

The problem with this simplistic approach is that it involves a division and if the Fourier transform of the IP has values near zero, then these will produce high values in the divided transforms. As we shall see in the Fourier domain, a peak results in a sinusoidal variation in the spatial domain, the so-called "ringing." To first order, an IP is a Gaussian function whose Fourier transform is also a Gaussian with low values at high frequencies. In the division, because the Gaussian has small values at high frequencies, this will magnify the high-frequency components of the stellar spectrum and introduce noise.

An alternative approach is to produce a deconvolved function using the maximum entropy method, which greatly reduces the noise and artifacts in the deconvolved spectrum. Maximum entropy has also been used to solve ill-posed problems like those encountered in Doppler imaging (Vogt et al. 1987).

Figure 6.10 shows a spectrum of β Geminorum that has been deconvolved from the IP using maximum entropy deconvolution. The lower panel shows the original

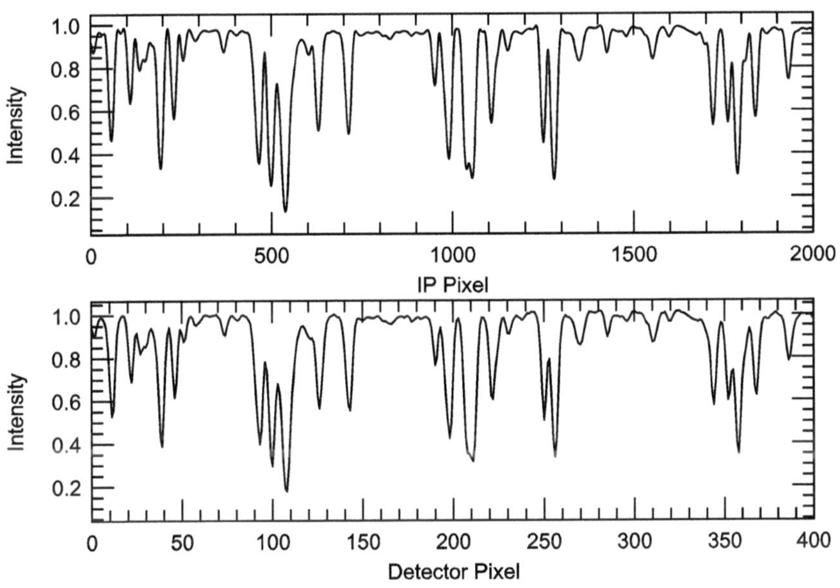

Figure 6.10. (Bottom) A section of the spectrum of β Gem shown in detector pixels. (Top) A deconvolved version of the same spectral region shown in IP pixels (five times oversampled). Note how the lines become sharper and blended features become better resolved.

spectrum as a function of "detector pixels," while the upper panel is the deconvolved spectrum as a function of oversampled "IP pixels."

6.5 Calculation of the Doppler Shift

The calculation of the RV requires solving the equation (Butler et al. 1996)

$$I_{\mathrm{m}} = k[T_{\mathrm{I}_2}(\lambda)I_{\mathrm{S}}(\lambda + \delta\lambda)] * \mathrm{IP}, \tag{6.3}$$

where I_S is the intrinsic (deconvolved) stellar spectrum, T_{I_2} is the transmission function of the I_2 cell, k is a normalization factor, $\delta\lambda$ is the wavelength (Doppler) shift, IP is the spectrograph instrumental profile, and * represents the convolution.

Because the IP can vary from spectral order to order and even across a single order, this equation must be solved in segments or chunks along each spectral order (Figure 6.11). Typically, one uses 10–40 chunks, each covering a wavelength range of about 1–4 Å. The total number of chunks depends on how rapidly the IP changes across a spectral order. If it is too wide in wavelength (too many chunks), the process may not adequately track the changes in the IP, too narrow (too few chunks) and you will get a poor solution to Equation (6.3) (no convergence) or there are just too few spectral lines in the chunk for a good RV measure.

In each of these spectral chunks, the reduction steps should:

1. Remove any slope in the continuum. Given the small size of the chunks, this can be done with a linear function. This requires two parameters.
2. Calculate the dispersion (angstroms/pixel) in the chunk. Again, due to the relatively small chunks it is sufficient to use a second order polynomial that has three parameters.
3. Calculate the IP containing 5–10 Gaussians (more can be used if needed). Each Gaussian (except for the central one) has variable positions, amplitudes, and widths.
4. Apply a Doppler shift to your template spectrum. This is one parameter.
5. Combine the FTS iodine spectrum and the Doppler-shifted template spectrum and convolve this with the IP produced in step 3. You compare this to your observed data by calculating the reduced χ^2. If you have not converged to the desired χ^2, go back to step 1 using the current values as your starting point and vary all the parameters.

So, in the end we are determining a total of 15 parameters when all we care about is one, the Doppler shift!

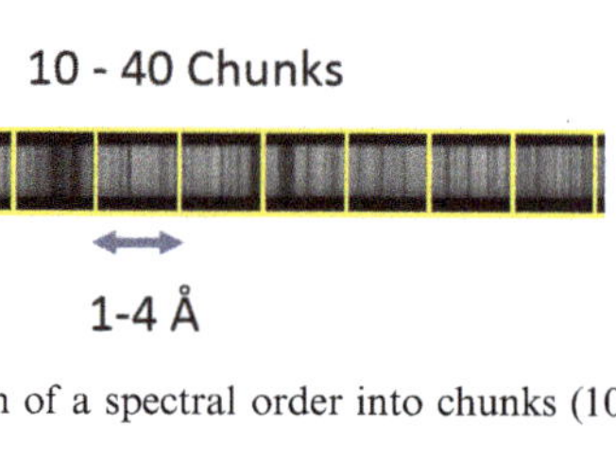

Figure 6.11. Schematic showing the division of a spectral order into chunks (10–40) each with width 1–4 Å.

In principle, allowing all the Gaussian parameters to vary may be too computationally intensive and the process may not converge. In practice, one fixes the location and widths of the satellite Gaussians, allowing only the heights to vary. It is a painstaking process to "crawl" through the χ^2 parameter, changing the location and widths of the Gaussians until a best solution is found. What defines a good solution? The best is to use a standard star which is constant in RV. A good target is τ Cet, which has been shown by most RV groups to have an RV constant to a few m s^{-1}. One varies the parameters of the IP Gaussians until one reduces the rms scatter. Alternatively, one can chose a known exoplanet host star with a single planet and a well-determined orbit. A good candidate is 51 Peg (Figure 6.4), for which after almost 25 years of study no additional companions have been found.

The standard procedure for calculating the error on the RV measurement is to take the rms scatter of all the chunks divided by the square root of the number of chunks. This should be weighted according to the rms scatter in the chunk or its spectral content (i.e., number of lines).

Is it possible to use Th–Ar emission lines or others from a hollow cathode lamp to get a measure of the IP? The short answer is no for two reasons. First, these emission lines are intrinsically broad and their shape contains no useful information about the IP. Second, there is a relatively low number density of these across the spectrum. Even if they were useful for measuring the IP, the sampling of this profile would be somewhat sparse.

In principle, the use of a laser frequency comb allows for the modeling of the IP because the emission "lines" of the comb are unresolved even for the highest resolving powers used for precise RV works.

6.6 Construction of an Iodine Cell

Of course, the procedure just described only works if you actually have access to an iodine cell. If one is not available for your spectrograph, it can easily be built. In fact, the iodine cell is without doubt the cheapest of all the available wavelength calibration methods.

A typical cell has a diameter of about 3 cm and a length of approximately 10 cm. The diameter is not critical, but should be large enough such that the clear window has a diameter large enough to accommodate the incoming beam while avoiding the edges of the window, which can be distorted during the manufacturing process. The length largely depends on the available space for the device. This author has constructed cells of length 3 cm up to 15 cm.[2] Fully constructed and filled iodine gas absorption cells can now be bought from commercial manufactures at a cost of ≈1500 Euros. However, you can build your own cell that functions well at a cost of a few hundred Euros.

The iodine cell is usually placed before the entrance slit (or fiber) of the spectrograph from the converging beam of the telescope. It is not necessary to use

[2] The cells in operation at McDonald Observatory, the Hobby-Eberly Telescope, the Tautenburg Observatory, and the Ondrejov Observatory were all constructed by the author.

it in a parallel beam, but keep in mind that the glass windows will change the focus of the telescope.

The construction of an iodine cell is straightforward, especially at a university which has a chemistry department with an in-house glass blower. All she or he needs to do is to take a glass tube of the appropriate diameter and length and to attach windows on each end. The materials can be purchased from any glass company. A narrow (≈3 mm) feed tube in the side of the cell allows iodine gas to enter.

Filling the cell is also simple. One first needs a glass manifold like the one shown in Figure 6.12, which can also be built by a glass blower. A cold trap ensures that the gas does not contaminate the vacuum pump. The cell's feed tube is attached to the manifold and then evacuated. After a good vacuum is achieved (after several minutes), the entrance valve is shut (if not all the iodine will go into the cold trap). It takes only a few minutes for the cell to fill with iodine gas. The amount of iodine in the cell can be gauged by eye. A torch is then used to pinch off the feed tube, allowing it collapse on itself due to the vacuum, thus leaving a permanently sealed cell. It is wise to construct several cells with different amounts of iodine gas. Testing each one on the spectrograph will determine the best device. Ideally, the iodine absorption lines should be deep enough, yet without making the light attenuation too high. A more sophisticated process includes wrapping the entire manifold with heating elements to ensure that the cell is filled at a relatively constant temperature.

It goes without saying that all glassware should be cleaned well to prevent contamination in the cell. If you chose to coat your windows with an antireflection coating, this should only be done on the outside surfaces to prevent contamination.

If this process seems a bit imprecise and not well calibrated, it does not make any difference so long as you use the same cell for your RV measurements. It is not so important what the absolute depths of the iodine lines are, only that they do not change with time.

The final step is to attach a temperature sensor (as close to the window as possible, if not on the window itself), wrap the cell with heating foil, and surround the cell tube with thermal insulating material. Temperature control can be done with a standard commercial temperature controller. Problems may arise in operating the cell at the

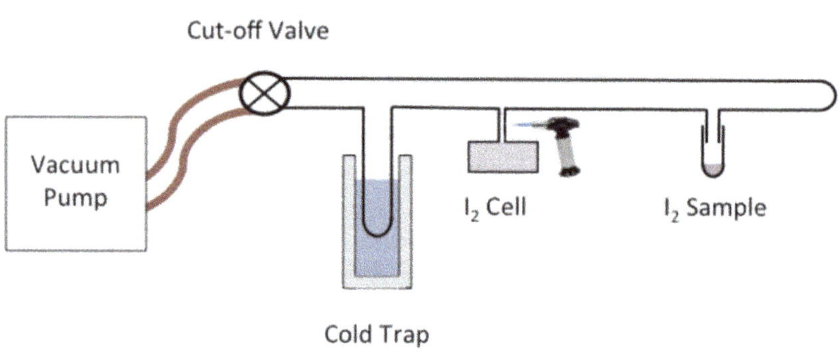

Figure 6.12. Schematic of the glass manifold used for filling the iodine cell. The cold trap with liquid nitrogen prevents iodine from contaminating the pump. A torch is used to pinch off the cell once enough iodine has been filled.

correct temperature since the value measured on the windows will be different than value measured on the tube which is covered in insulation (Wang et al. 2019).

The cell should always be operated at a temperature above which it was filled. If the cell is filled at roughly room temperature, the operating temperature should be 50 °C–80 °C. If the cell is located in a place where drafts of air can blow across the windows, it is best to use higher operating temperatures to ensure that no iodine condenses on the windows. Once above the fill temperature, the spectrum of molecular iodine should be insensitive to variations of the cell temperature by several degrees.

6.7 Closing Remarks

The iodine cell method has been in use for the past 30 years. It may not hold the record for providing the best RV precision that is possible compared to, say, a laser frequency comb, but there are several things to consider if a researcher wants to use a cell for precise RVs.

Although the method has not achieved the precision of stabilized spectrographs such as HARPS, it is worth noting that it comes close. For many spectrographs (e.g., HIRES at the Keck Observatory or UVES at ESO's VLT), the iodine method has produced an RV precision of 2–3 ms or only a factor of ≈2 worse compared to stabilized instruments. The fact that this was achieved with spectrographs that were general-purpose instruments (e.g., with moving parts) not designed for precise RV work is a testament to the power and utility of the method. If one considers the "working precision," i.e., the actual precision one achieves on a sample of real stars, then the iodine method fares reasonably well (see Chapter 4).

The iodine cell method is low cost and low maintenance. If one wants to perform RV measurements at a remote, robotic telescope, an iodine cell is an attractive choice for wavelength calibration. If costs of the instrument are the main concern, then it may be the only viable choice. Also, if the goal is to perform RV measurements over a long time, e.g., searching for planets in long-period orbits, then the iodine cell may be the only choice. Of all the wavelength calibration methods, it is the only one that has a proven track record of stability spanning several decades (e.g., Hatzes et al. 2006; Blunt et al. 2019).

Currently, the iodine method is the only one that can provide a simultaneous measurement of the instrumental profile. If one wants to achieve RV precision at the level of a few cm s^{-1}, then one must monitor and model changes in the IP. Laser frequency combs may hold some promise to do this, but as of this writing, no attempts have been made to do so. If one is considering building an "ultra-precise" RV machine, then using an iodine cell, possibly illuminating a fiber optic, should be considered.

Finally, and most importantly, the iodine cell method currently is the *only* method that provides an in situ measurement of the wavelength calibration at the same time and at the same place on the detector as the stellar observations. You have to have faith that the methods that provide the calibration adjacent to the stellar spectrum give the same wavelength solution as for the stellar spectrum.

Clearly, this has not been a problem with an RV precision of ~m s^{-1}, but this may be an issue for achieving a precision of ~cm s^{-1}.

References

Blunt, S., Endl, M., Weiss, L. M., et al. 2019, AJ, 158, 181
Butler, R. P., Marcy, G. W., Williams, E., et al. 1996, PASP, 108, 500
Endl, M., Kürster, M., & Els, S. 2000, A&A, 362, 585
Hatzes, A. P., Cochran, W. D., Endl, M., et al. 2006, A&A, 457, 335
Jansson, P. A. 1997, Deconvolution of Images and Spectra (San Diego, CA: Academic)
Valenti, J. A., Butler, R. P., & Marcy, G. W. 1995, PASP, 107, 966
Vogt, S. S., Penrod, G. D., & Hatzes, A. P. 1987, ApJ, 321, 496
Wang, S. X., Wright, J., MacQueen, P., et al. 2019, arXiv: 1910.10756

AAS | IOP Astronomy

The Doppler Method for the Detection of Exoplanets

A P Hatzes

Chapter 7

Frequency Analysis of Time Series Data

7.1 Introduction

In 1807, the French mathematician and physicist Jean-Baptiste Joseph Fourier demonstrated that any continuous function could be represented by a trigonometric series (sines and cosines). The basis of the Fourier transform was born, and to this day, it serves as an indispensable tool in searching for periodic signals in time series data. The reason is evident—if you have a continuous function consisting of a pure sine wave, then in the Fourier domain this appears as a δ-function at the frequency of the sine wave and with a height corresponding to the amplitude of the sine. Detecting periodic signals due to planets in your RV data is now reduced to finding peaks in your Fourier transform. Searching for periodic signals in a time series is not restricted to the radial velocity (RV) method, but is found in many branches of astronomy (and physics), most notably the study of stellar oscillations.

Unfortunately, identifying peaks in the Fourier transform due to planetary signals is often not simple. You have uneven time sampling as well as gaps in your data, due to the diurnal motion of Earth, seasons, and weather. These along with the presence of noise can make the interpretation of features in the Fourier transform challenging. Exoplanets reported in the literature have often lived and died by how well, or poorly, researchers interpret the Fourier transform or periodogram.

In this chapter, we look into the frequency analysis of time series data. Emphasis will be given on how the sampling window affects the periodogram, common misconceptions, and possible pitfalls when it comes to interpreting the Fourier transform.

Searching for periodic signals in unevenly spaced time series data is a long-standing problem in astronomy and other disciplines, and one can devote entire textbooks to the problem. There are a variety of methods that have been employed. One can simply use least-squares fitting of sine functions to your data, and the period that produces the best fit to your data (as measured by the reduced χ^2) is the signal that is in your data. Other approaches are based on the discrete Fourier transform (DFT) and is often called the "classic" periodogram.

Most researchers searching for exoplanets with RV measurements use a variant called the Lomb–Scargle (LS) periodogram, or the improved generalized LS (GLS) periodogram. This author uses all of them, choosing one or the other depending on the problem at hand. The reader will thus see all types of periodograms (DFT, LS, and GLS) so as to get a better feel of their similarities and differences. You should be comfortable dealing with all types. All of these will be referred to as as "periodograms," but keep in mind that all of these can be treated simply as a Fourier transform.

7.2 The Discrete Fourier Transform

The Fourier transform is the classic method for finding periodic signals in your time series data. Because with experimental data we are always dealing with discrete time series, we use the DFT defined as

$$\mathrm{DFT}_X(\omega) = \sum_{i=1}^{N} X(t_j)e^{-i\omega t_j}, \tag{7.1}$$

where $e^{i\omega t}$ is the complex trigonometric function $\cos(\omega t) + i\,\sin(\omega t)$, N is the number of data points sampled at times t_j, and ω the frequency.[1] The DFT is often called the classic periodogram.

The power is defined by

$$P_X(\omega) = \frac{1}{N}|FT_x(\omega)|^2 = \frac{1}{N}\left[\left(\sum_{j=1}^{N} X_j \cos \omega t_j\right)^2 + \left(\sum_{j=1}^{N} X_j \sin \omega t_j\right)^2\right]. \tag{7.2}$$

The uncertainty in the determination of the frequency is (Kovacs 1981)

$$\delta\omega = \frac{3\pi\sigma_N}{2(N_0)^{1/2}TA}, \tag{7.3}$$

where $\delta\omega$ is the uncertainty in the frequency ($\omega = 2\pi/P$, P = period), A is the amplitude of the signal, σ_N^2 its variance after subtracting true signals, N_0 the number of measurements, and T the length of the data set in time.

The Fourier transform is often plotted as power versus frequency. Here when I use the DFT, I will always plot the amplitude versus the frequency—the so-called amplitude spectrum. The reason is that the amplitude translates into a more useful unit, which in our case is m s^{-1}.

The Fourier transform is essentially the basis for all periodograms. Understanding the behavior of the Fourier transform of a function (the so-called "Fourier domain") is essential for interpreting features in the periodogram.

[1] Frequency is often measured as angular frequency, which is related to the period by $\omega = 2\pi/P$. Throughout this paper, when I refer to a frequency it is merely the inverse of the period, or day^{-1}.

It is beyond the scope of this book to go into the details of Fourier analysis. Instead, the reader is referred to the excellent textbook *The Fourier Transform and Its Applications* by Bracewell (1978). Here we will cover the basics in order for the reader to interpret better periodic signals in time series data.

To interpret DFTs and periodograms, it is useful to keep in mind two key properties of the Fourier transform:

1. The time domain, t, maps into $1/t$ or the frequency in the Fourier domain. If your time stamps are days in normal space, they will be inverse days in Fourier space. Because of the inverse mapping of t into $1/t$, a function that is broad in the time domain will be narrow in the Fourier domain, and vice versa. The extreme of this is a constant function (infinitely broad), which maps onto a δ-function at the origin (infinitely narrow).
2. The convolution theorem, which states that the convolution of two functions in the time domain is the product of the individual Fourier transforms in the Fourier domain.

7.2.1 Convolution

The convolution of two functions $f(t)$ and $g(t)$ is defined as

$$f*g = \int_{-\infty}^{+\infty} f(\tau)g(t - \tau)d\tau. \tag{7.4}$$

To calculate the value of the convolution function at x, you reverse the function g, move it to a position t, and measure the area (integral) under both functions and then move to the next position τ. That is, you reverse g and slide it across f, while integrating under both functions. It is often called a smoothing function because if you convolve a time series, or spectral data, by, say, a box of width N, you are producing a running average of points within that box. The convolution theorem states that this complicated mathematical process reduces to a product of the Fourier transforms of the two functions:

$$f(t)*g(t) \equiv F(\omega) \cdot G(\omega). \tag{7.5}$$

It is important to note that the convolution theorem works both ways—if you multiply two functions in the time domain, that is the same as convolving the individual Fourier transforms in the Fourier domain. As we will see, this fact can produce some rather complicated periodograms when looking for periodic signals in RV data.

Convolution versus Cross-correlation

It is of interest to compare the differences between convolution and cross-correlation. You will note that Equations (7.4) and (5.1) are nearly identical, except for a minus sign. Figure 7.1 highlights the differences. Cross-correlation is a measure of the similarity between two signals. It is a "match signal detector," and the more similar the two signals are to each other, the higher the contrast of the CCF. And in

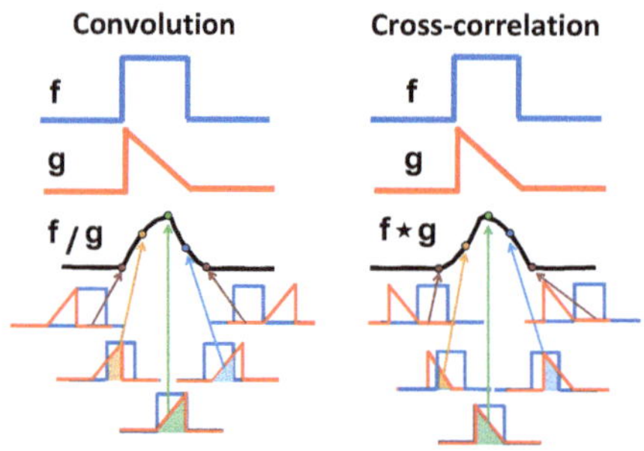

Figure 7.1. A schematic of the convolution of two functions, *f* and *g* (left), and their cross-correlation (right).

the case of Doppler measurements, it tells you how much you have to shift your signal in velocity space to get a match.

On the other hand, convolution is a measure of the effect of one signal on another, or the system impulse response. It is a blurring process of one signal on another. We have seen this in Chapter 6, where the instrumental profile (IP) is the response of your spectrograph to your spectrum, i.e., the observed stellar spectrum is the intrinsic spectrum convolved with the IP.

Convolution is a process by which you smooth data. For example, suppose you have a noisy spectrum and you want to make it look nicer by smoothing it with a box function. The brute force method is to take your box of some width, place it on a pixel, and replace the value by the mean value of all pixels in the box. You then slide the box over to the next pixel and thus compute the running average. This is convolution.

The elegant method is to exploit the convolution theorem. You take the Fourier transform of your spectrum and then the Fourier transform of your box function.[2] You multiply the two together and then take the inverse transform. You will arrive at the same answer as computing the moving average.

7.2.2 Visualizing Fourier Transforms

The best way to get an intuitive understanding of Fourier transforms is to visualize these, or as a former colleague of mine once said, "you need to think in Fourier space." Following the example of Bracewell's "pictorial atlas" of Fourier transforms, Figure 7.2 shows some common Fourier transforms that are useful for interpreting periodograms. Note that these figures show the DFT as a positive amplitude spectrum. In reality, the DFT has both positive and negative values, as well as negative frequencies. Most programs either give the amplitude, or the power (amplitude squared) and only positive frequencies. As we will shortly see when we discuss the spectral window, it is important to remember that there are negative frequencies as well.

[2] If want to save time and are clever, you can exploit the fact that you know the Fourier transform of the box function, which is a sinc.

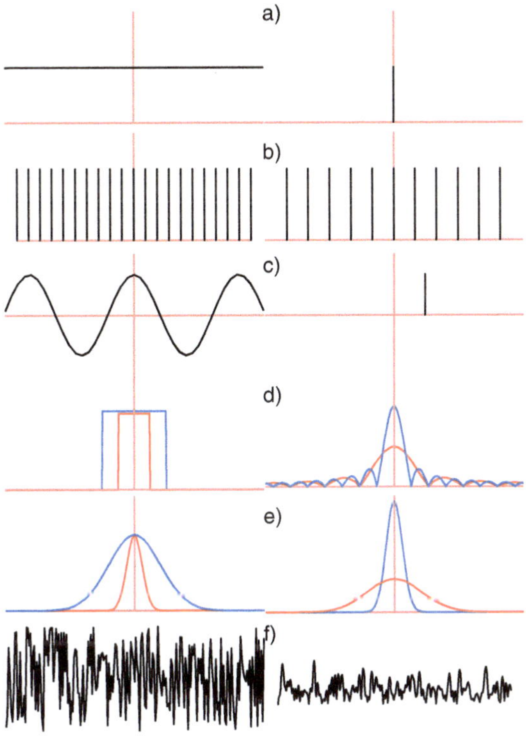

Figure 7.2. Pictorial Fourier transforms of some useful functions. The functions are shown in the time domain at the left and Fourier domain at the right. (a) A constant function, (b) a sampling function, (c) a cosine function, (d) a slit function, (e) a Gaussian, and (f) random noise. See text for more details.

A constant-valued function (panel a) has a Fourier transform consisting of a δ-function at the origin. This is easy to understand given the reciprocal nature of the time and Fourier domain. A constant function is infinitely broad, which means it must be infinitely narrow in the Fourier domain. Because there is no characteristic frequency, the δ-function must be at the origin.

A regular sampling function (i.e., series of δ-functions) with a spacing ΔT transforms into a series of δ-functions, but with a spacing of $1/\Delta T$ (panel b). A cosine (and sine) transform to a δ-function (panel c) at the appropriate amplitude and frequency (both negative and positive) and is the main reason that we use the Fourier transform to search for periodic signals in time series data. Note that because a sine function is odd ($\sin(-x) = -\sin x$), the negative frequency actually has a negative amplitude δ-function. Here we show only positive frequencies, the output of all periodgram programs.

A box or slit function (panel d) transforms into a sinc function and a Gaussian (panel d) into another Gaussian function. Note that a wide (narrow) box/Gaussian has a narrow (wide) function in the Fourier domain.

Finally, random noise has a Fourier transform that is again random (panel f). However, if σ_T is the rms scatter in the time domain, the scatter of the amplitude

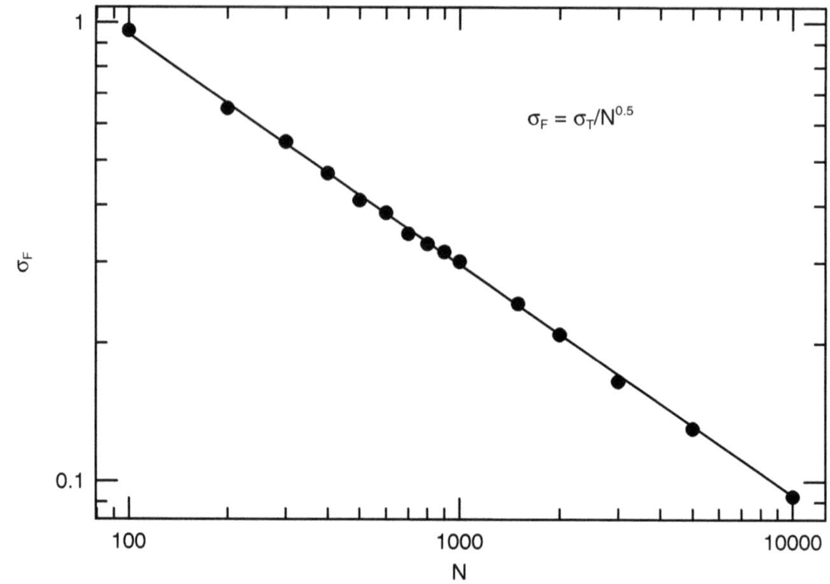

Figure 7.3. The standard deviation, σ_F, of the Fourier amplitude spectrum of a random noise time series with standard deviation of σ_T as a function of the number of points in the time series, N.

spectrum is $\sigma_T/\sqrt{N}$, where N is the number of measurements. This is shown in Figure 7.3, which was produced using simulated time series with even sampling of random noise having a standard deviation σ_T. Shown in the figure is the standard deviation of the DFT amplitude spectrum, σ_F, as a function of the number of measurements. If you want to decrease the Fourier noise level in your amplitude spectrum, you simply have to take more measurements. The results shown in Figure 7.3 are useful because they can be used to estimate how many measurements we need to detect a signal in our data.

7.3 The Lomb–Scargle Periodogram

A large part of the astronomical community, especially exoplanet researchers, use an alternative form of the DFT, namely the LS periodogram (Lomb 1976; Scargle 1982). This is defined by a bit more complicated expression than the DFT:

$$P_X(\omega) = \frac{1}{2}\left\{\frac{[\sum_j X_j \cos\omega(t_j - \tau)]^2}{\sum_j \cos^2\omega(t_j - \tau)} + \frac{[\sum_j X_j \sin\omega(t_j - \tau)]^2}{\sum_j \sin^2\omega(t_j - \tau)}\right\}, \tag{7.6}$$

where τ is defined by

$$\tan(2\omega\tau) = \sum_j \sin(2\omega t_j)/\sum_j \cos(2\omega t_j).$$

The periodogram defined in this way has useful statistical properties that enables one to determine the statistical significance of a periodic signal in the data. One of the main problems of time series analysis is finding a periodic signal that is real and not due to noise. The LS periodogram gives you an estimate of the significance of such a signal.

The DFT and the LS periodogram are intimately related. In fact, Scargle (1982) showed that the LS periodogram is the equivalent of sine fitting of data, essentially a DFT. It is worth mentioning, however, the differences between the DFT and the LS periodogram. In the DFT, the power at a frequency is just the amplitude squared of the periodic signal. The amplitude (or power) of the signal can be read directly from the DFT amplitude (power) spectrum. On the other hand, the power in the LS periodogram is related in a nonlinear way to the statistical significance of the signal, as we shall soon see.

7.4 The Generalized Lomb–Scargle Periodogram

Although the LS periodogram has proven to be a powerful and versatile tool for the frequency analysis of uneven time series, it suffers from two major drawbacks that limit its performance:

1. It does not take measurement errors into account.
2. In the analysis, a simple mean of the data is subtracted.

Regarding the first point, it is standard practice to take into account the error in the measurements by applying weights. Clearly, you want to give higher weight to the higher quality data. In its initial implementation, the LS applied no weights so all data, good and bad, made equal contributions to the periodogram.

In using the LS or the DFT, for that matter, one must subtract the mean value of the data. The reason for this is clear as a constant offset will produce a δ-function at the origin (zero frequency) which can influence the periodogram. LS and DFT simply subtract the mean value of the time series. It is more appropriate to fit the offset value, so instead of solving for $y = a \cos \omega t + b \sin \omega t$ as is done for LS, we solve for

$$y = a \cos \omega t + b \sin \omega t + c. \tag{7.7}$$

where $\omega = 2\pi/P$.

Cumming et al. (1999, p. 1) recognized that this made LS "non-robust when the number of observations is small, the sampling uneven, or for periods comparable or greater than the duration of the observations." They thus introduced the "floating mean periodogram," but without an analytical solution. This was later done by Zechmeister & Kürster (2009).

We therefore have to minimize the squared difference (χ^2) between the model function $y(t)$ and the data y_i:

$$\chi^2 = \sum_{i=1}^{N} \frac{[y_i - y(t_i)]^2}{\sigma_i^2} = W \sum [y_i - y(t_i)]^2, \tag{7.8}$$

where the normalized weights are

$$w_i = \frac{1}{W}\frac{1}{\sigma_i^2}\left(W = \sum \frac{1}{\sigma_i^2};\ \sum w_i = 1\right). \tag{7.9}$$

The power can be written as

$$p(\omega) = \frac{\chi^2_{\text{const}} - \chi^2_{\sin}(\omega)}{\chi^2_{\text{const}}} \tag{7.10}$$

$$= \frac{1}{YY \cdot D} \cdot \frac{SS \cdot YC^2 + CC \cdot YS^2 - 2CS \cdot YC \cdot YS}{CC \cdot SS - CS^2}, \tag{7.11}$$

with

$$D(\omega) = CC \cdot SS - CS^2 \tag{7.12}$$

and the following abbreviations for the sums (the hats indicate the sums of the classical LS periodogram):

$$YY = \hat{YY} - Y \cdot Y \quad \hat{YY} = \sum_i w_i y_i^2 \quad Y = \sum_i w_i y_i \tag{7.13}$$

$$YC(\omega) = \hat{YC} - Y \cdot C \quad \hat{YC} = \sum_i w_i y_i \cos \omega t_i \quad C = \sum_i w_i \cos \omega t_i \tag{7.14}$$

$$YS(\omega) = \hat{YS} - Y \cdot S \quad \hat{YS} = \sum_i w_i y_i \sin \omega t_i \quad S = \sum_i w_i \sin \omega t_i \tag{7.15}$$

$$CC(\omega) = \hat{CC} - C \cdot C \quad \hat{CC} = \sum_i w_i \cos^2 \omega t_i \tag{7.16}$$

$$SS(\omega) = \hat{SS} - S \cdot S \quad \hat{SS} = \sum_i w_i \sin^2 \omega t_i \tag{7.17}$$

$$CS(\omega) = \hat{CS} - C \cdot S \quad \hat{CS} = \sum_i w_i \cos \omega t_i \sin \omega t_i. \tag{7.18}$$

When replacing t_i with $\tau_i = t_i - \tau$ (because of the time-translation invariance, it will not affect the χ^2 of the sine fit) and choosing the parameter τ as

$$\begin{aligned} \tan 2\omega\tau &= \frac{2CS}{CC - SS} \\ &= \frac{\sum_i w_i \sin 2\omega t_i - 2\sum_i w_i \cos \omega t_i \sum_i w_i \sin \omega t_i}{\sum_i w_i \cos 2\omega t_i - \left[\left(\sum_i w_i \cos \omega t_i\right)^2 - \left(\sum_i w_i \sin \omega t_i\right)^2\right]}, \end{aligned} \tag{7.19}$$

the interaction term in Equation (7.11) disappears such that $CS_\tau = \sum w_i \cos\omega(t_i - \tau)\sin\omega(t_i - \tau) - \sum w_i \cos\omega(t_i - \tau)\sum w_i \sin\omega(t_i - \tau) = 0$. Calculating this parameter $\tau(\omega)$ for each frequency, the periodogram in Equation (7.11) simplifies to

$$p(\omega) = \frac{1}{YY}\left[\frac{YC_\tau^2}{CC_\tau} + \frac{YS_\tau^2}{SS_\tau}\right]. \tag{7.20}$$

The form of this equation is similar to Equation (7.6) of the classical LS periodogram, but now accounting for measurement errors (weights w_i in all sums) and a floating mean (additional terms in all sums and $\tan 2\omega\tau$). Due to the normalization to unity, the GLS power $p(\omega)$ lies in the range of $0 \leqslant p \leqslant 1$, with $p = 0$ ($\chi^2_{\rm sin} = \chi^2_{\rm const}$) showing no improvement of the fit and $p = 1$ ($\chi^2_{\rm sin} = 0$) indicating a "perfect" fit.

7.5 The Bayesian Generalized Lomb–Scargle Periodogram

One drawback of the LS and GLS is that the power is in arbitrary units, which make it difficult to assess the relative probability of two peaks. For this reason, Bretthorst (2001) generalized the LS periodogram by using Bayesian probability theory. Mortier et al. (2015) further extended this formalism. The mathematical description can be found in that paper and will not be repeated here.

Moriter et al. (2015) presented an example which nicely highlights the differences between LS and GLS. The authors considered a sinusoidal signal with a period of 105 days and an amplitude of unity. The synthetic curve was sampled unevenly 100 times. Half of the points had a mean error of 0.4 (hereafter the "low-noise" data) while the other half had an error of 1.1 (the "high-noise" data). The data are shown in the left panel of Figure 7.4.

The right panel shows the LS, GLS, and BGLS periodograms. For BGLS, the power is normalized so that the highest peak has a 100% probability. With the BGLS periodogram, one can assess the relative probabilities of other peaks, which in this case are exceedingly small. Both GLS and BGLS find the correct period at 105 days, but the LS selects a peak near 50 days, which also seems to be significant. This peak is not an alias of the dominant peak. I should note that the simple DFT produces the same result as the LS and in what follows, the discussion for the LS also holds for the DFT.

It is worth investigating the reason for this difference by taking a closer look at the data, in particular by examining the high- and low-noise data separately. The upper-left panel of Figure 7.5 shows the LS periodogram using only the low-noise data. It easily finds the correct period. The LS periodogram of the high-noise data, however, indeed finds a peak at 50 days (lower left panel), which is significant, having a false-alarm probability (FAP) of $\approx 10^{-6}$ determined through the bootstrap procedure (see below). The phasing of the data to the 50 day period (lower-right panel of Figure 7.5) shows clear variations at this period. In this example, even though the errors in the high-noise data are presumably random, they seemed to have conspired in a not so obvious way with the 105 day period to produce variations in the data at the shorter period. A DFT of the high-noise data yields a K-amplitude of 1.5, or 50% higher than the amplitude of the 105 day period.

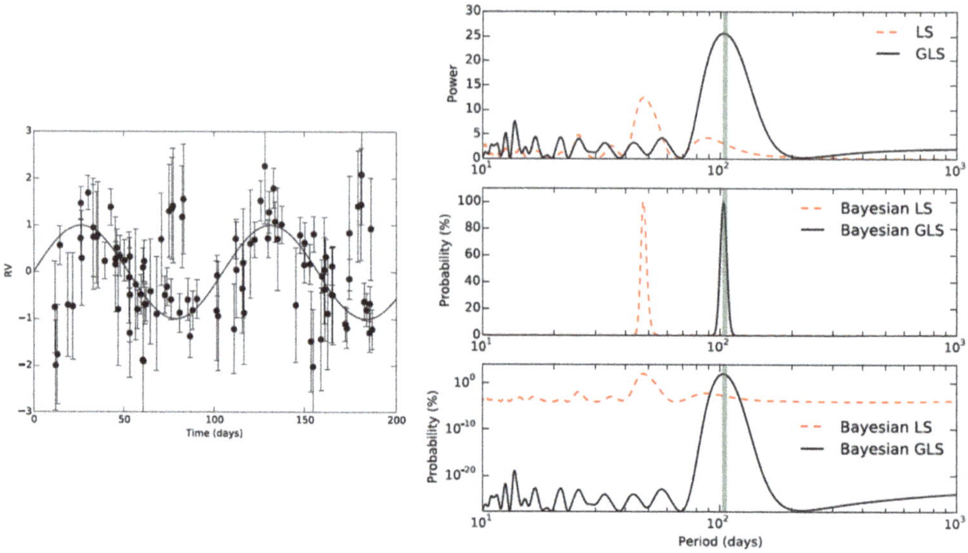

Figure 7.4. (Left) Simulated data from Mortier et al. (2015) of a sine function with a period of 105 days and an amplitude of 1.0. Half of the data have an error of 0.4 and the other half 1.1. (Right, top) Comparison of periodograms from Lomb–Scargle (LS) and generalized Lomb–Scargle (GLS). (Right, middle) The Bayesian LS periodogram. In this instance, power is a measure of the relative probability of the signal. (Right, bottom) Same as the middle panel but on a logarithm scale. The LS finds a period of 50 days whereas GLS and BGLS find the correct period of 105 days.

This may explain why the LS chooses the 50 day period. The high-noise data appears to have a periodic signal present at 50 days and with an amplitude that is 50% higher than that of the true 105 day period, which is clearly evident in the low-noise data. Because the LS assigns equal weights (i.e., no weights) to both the low- and high-noise data, it merely chooses the signal with the highest amplitude, which is at a period of 50 days. Both the GLS and BGLS weigh the data according to the errors, which is dominated by the higher quality data having the 105 day period. In this case, applying weights has a significant impact.

It seems that the noise in the data used by Mortier et al. (2015) has a peculiar structure. As a test, another set of synthetic data using a 105 day period was made according to the time stamps of the low- and high-noise data and the appropriate noise. That is, for the high-noise points, the random noise that was added had an rms scatter 2.65 times higher than that for the low-noise data points. In this case, the LS easily finds the correct period (Figure 7.6). For a different set of noise characteristics, the LS works fine. It is generally the case that if one has a large number of data points the LS and GLS should arrive at the same answer.

So, the lesson learned is that if the uncertainties of most of your data are comparable, the LS should give consistent results. If you have a wide range of uncertainties, particularly if a large fraction of the data has much larger errors, you should trust GLS results over those of the LS. In general, you should use periodograms that weight the data values accordingly.

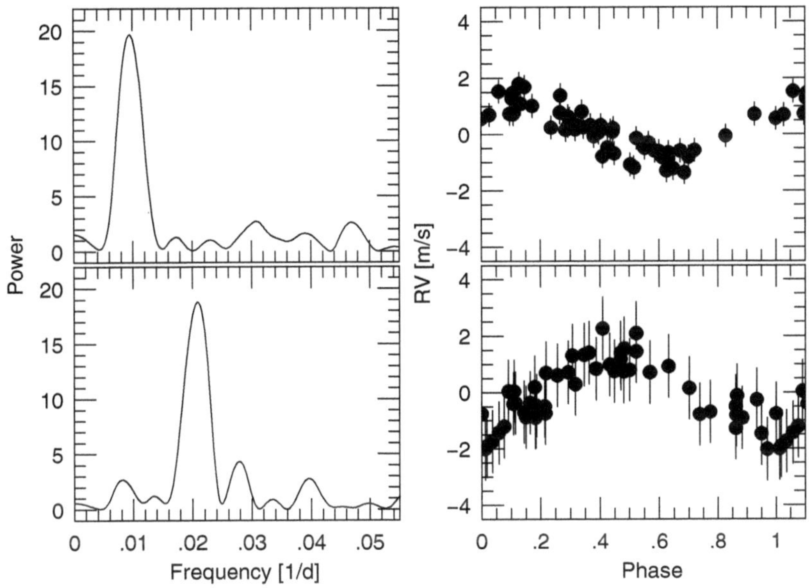

Figure 7.5. (Top left) The LS periodogram of the simulated data shown in Figure 7.4 but for only the data points with the smaller errors. It finds the correct peak at a period of 105 days ($\nu = 0.0095$ day^{-1}). (Top right) The data phased to the 105 day period. (Bottom left) The LS periodogram of the simulated data but only using the data points with the larger errors. In this case, the LS finds a period of 50 days. (Bottom right) The large error data phased to the 50 day period. In this, the noise characteristics of the large-error data mimics a shorter period signal.

7.6 Comparison of the Types of Periodograms

Before we leave our discussion on periodograms it is of interest to compare the various types. We have presented the classic (DFT), LS, and GLS periodograms. We have not discussed in detail the least-squares sine fitting, but have hinted that it produces results similar to the other periodograms. As reminder, the least-squares sine fitting merely fits a function of the form $y = A\sin(\omega t + \phi) + c$ to your data. You chose a test frequency (period) and find the best fit to the data, varying amplitude and phase. The quality of the fit, as measured by χ^2, as a function of frequency is your "periodogram." You now search for a minimum in the χ^2.

Figure 7.7 shows a comparison of the various periodograms, DFT, LS, and GLS, and sine fitting as applied to the public RV data for 51 Peg. To facilitate the comparison between the periodograms, we plot the power of the DFT and not the amplitude because the LS and GLS display the power. We also flip the y-axis for the sine-fitting periodogram so that χ^2 decreases upwards. In this way, it can be compared directly to the peaks in the other periodograms. The take-home message is that given sufficient measurements and with uncertainties that are comparable for each data point, all methods found the correct signal at $\nu = 0.236$ day^{-1}.

It is interesting to note that the DFT, LS, and GLS look remarkably the same. On the other hand, the least-squares sine fitting shows a different side-lobe structure (caused by the spectral window discussed below). The one at lower frequencies (an

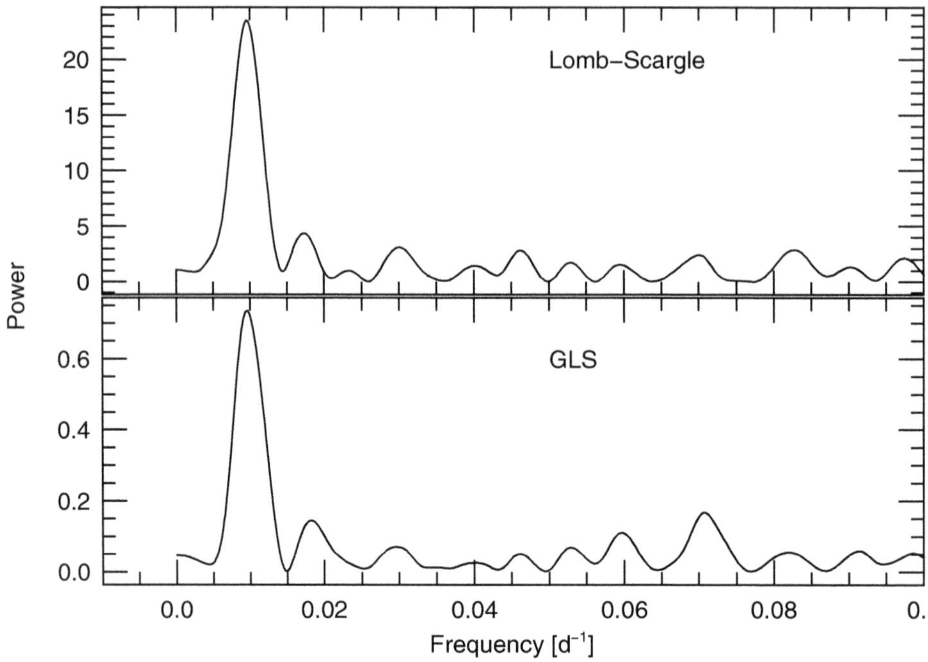

Figure 7.6. Periodograms of the simulated data shown in Figure 7.5 (i.e., period = 105 days, amplitude = 1), using different random noises, but again with half the data having $\sigma = 0.4$ and the other half with $\sigma = 1.0$. In this case, the LS and GLS find the appropriate period.

alias) has a much higher amplitude than the one at higher frequencies. In the other periodograms, the peaks of the side lobes have equal power. The least-squares sine fitting does, however, find the correct dominant frequency. One should note that χ^2 is not strictly a power, which makes a direct comparison more difficult.

7.7 The Spectral Window

The spectral or sampling window is a periodogram of the function that has values of unity at the time stamps of your measurements and zero elsewhere. It gives you important information on how the sampling of your data affects your periodogram. For ground-based astronomical observations, the spectral window can be quite complex, due to gaps introduced by nightly, monthly, and yearly sampling. These can result in peaks in the periodogram which at best merely complicate its interpretation and at worse make one choose spurious or alias signals as being real. Many researchers can be lazy and neglect to look at your window function, but it is essential for interpreting the features that you find in your periodogram.

The spectral window is often poorly understood by students first embarking into frequency analysis and is a source of misinterpretation and misconceptions. The top panel of Figure 7.8 shows a typical spectral window for an astronomical time series that is given by virtually all DFT or periodogram programs. It has a peak value of unity at zero frequency, and at higher frequencies one sees features corresponding to

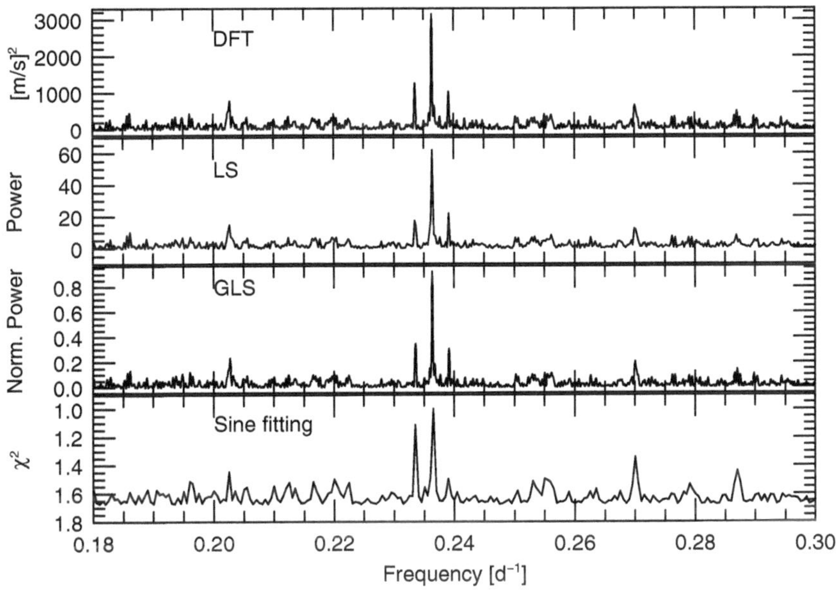

Figure 7.7. The various periodograms of the RV data for 51 Peg. (Top) The DFT power spectrum. (Second from the top) The LS periodogram. (Third from the top). The GLS periodogram. (Bottom) Least-squares sine fitting (reduced χ^2 versus frequency). The value of χ^2 is shown in reverse (larger to smaller values) in order to make a more direct comparison to the other periodograms.

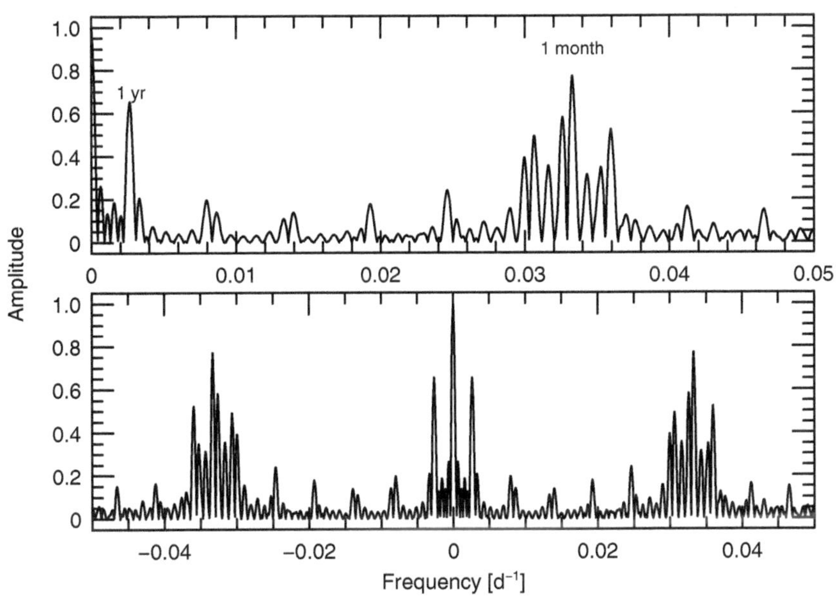

Figure 7.8. (Top) The sampling (spectral) window as an output of most programs producing periodograms or DFT. (Bottom) The true representation of the spectral window showing negative and positive frequencies.

the one-year and one-month sampling. A common misconception is that the peaks in the spectral window will appear exactly at the same frequencies in the periodogram, but this is not the case. To understand where features due to the spectral window appear in the data periodogram, one needs to keep in mind (1) the true representation of the spectral window and (2) the convolution theorem.

The first step in understanding the effects of the spectral window is to realize that the top panel only represents half of the full spectral window. The true representation of the spectral window is shown in the lower panel of Figure 7.8. It has both positive and negative frequencies (remember that the Fourier transform has both) and is symmetrical about the origin. For practical purposes, all periodogram programs only give you positive frequencies, which is sufficient given the symmetry of the function.

The second step to understanding the window pattern is to recall the convolution theorem: a multiplication of two functions in the time domain is the convolution of the individual transforms in the Fourier domain. Figure 7.9 shows this process in the time domain (left panels) and Fourier domain (right panels). Suppose a star shows periodic RV variations (e.g., a planet in a circular orbit), as shown in the top-left panel. We do not measure a continuous function, but rather the orbit in a sampling

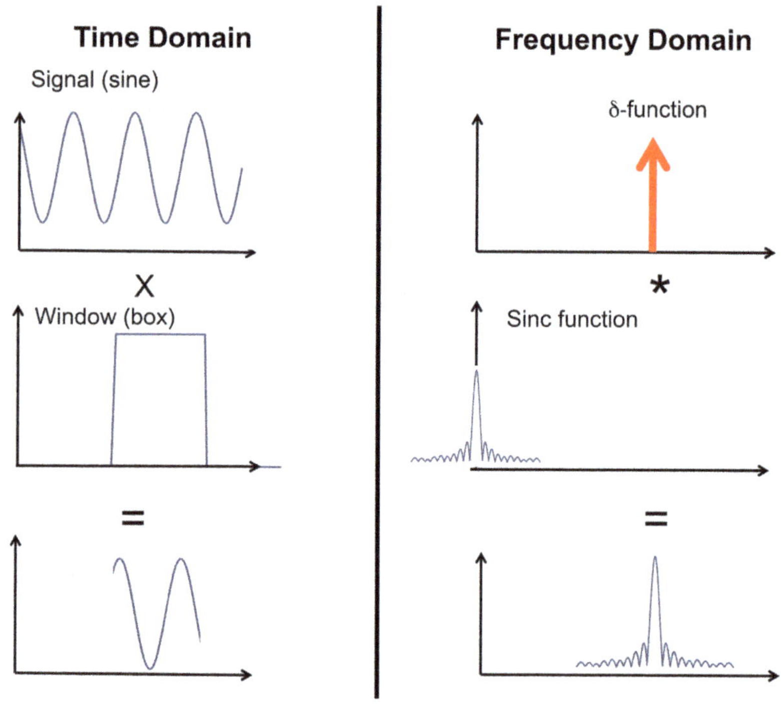

Figure 7.9. (Left panels) In the time domain, the observations consist of an infinite sine wave multiplied by the sampling window, in this case a simple box function. The observed data consist of a truncated sine function. (Right panels) In the Fourier domain, the product of the sine with the box functions is a convolution of the Fourier transform of the sine (i.e., δ-function) with the Fourier transform of the box function (i.e., sinc). This sinc function appears at the location of the frequency of the δ-function with the same amplitude.

window represented by the box function in the middle. (For simplicity we will assume that within this box we have continuous sampling.) Our time series is truncated—we observe the star for a time and then stop. This is the product of the box (sampling length) and the sine functions, i.e., a truncated sine function (lower-left panel).

In the periodogram (Fourier domain), the infinite sine wave will appear as a δ-function (right side of Figure 7.9). Our box function is a sinc-function in the periodogram (middle-right panel). According to the convolution theorem, the product of the box and sine functions in the time domain will appear as a convolution of the δ-function, with the sinc function in the Fourier domain. The periodogram will appear as a sinc with the appropriate amplitude at the frequency of the sine wave.

Figure 7.10 shows the DFT amplitude spectrum of a sine function with a period of 7.35 days and an amplitude of 100 m s^{-1}, and a realistic spectral window (inset in figure). One can clearly see the spectral window pattern appears at the expected frequency of 0.136 day^{-1}.

It is important to note that because the spectral window is convolved with every δ-function in the periodogram, this window structure will appear at the frequency of every periodic signal that is present in the time series data. If you have two sine waves in the data, the window function will appear twice (Figure 7.11). This can make for a very complicated-looking periodogram with the peaks of the window function at one signal overlapping with others. There will be a myriad of peaks making it difficult to find the few that are related to true periodic signals in the data. For this reason, it is

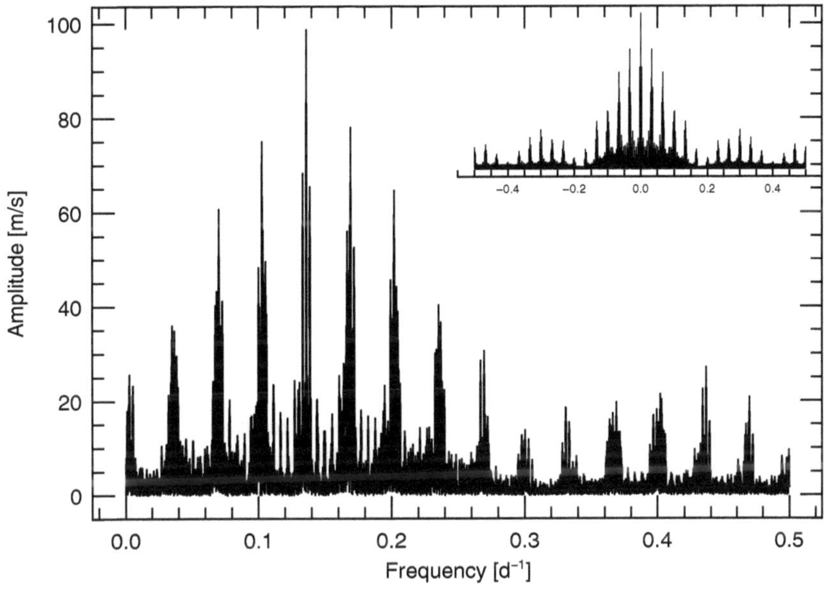

Figure 7.10. The DFT amplitude spectrum of a sine function with a period of 7.35 days and an amplitude of 100 m s^{-1} sampled every night for five nights once a month for three years. The dominant signal is at the correct frequency (0.136 day^{-1}), but a large number of other peaks due to the sampling window are present.

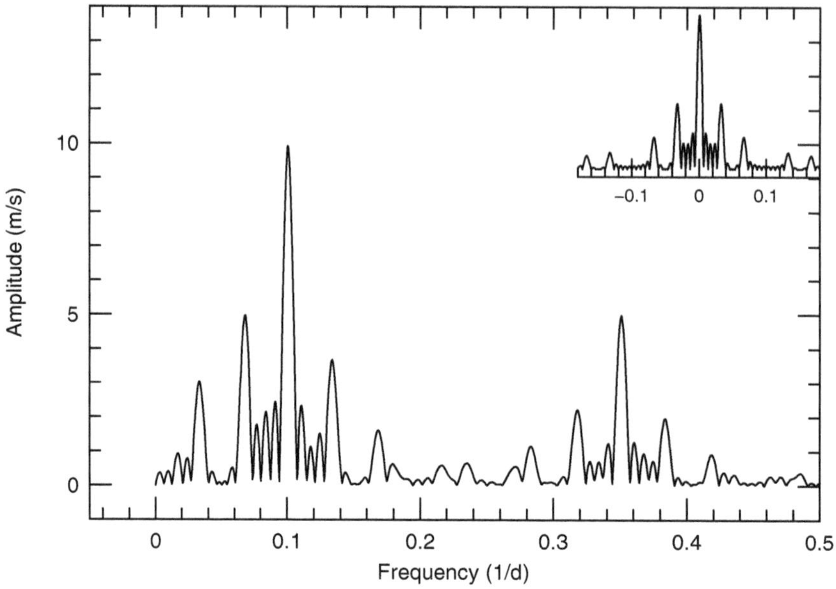

Figure 7.11. The sampled time series consisting of two sine functions with periods of 10 days (ν = 0.1 day^{-1}) and 2.85 days (ν = 0.35 day^{-1}). The DFT of the window function is shown in the upper panel. Because the observed DFT is a convolution of the Fourier transform of the window function (spectral window) with the data transform (two δ-functions), the spectral window appears at each signal frequency.

very dangerous to interpret features in the raw periodogram without checking the spectral window. You first have to remove the periodic signal you believe is in the data, which also removes all of the peaks due to the window function. We will return to this point at length later on when we discuss the procedure of prewhitening.

Another common misconception regarding the spectral window is that it will introduce peaks in your periodogram even in the absence of a true signal. You cannot identify a peak in the data periodogram and then demonstrate that it is not due to the window function by simply producing a time series of random data, sampling it like the real data, and then looking at the periodogram. You are guaranteed to find no significant peak in the simulated data, yet you have not confirmed that your signal is real. This can be seen if you replace the sine function in Figure 7.9 by random noise. Its Fourier transform will again be random noise, and the convolution of random noise with a sinc function is once again random noise with no significant peaks. All that you have shown is that a periodogram of random data is also random data! *For the spectral window to produce peaks in the periodogram, there must already be a periodic signal in the data.* This signal can be from the star, systematic errors, etc.

7.8 The Nyquist Frequency and Aliasing

The spectral window demonstrates a problem that plagues the period analysis of time series. Ground-based astronomical observations that have data gaps will produce false alias signals in your periodograms. An alias period is due to the

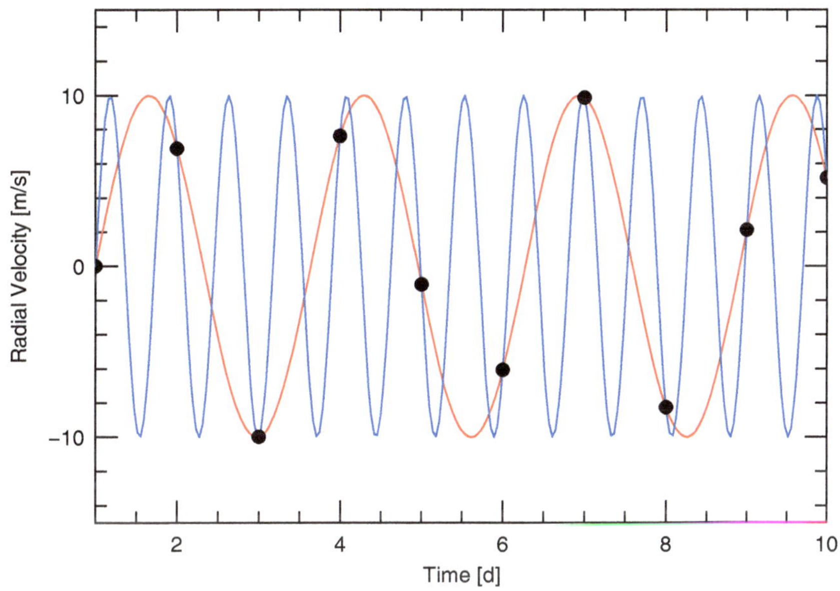

Figure 7.12. A sine wave with a period of 0.735 days (blue curve) that is undersampled with only one observation per night (points) is indistinguishable from a sine wave with a period of 2.77 days.

undersampling of the time series that causes a short period signal to appear as a much longer one. This is demonstrated in Figure 7.12, which shows two sine functions of different periods. If the short-period sine wave was actually in the data and you sampled it at the rate shown by the dots, you would not be able to distinguish which signal was in your data—both fit your measurements.

To avoid aliases, one must satisfy the Nyquist sampling criterion. If your sampling rate is δT, this corresponds to a sampling frequency of $f_s = 1/\delta T$. If you want to detect frequency signals higher than f_c, then the Nyquist criterion states

$$f_s \geqslant 2f_c. \tag{7.21}$$

For example, if you observe a star once a night, your sampling frequency is 1 day^{-1}. The Nyquist frequency (f_c) in this case is 0.5 day^{-1}. Thus, you will not be able to detect a periodic signal with a frequency higher than $2f_s$, or equivalently, a period shorter than 2 days. If shorter periods were in the data, these will appear at a longer period in your Fourier spectrum. You will see both the true period and its alias in the periodogram.

If f_d is the frequency of a signal present in your data and you make observations at intervals f_s, then alias frequencies, f_a, will appear at

$$f_a = f_d \pm nf_s \tag{7.22}$$

where $n = 1, 2, 3\ldots$.

For example, if you have a signal at 2.3 days ($f_d = 0.43$ d^{-1}), then the aliases due to a 1 day sampling interval will appear at $f_a = 1.43$ day^{-1}, 2.43 day^{-1}, etc.

It is important to realize that the Nyquist frequency represents the limits of your "frequency coordinate system." All amplitudes beyond the Nyquist frequency just repeat frequencies already present in the data. In practice, you only perform your periodogram out to the Nyquist frequency because at higher frequencies, you have no new information. It is also important to realize that the periodogram also has negative frequencies that in practice are only seen beyond the Nyquist frequency. Essentially, the full DFT amplitude spectrum or periodogram is a periodic pattern out to the Nyquist frequency that just repeats as you go to higher frequencies.

Figure 7.13 demonstrates this more clearly. It shows the DFT amplitude spectrum of a single sine function with a frequency of f_0 = 0.379 day^{-1} with one-day sampling. The amplitude spectrum is shown out to a frequency of 2 day^{-1} as well as −1 day^{-1}—frequencies that are usually not shown by periodogram programs. One can clearly see the 1 day aliases due to f_0 ($f_0 \pm 1$), but also an alias peak at $1 - f_0$. Where does this come from?

It is important to know that not only do frequencies essentially "repeat" past the Nyquist frequency, $f_{\mathrm{Nyp}}\left(= 0.5c\ d^{-1}\right)$, but that now, the negative frequencies of the periodogram come into view. The amplitude spectrum coordinate system is centered on the origin, going from $-f_{\mathrm{Nyq}}$ to $+f_{\mathrm{Nyq}}$ (white region). This coordinate system is

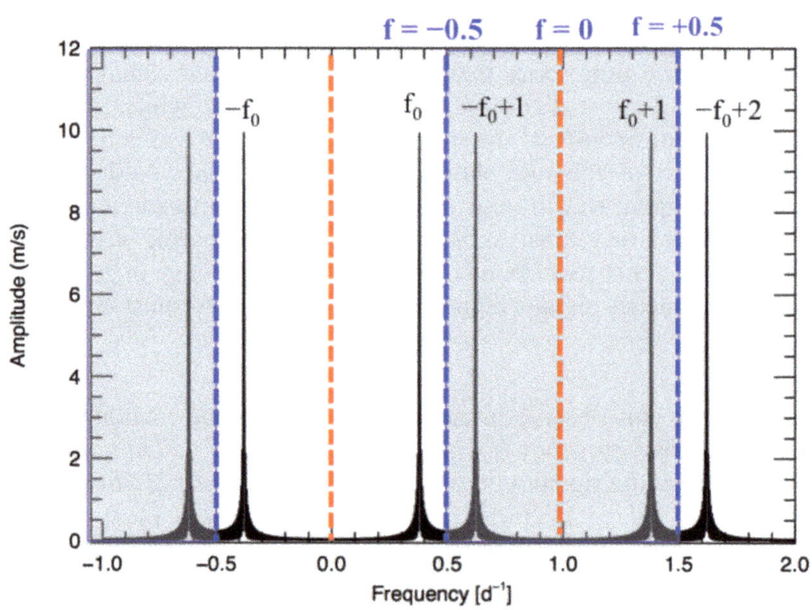

Figure 7.13. Cyclic nature of the periodogram using the Fourier amplitude spectrum (DFT) of a 2.64 day sine wave ($f_0 = \nu = 0.379$ day^{-1}). This is shown from a frequency of −1.0 day^{-1} to 2.0 day^{-1}, well beyond the Nyquist frequency of $f_{\mathrm{Nyquist}} = 0.5$ day^{-1}. One can only unambiguously detect signals in the frequency "world coordinate system" spanning $-f_{\mathrm{Nyq}}$ to $+f_{\mathrm{Nyq}}$ and shown by the dashed blue lines. In this range, we see f_0 and $-f_0$. This coordinate system is repeated every 1 day^{-1}, shown as alternating white and blue regions and with the origin (zero frequency, red dashed line) repeated at intervals of twice the Nyquist frequency (the 1 day alias). In plotting, it is only at positive frequencies in the range 0 to 1 day^{-1} does one see f_0 and $-f_0 + 1$ straddling the origin. The peak to the right of the Nyquist frequency is thus the alias of the negative value of f_0.

then repeated at every interval of 1 day^{-1} (2 f_{Nyq}) in frequency, alternating between blue and white regions for clarity. The red dashed line marks the origin (zero frequency) which repeats every day^{-1}. The peak seen at $1 - f_0$ is merely the 1 day alias of the negative signal frequency. Remember, you will see 1 day aliases of both the positive and negative values of a signal frequency.

A famous case of aliasing and neglecting to consider the Nyquist frequency is the planet 55 Cnc e. This was one of the first hot Neptunes discovered with the RV method. The planet was reported to have a 2.85 day orbital period corresponding to a minimum mass of 17.7 $M_\odot$ (McArthur et al. 2004). It was later demonstrated that the alias period of 0.73 day was the true one. In this case, the investigators missed out on an even more important discovery because it turned out that 55 Cnc e was the first transiting hot Neptune!

Figure 7.14 shows the DFT power spectrum of the RV measurements for 55 Cnc plotted beyond the Nyquist frequency of ≈0.5 day^{-1} and demonstrates the ambiguity of the true signal due to aliasing. One sees a peak at the "discovery" orbital frequency at 0.35 day^{-1} (P = 2.85 days). However, there is a slightly higher peak at the alias frequency of 1.35 day^{-1}. There is also another, most likely alias at 0.65 day^{-1}, or about one-half the higher frequency.

One can easily check that there is only one signal in the data by fitting and removing sine functions to the peaks individually. If you fit the 0.35 day^{-1} signal and remove it, the ones at 1.35 day^{-1} and 0.65 day^{-1} disappear. Likewise, by fitting sine waves using 1.35 day^{-1} and removing it, you eliminate the other aliases

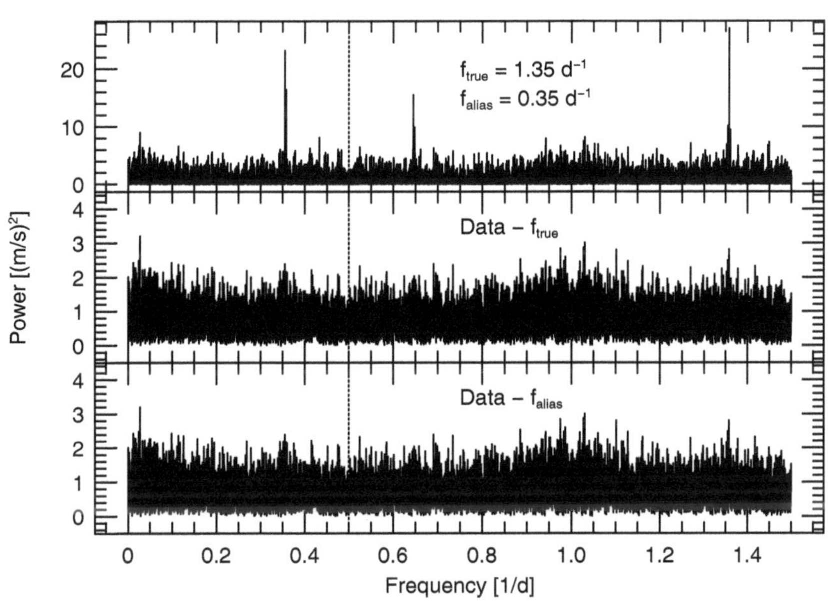

Figure 7.14. (Top) DFT power spectrum of the RV measurements for 55 Cnc. The vertical dashed line represents the Nyquist frequency. (Center) DFT power spectrum of the RVs after removing the true orbital frequency of the planet at 1.35 day^{-1}. (Bottom) DFT power spectrum of the RVs after removing the signal of the alias frequency at 0.35 day^{-1}.

(Figure 7.14). You simply do not know which is the true frequency that is in the data.

The only way to distinguish between an alias and a real frequency is to increase the Nyquist frequency. For example, if you suspect that the true period in your data is 0.85 days, it is a poor strategy to observe the star just once per night. Rather, the intelligent strategy would be to take several observations per night, or preferably observe the star throughout the night for several consecutive nights.

7.9 Frequency Resolution

We have seen the shape of the DFT of a pure sine wave is that of a sinc function. The width of this is determined by the length of time of your measurements—the longer you observe, the narrower the peak (top panel of Figure 7.15). If your data have a total time length Δt, the full width of your peak in frequency space will be $1/\Delta T$. This defines the frequency resolution of your time series.

If you have two very closely spaced frequencies that you wish to resolve, it is essential that you have a sufficiently long time string (lower panel of Figure 7.15). A typical application is trying to extract a planet signal with a period that is only slightly different from the rotational period of the star. If the orbital frequency of the planet is f_{p} and the rotational frequency of the star is f_{s}, then to first order, if you want to be confident of separating these two signals, you need measurements spanning a time span ΔT defined by

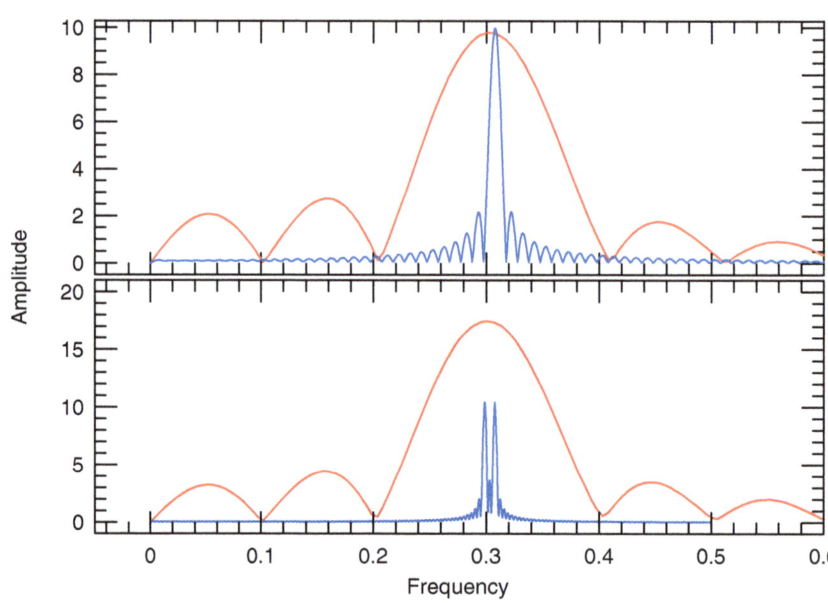

Figure 7.15. (Top) DFT amplitude spectrum of a periodic signal at $P = 3.25$ days for a 10 day time window (broad red curve) and a 100 day time window (narrow blue curve). (Bottom) DFT amplitude spectrum from a series consisting of two closely spaced sine functions with $P = 3.24$ days and $P = 3.35$ days and the same amplitude. The red (broad) curve is for a 10 day time series and the blue (broad curve) is for a 300 day time series.

$$\Delta T \approx \left| \left(\frac{1}{f_{\mathrm{p}}} - \frac{1}{f_{\mathrm{s}}} \right)^{-1} \right|. \tag{7.23}$$

The length of your observing window, ΔT, corresponds to a frequency of $1/\Delta T$, and this must be much less than the frequency separation of the two signals you are trying to detect. For example, if you have a planet with an orbital period of 8.9 days and a star that has a dominant rotational period at 11.2 days, then you would have to obtain measurements of the star over about 40 days in order to resolve the two.

In practice, by identifying and removing dominant frequencies to find secondary signals (prewhitening), one can separate closely spaced frequencies using data covering a much shorter time span than indicated by Equation (7.23).

To demonstrate this, let us take two signals with periods of P_1 = 11.2 days and P_2 = 8.9 days and amplitudes of $K_1 = 10$ m s^{-1} and $K_2 = 5$ m s^{-1}, respectively. This could be a typical case of trying to detect the RV signal of a transiting planet (say with period P_2) from the presence of rotational modulation due to activity (P_1). We consider a time string of these combined signals of increasing time length. To minimize the effects of the window function, we take the idealized case of regular sampling with no time gaps. For good measure, we add random noise at a level of $\sigma = 2$ m s^{-1}.

The left panels of Figure 7.16 show the amplitude spectrum of the time series with lengths of 10, 20, 30, and 40 days. The vertical blue lines indicate the frequencies found by prewhitening. The right panels shows the respective time strings and the

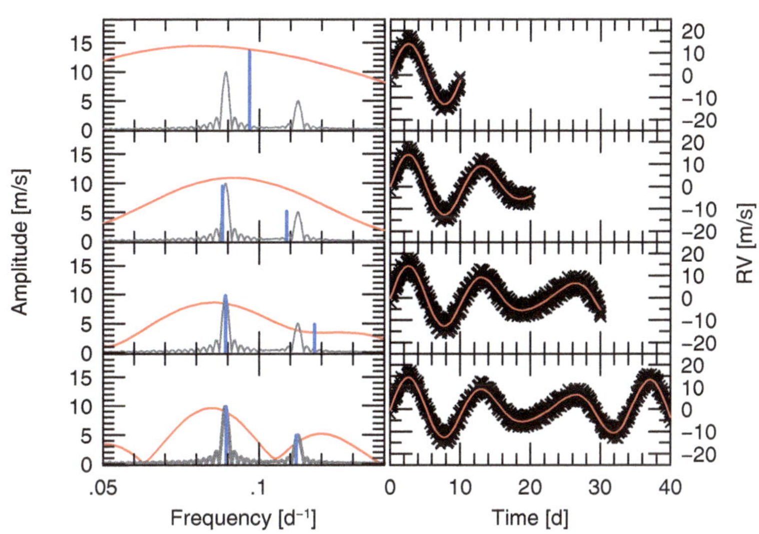

Figure 7.16. Resolving two closely spaced frequencies at ν = 0.0892 day^{-1}, 0.1123 day^{-1} (P = 8.9 days and 11.2 days). (Left panels) Prewhitening results (red line) for simulated data strings of 10, 20, 30, and 40 days (top to bottom). The gray line is for a time string of length 500 days shown as a reference curve. The blue vertical lines represent the frequency and amplitudes found by doing a simultaneous fit to the frequencies found by prewhitening. (Right panels) The fits to the RV data using the best-fit frequencies and amplitudes.

resulting fits from the frequencies that were found. For clarity we zoom into the peak of interest so you only see a narrow region centered on the peak, but if you plot the DFT out to higher frequencies, the sinc function is apparent.

After 10 days, only one periodic signal is found with a much higher amplitude than is in the data. Over a short time span, an 11.2 day signal is indistinguishable from an 8.9 day signal. A single period signal with $P = 12.4$ days is found. After 20 days, the DFT amplitude spectrum still shows no evidence for a second signal. However, by removing the dominant peak, a second peak is found in the residuals, and a simultaneous fit to the two sine functions does a good job of reconstructing the input signals. After 30 days, the recovery of the input signals is excellent. So, in practice, it suffices to have roughly half the time span indicated by Equation (7.23).

One final remark, although prewhitening and simultaneous sine fitting can recover the input signals, simply looking at the peaks in the periodogram will not tell you the true frequency of the signals that are present. Interference between the two signals and the effects of the spectral window may deceive you. You need to perform a simultaneous sine fit to the data, remove the next dominant signal, and look for additional ones. A simultaneous fit should then be performed using all of the signals that were found.

7.10 Assessing the Statistical Significance

If one has noisy data and sparse sampling, it is relatively easy to find a periodic signal that fits the data. For example, Figure 7.17 shows 10 simulated RV measurements with noise. A periodogram finds the highest peak at $P = 0.59$ days,

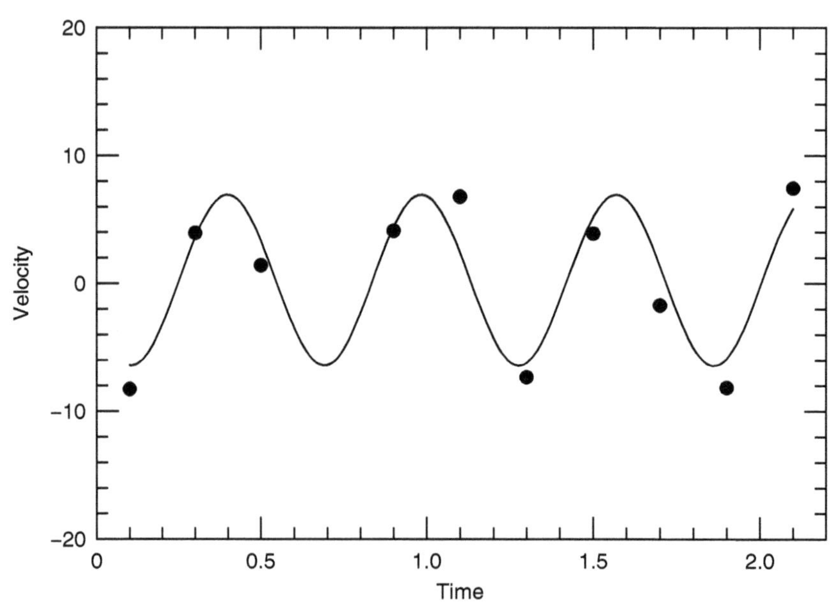

Figure 7.17. A sine fit (curve) to 10 simulated RV measurements consisting only of random noise.

which was seen to fit the data quite well. The only problem is that the data were generated using pure random noise—there is no periodic signal in the data. Of course, I have cheated a bit by not showing the error bars of the measurements—the sine fit would be less convincing (which is why you should always plot error bars). If you have noisy data of limited time span and sparse sampling, you almost *always* will be able to fit a sine curve to your data. Even if you have lots of data and good sampling, noise can often appear as a real signal. So, the question is, how do you know you have a real signal in your data that is not due to noise? Here we provide tools to answer this.

7.10.1 Using the Lomb–Scargle Periodogram

The LS periodogram was largely developed to help researchers assess whether a peak in a periodogram was real because the power of a peak in the LS periodogram is related to the statistical significance of a signal.

In assessing the statistical significance of a peak in the LS periodogram, there are two cases to consider. The first is when you want to know if noise can produce a peak with power higher than the one you found in your data, the so-called false alarm probability (FAP) over a wide frequency range. The second case is if there is a known periodic signal in your data. In this case, you want to ask what the FAP is *exactly* at that frequency.

If you want to estimate the FAP over a frequency range f_1 to f_2, then for a peak with a certain power, the FAP is (Scargle 1982)

$$\mathrm{FAP} = 1 - (1 - e^{-z})^{N_\mathrm{i}}, \tag{7.24}$$

where z is the power of the peak in the LS periodogram and N_i is the number of independent frequencies. It is often difficult to estimate the number of independent frequencies, but as a rough approximation, one can just take the number of data measurements. Horne & Baliunas (1986) gave an empirical expression relating the number of independent frequencies to the number of measurements, namely

$$N_\mathrm{i} = -6.362 + 1.193 + 0.00098 N_0^2, \tag{7.25}$$

where N_i is the number of independent frequencies and N_0 is the number of measurements. For large N, the FAP $\approx Ne^{-z}$.

The LS power thus gives you a first indication as to whether you should believe a signal is real or due to noise. It has been my experience that:

z = 6–10: probably due to noise.
z = 8–12: borderline case, but most likely noise.
z = 12–15: interesting, most likely real, but in some cases can be due to noise.
z = 15–20: with high probability real, but nature can still fool you.
z > 20: definitely a real signal.

Of course, the situation is different if you are trying to assess the FAP for a peak at a signal known to be in your data. This is the case if you are using RV data to detect the orbital motion of a known transiting planet that was found by a photometric survey. You are not interested in assessing the FAP that a peak can have a certain height over a broad frequency range, but rather the FAP of a peak at a precise frequency. In this case, you have only one independent frequency, or $N = 1$. Thus, Equation (7.24) reduces to

$$\text{FAP} = e^{-z}. \tag{7.26}$$

In such instances, the significance of the signal can be much higher than is indicated by Equation (7.24).

As an example consider a time series of 100 RV measurements of a star where you are looking for a periodic signal in the data. You perform a periodogram out to the Nyquist frequency, and your highest peak has $z = 7$. According to Equation (7.24), your FAP is an unconvincing 9%. On the other hand, if the star has a known transiting planet with an orbital frequency of f_{orb}, you find a peak at f_{orb} with $z = 7$. In this case, the FAP has a more convincing value of 0.09%, or two orders of magnitude smaller.

7.10.2 Using the Fourier Amplitude Spectrum

It is possible to get an estimate of the FAP from the Fourier amplitude spectrum. This is done by comparing the height of a peak (i.e., signal S) to the mean height of surrounding peaks (i.e., noise, N). The signal-to-noise ratio, S/N, increases with the significance of the signal.

Monte Carlo simulations (Kuschnig et al. 1997) established that if a peak in the amplitude spectrum has a height approximately 3.6 times higher than the mean peaks of the noise that surrounds it, then this corresponds to an FAP ≈ 0.01.

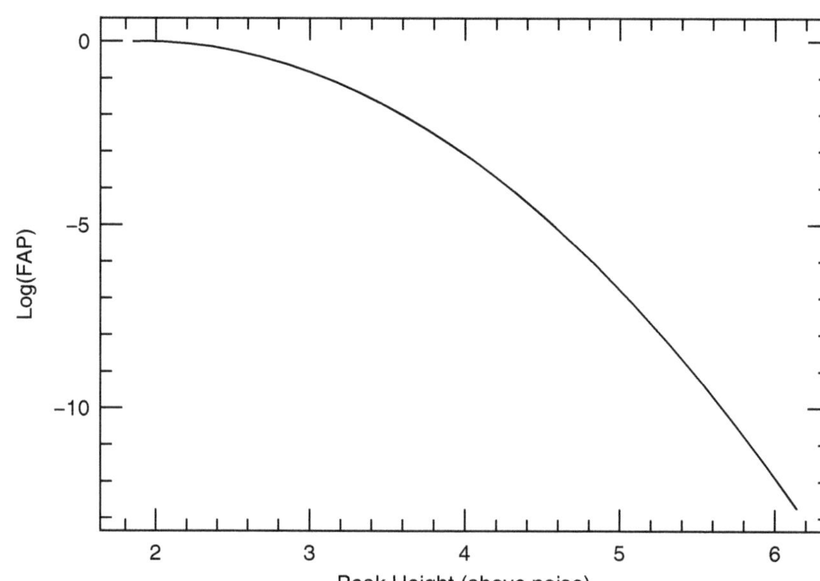

Figure 7.18. The FAP for a peak in the DFT amplitude spectrum as a function of the height above the mean amplitude of the surrounding peak.

Figure 7.18 shows the FAP as a function of the peak height above the noise. This was generated using Figure 4 in Kushnig et al. (1997).

This behavior of the FAP with S/N can be fit with a polynomial of the form

$$\log(\text{FAP}) \approx -2.50 + 2.62(\text{S/N}) - 0.69(\text{S/N})^2. \tag{7.27}$$

Again, this expression should only be used to get a rough estimate of the FAP. A proper bootstrap method (see below) should be used if you suspect a signal to be statistically significant.

It is instructive to compare how a significant signal appears in the classic and LS periodograms. The left panel of Figure 7.19 shows the LS periodogram of a portion of a time series consisting of a periodic signal in the presence of noise. The peak is marginally significant, having a power $z \approx 9$, or FAP $\approx 2\%$. The top-right panel shows the LS using a longer segment of the time series. The peak has become more significant with $z \approx 24$ (FAP $\approx 10^{-10}$). The significance of the signal in this instance is measured by the actual power of the peak.

The lower panels show the DFT amplitude spectra of the same stretch of the time series. The DFT of the short segment (lower left) shows a peak at the correct

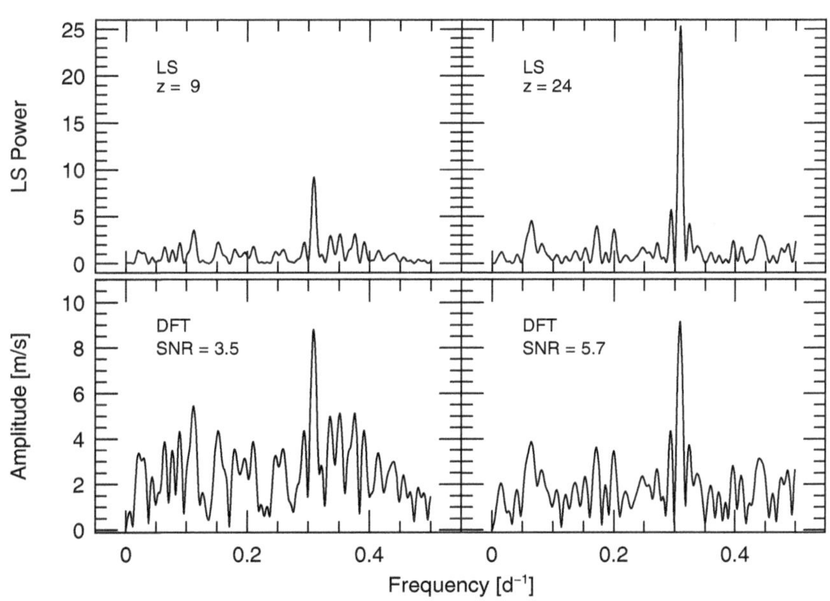

Figure 7.19. Comparison of the measure of significance of a peak with the LS (top panels) and DFT (lower panels) using a synthetic time series consisting of a sine function with a period of 3.35 days (ν = 0.298 day^{-1}) and an amplitude of 10 m s^{-1}. Noise is also present. (Top right) LS periodogram of a short segment of the time series. The signal is marginally significant. (Top left) LS periodogram of a longer segment of the time series. The significance of the detection has become higher, thus the LS power has increased. (Lower left) The DFT amplitude spectrum of the short time series. (Lower right) The DFT of the longer time series. Unlike the LS, the amplitude (and power) of the signal remains constant, but the significance is indicated by the lower mean amplitude of the surrounding peaks. In the DFT case, the significance of the signal is measured by the height of the peak with respect to the surrounding noise level.

frequency, but with low significance. After a longer time series, the peak becomes more pronounced, yet maintains the same amplitude. In this case the DFT only tells you the amplitude of the signal and "an amplitude is an amplitude"—it should not change. It does not tell you directly the statistical significance. Note, however, that the mean amplitude of the surrounding peaks (the Fourier noise level) has dropped. In this case, the significance of the peak is indicated by the ratio of the main peak with respect to the mean level of the surrounding noise peaks.

In short, a significant signal in the LS periodogram translates into increased power; for the DFT or classic periodogram, it translates into a decreased noise floor.

Estimating the Number of Measurements You Need

The ratio of the peak in the DFT to the surrounding mean level is a measure of the S/N of your signal, and it can be used to estimate how many more measurements are required to make a signal significant. Suppose that after 40 RV measurements you have detected a signal with S/N $\approx$ 3.3, or FAP $\approx$ 5%. You would like to know how many more measurements will be required to decrease the FAP to $\approx$0.1%, which requires S/N $\approx$ 4. This means you need to decrease the Fourier noise level by a factor of 1.2. The noise level decreases as $\sqrt{N}$ (Figure 7.3). Therefore, you would need $1.2^2 \times 40$, or approximately 60 measurements, with roughly the same sampling pathology to make the detection significant.

7.10.3 Bootstrap Randomization

The FAP estimated using the above expressions should just be taken as a rough estimate or a "quick look" at the FAP. To get a more accurate assessment of the FAP, one should always use a so-called bootstrap analysis using one of two methods. Both are based on randomly shuffling your data.

Method 1: Random Noise

The first method is to create random noise with a standard deviation having the same value as the rms scatter in your data. One calculates the LS periodogram and then finds the highest peak in the periodogram. This is done a large number (10,000–100,000) of times for different random numbers. The fraction of random data sets having LS power higher than that in your data is the FAP.

Method 2: Bootstrap

Method 1 of course assumes that your noise is Gaussian and that you have a good handle on your errors. What if the noise is non-Gaussian, or you are not sure of your true errors? The most common form of the bootstrap is to take the actual data and randomly shuffle it, keeping the time values fixed (Murdoch et al. 1993; Kuerster et al. 1997). You calculate the LS periodogram, find the highest peak, then reshuffle the data. The fraction of the shuffled-data periodograms having power larger than the original data gives you the FAP. This method more or less preserves the statistical characteristics of the noise in your data. Of course, if you still have a periodic signal in your data, then this will create a larger rms scatter than what you

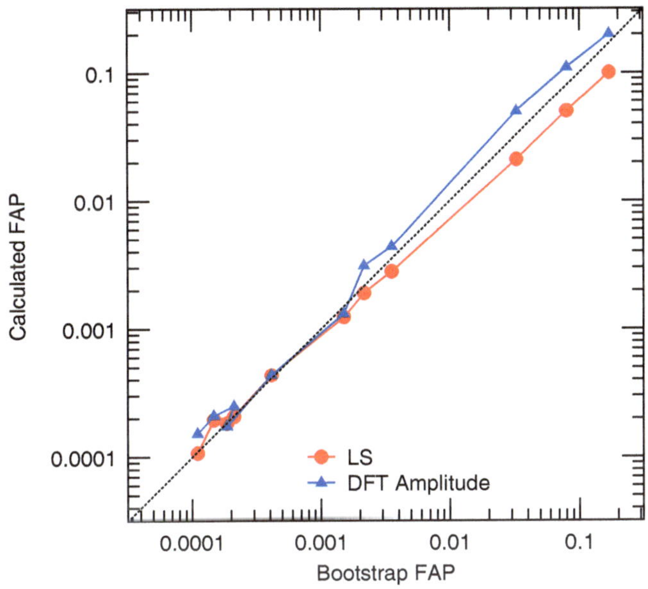

Figure 7.20. The false alarm probability (FAP) calculated using the LS periodogram and Equation (7.24) (red points) and the DFT amplitude and Equation (7.27) (blue triangles) versus results from the bootstrap (Method 2). The dashed line has a slope of unity.

would expect due to measurement uncertainties. The bootstrap thus represents a conservative estimate of the FAP.

How well do the analytical expressions for FAP (Equations (7.24) and (7.27)) agree with the results of the bootstrap process? Figure 7.20 shows the numerical results comparing the analytical FAP to those of the bootstrap (Method 2). An input sine signal using realistic sampling was used, and the noise level at different levels was added. One can see that the analytical expressions give a reasonably good approximation of the FAP.

Using the bootstrap to estimate the FAP for peak at a known frequency (e.g., a transiting planet with orbital frequency f_{orbit}) can be more problematic. The bootstrap should be done on a narrow frequency window centered at about f_{orbit}, but how narrow? If it is too wide, the FAP is overestimated, too small and it may be underestimated or give bad results.

One approach is to use a windowing technique. You first take a relatively large frequency window (at least 10 times the frequency resolution) centered on your frequency of interest, in this case f_{orbit}, and perform the bootstrap. You then do this for windows of ever smaller width ("winnowing window") and plot the bootstrap FAP versus the width of the window. The intercept (zero window width) gives you the FAP. This process is shown for RV data taken for the transiting planet GL 357 b (Luque et al. 2019). An LS periodogram of the RV data reveals a peak with $z = 6.79$, which corresponds to an FAP ≈ 0.0011 according to Equation (7.26). Figure 7.21 shows the results of the "windowing" bootstrap process for the peak found at the orbital frequency of 0.2544 day^{-1}. Points show the FAP as determined by a bootstrap as a

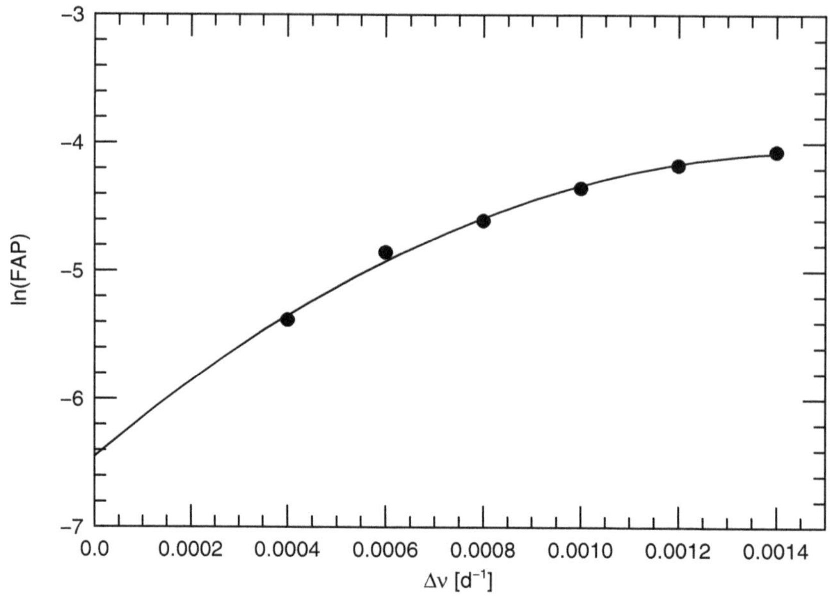

Figure 7.21. The natural logarithm of the false alarm probability (FAP) for the RV signal due to the transiting planet GL 357 b. The bootstrap randomization method was used to calculate the FAP over a narrow frequency window, $\Delta\nu$, centered on the orbital frequency of the planet and for various window widths (points). The extrapolation of the polynomial fit (curve) to $\Delta\nu = 0$ yields the FAP of the RV signal at the orbital frequency of the planet.

function of different window sizes, $\Delta\nu$, centered on the orbital frequency. Extrapolating the polynomial to $\Delta\nu = 0$ gives the FAP for the peak at the orbital frequency. This yields ln(FAP) = −6.44 or FAP 0.0016, entirely consistent with the analytical result.

7.11 Finding Multiperiodic Signals in Your Data

In astronomical time series, data are often multi-periodic. This is certainly true for exoplanets, which are often found in multiple systems. For stellar oscillations, there can be tens to hundreds of oscillation modes.

As we have seen, the spectral window can often introduce more peaks in your data than the actual signal (Figure 7.8), so it is pointless to try to identify real peaks in the time series just from the raw periodogram. One must first fit and remove the dominant peak in your data, and by doing so, you remove all of its alias peaks. What remains is the peaks of additional signals, and of course, their alias peaks.

This is demonstrated in the left panels of Figure 7.22, which show the DFT amplitude spectrum of a time series consisting of two sine functions with $P_1 = 7.35$ days ($f_1 = 0.136$ day^{-1}), $K_1 = 100$ m s^{-1}, and $P_2 = 3.75$ days ($f_2 = 0.267$ day^{-1}), $K_2 = 75$ m s^{-1}. The time series was sampled in the same manner as the data for Figure 7.10, and random noise at a level of $\sigma = 5$ m s^{-1} was also added.

The top left panel shows the initial amplitude spectrum. The middle panel is the amplitude spectrum of the residuals after fitting and removing the dominant peak at

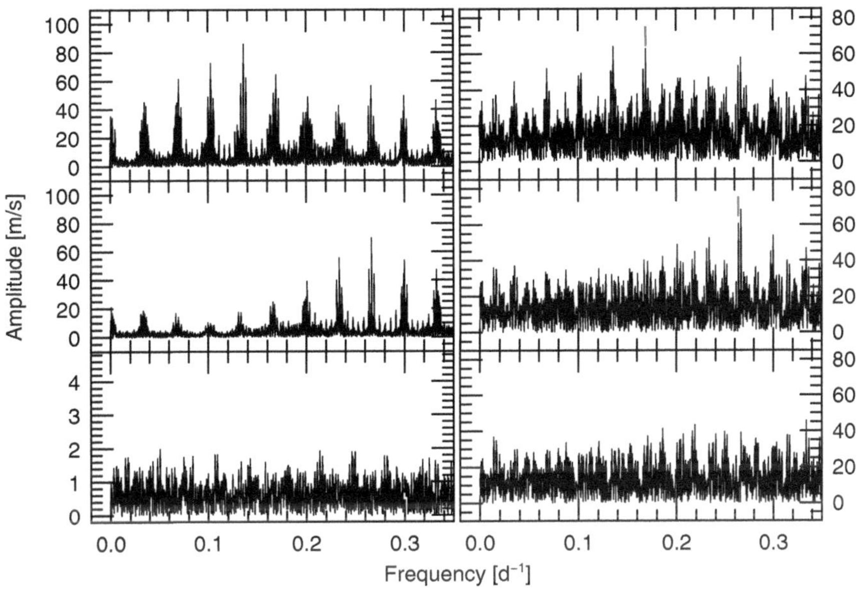

Figure 7.22. (Top left) The DFT amplitude spectrum of two sine functions (P_1 = 7.35 days, K = 100 m s^{-1}; P_1 = 3.75 days, K = 65 m s^{-1}) sampled in the same manner and with the window function show in Figure 7.10. The noise level is 5 m s^{-1}. (Middle left) The DFT amplitude spectrum after removing the 7.35 day period. (Bottom left) DFT amplitude spectrum after removing the 3.75 day period. (Top right) DFT amplitude of the same time series but with 20 times the noise level (σ = 100 m s^{-1}). (Middle right) DFT after removing the alias frequency at 5.91 days (f = 0.169 day^{-1}) indicated by the vertical red line in the top panel. (Bottom right) DFT after removing the alias period at 3.78 days (ν = 0.0.264 day^{-1}) shown by the red vertical line in the middle panel. Despite removing the wrong frequencies, the DFT of the residuals is at the noise level.

$f_1 = 0.136$ day^{-1}. Notice that this has also removed all of its associated peaks due to the spectral window. After fitting and removing the dominant peak in the residuals ($f_2 = 0.267$ day^{-1}), we arrive at the level of the Fourier noise. So, despite more than 10, apparently significant, peaks appearing in the original amplitude spectrum, there are in fact only two true signals present.

Things get more complicated once the noise level in the time series increases. The noise and its frequency characteristics may cause an alias peak to have a higher amplitude than the true one. You can thus choose and fit the alias frequency, but because it is an alias, you will still have an adequate fit to the data.

This is demonstrated in the right panels of Figure 7.22, where we have taken the same time series used for the left panels, but in this case added noise at a level of $\sigma = 100$ m s^{-1}, i.e., comparable to the amplitude of the dominant signal. You can see in the amplitude spectrum (top panel) that much of the structure of the spectral window is now lost in the noise. One can also see that one of the alias peaks of f_1, which is at f = 0.169 day^{-1} (shown by the red line), now has an amplitude comparable to the true frequency. For demonstration purposes, we fit and removed this alias peak. The residuals in the middle panel show that, despite choosing the alias, we have still removed the associated peaks from the spectral window.

In this case, the dominant peak in the amplitude spectrum of the residual data (middle right panel) occurs at the proper frequency of f_2. However, there is a nearby, high-amplitude peak at $f = 0.264$ day^{-1}. If you look carefully at the low-noise DFT in the top-left panel, you will see that every peak consists of a "triplet" of peaks due to the sampling window. The noise has boosted the amplitude of the satellite peak. Fitting and removing this alias peak (marked by red line in the panel), even though it is not the dominant one, we arrive at the level of the noise (lower-right panel of Figure 7.22). So, in this example, we have chosen the wrong frequencies in fitting the data, but because these are aliases, we still have an adequate fit to the data.

Is there any way to check if you selected the correct peak? In most cases, the true frequency tends to be the highest peak in the periodogram, but because of noise, that may not always be the case. You can get some indication as to which peak is the true one via simple simulations. Generate a sine wave using the frequency and amplitude of the dominant peak in the periodogram. Sample this in the same manner as the data and add noise at a level appropriate for the real data. Look at the periodogram to see how often the maximum peak occurs at the true one or the alias. Now do this a large number of times using different random noise. You then perform the same procedure, but instead using a sine fit to the alias peak. In this way you get a rough "probability" that the highest peak is in fact the correct one.

The procedure of finding multiperiodic signals we have just performed is called prewhitening, and it is frequently used for finding multiperiodic signals in time series data. The name prewhitening stems from the fact that time series data consisting of only random noise will have a frequency power spectrum that is flat or "white." That is to say, you have roughly equal contributions from all frequencies. This is opposed to data having a signal of a particular frequency or "color." By removing a periodic signal (peak in the periodogram), you are making the power spectrum "whiter" for the next step.

Prewhitening is a simple procedure with the following steps:

1. Perform a periodogram of your data.
2. Find the highest (significant) peak.
3. Fit a sine function (amplitude and phase), keeping this frequency fixed.
4. Remove this sine fit from your data. Removing this signal will remove its alias frequencies as well.
5. Perform a periodogram on the residuals.
6. If there are no more significant peaks in your data, stop. If not, go to step 2.

Figure 7.23 shows the process applied to RV data from the planet-hosting star 55 Cnc.[3] By virtue of the large data set pre-whitening works well in this case and it finds all the published planets. With the proper tools, finding all the planets of 55 Cnc took a matter of minutes.

[3] A convenient program for performing a DFT and pre-whitening analysis on time series data is Period04 (Lenz & Breger 2005, 2014). As of this writing the code can be downloaded at https://www.univie.ac.at/tops/Period04/.

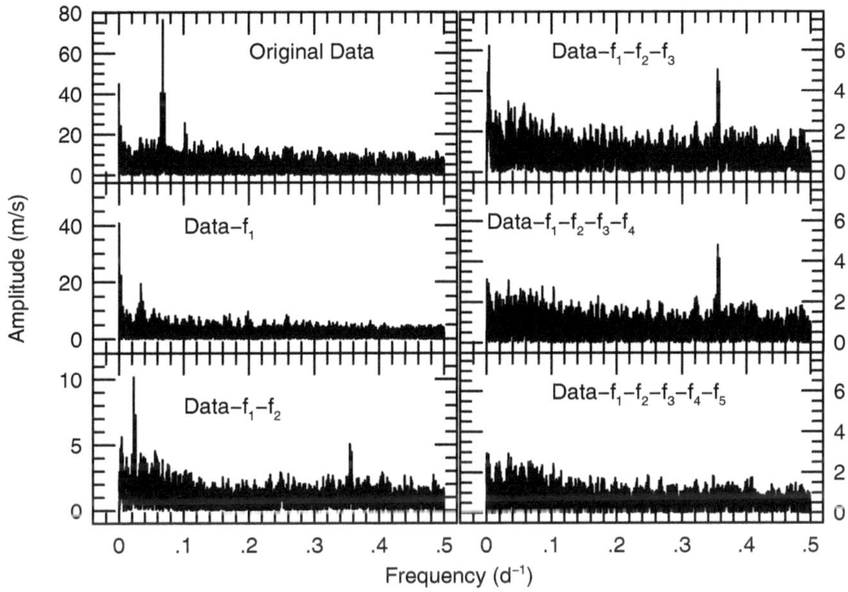

Figure 7.23. The prewhitening procedure performed on radial velocity measurements of the planet-hosting star 55 Cnc. $f_1 = 0.068$ day^{-1}, $f_2 = 000194$ day^{-1}, $f_3 = 00225$ day^{-1}, $f_4 = 000038$ day^{-1}, $f_5 = 1.357$ day^{-1}. Note that for f_5 we chose the true frequency of the hot Neptune rather than the alias of the discovery paper.

With prewhitening you will always find a highest peak that you can fit, but you should not be doing that to noise. Therefore, the process should stop when there are no longer any significant peaks in the periodogram. The definition of "significance" is often a matter of personal preference, but this should never be higher than an FAP of about 1%, otherwise one is most likely only fitting noise. If one is dealing with DFT amplitude spectra, this means that a significant peak must have a height at least 3.6 times the surrounding noise peaks.

When dealing with Keplerian orbits, it is wise to use a refined version of the prewhitening procedure. Rather than using sine functions (i.e., circular orbits), it is best to fit the data with the elements of a true Keplerian orbit including the eccentricity (replace the sine function in step 3 with a Keplerian orbit). As we shall later see, eccentric orbits can produce additional peaks in the periodogram that are harmonics of the orbital frequency. Fitting and removing a pure sine function will not remove these harmonics.

I should caution the reader that the prewhitening procedure will tell you all significant periodic signals in your data. What it will not tell you is the nature of these signals. A periodogram is essentially a Fourier transform of your time series. The prewhitening procedure just finds the dominant Fourier components of the sine series representing the data.[4] If you think in the Fourier domain, all peaks are in a sense "real" as they are needed reproduce the observed time series. (Take the inverse

[4] For these reasons, this author always prefers to call the prewhitening process Fourier component analysis or Fourier filtering.

transform and you get back the time series you started with.) Interpreting the nature of the individual signals in the periodogram is the task of the researcher and is often the most difficult part of frequency analysis. It can also lead to false conclusions.

7.12 Required Number of Observations

In planning observational strategies for the detection of periodic signals in time series data, the question naturally arises as to how many observations one needs for a significant detection. First, what is a significant detection? Cochran & Hatzes (1996) proposed an FAP < 0.001 as a significance criterion for the safe announcement of a planet detection. We note that an FAP < 0.001 only establishes that a periodic signal is in your data. It does not establish the nature of this signal or if it is actually due to a planetary companion.

Second, how many measurements are required? Cochran & Hatzes (1996) investigated this using the LS periodogram They defined a power S/N as $\xi = X_0/2\sigma_R$, where X_0 is the amplitude of the signal and σ_R is the measurement uncertainty. There are two ways of improving the detection efficiency. One can lower the measurement uncertainty, and great efforts over the past few decades have seen a remarkable improvement in the RV precision. However, current techniques have largely "hit a wall" of $\sigma_R \approx$ 1–2 m s^{-1}. The limitation is not so much due to instrumentation, but rather the intrinsic variability of the star (see Chapters 9 and 10).

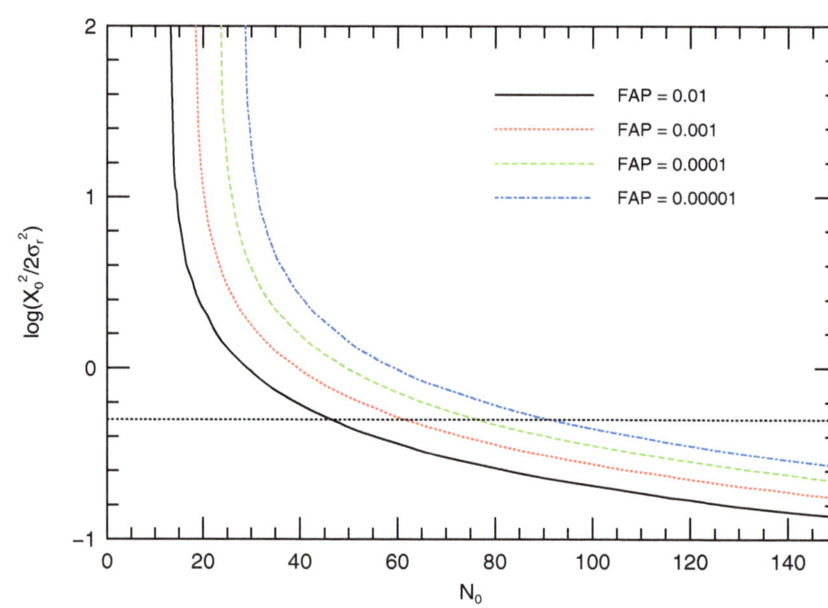

Figure 7.24. The power signal-to-noise ratio as a function of the number of independent observations N_0 spaced over the orbital phase for a variety of false alarm probabilities, FAPs. (From Cochran & Hatzes 1996. With permission of Springer.) The horizontal dashed line marks $X_0 = \sigma_r^2$, i.e., an RV amplitude equal to the measurement error.

Alternatively, one can increase the detection efficiency for low RV amplitude planets by increasing the number of measurements.

Figure 7.24 shows the power S/N, ξ, as a function of the number of independent measurements, N_0. Also shown are curves for detections at FAPs of 0.01, 0.001, 0.0001, and 0.00001. These curves were computed taking observations spread over a full orbital phase, i.e., with good phase coverage. As a rule of thumb, one needs at least $N_0 = 30$ before one gets any meaningful information about the presence of a periodic signal in the data. If $X_0 = \sigma_R$, then one needs about 40–45 measurements for FAP = 0.01, and $N_0 = 90$ to get the FAP to the very low value of 0.00001.

Of course, this simulation was made assuming only one planet was in the system, and we know that planets most likely come in multiple systems. In particular, the *Kepler* mission has shown that multiple, compact planetary systems with up to five to seven planets with periods less then tens of days may be common around stars, particularly low-mass ones. These are the types of systems that space-based transit missions like *PLATO*[5] will find. The large number of periodic signals in the data will require more observations to detect all planets in the system reliably .

As a simple simulation, let's take the Trappist-1 system as a "typical" compact system. Trappist-1 is an M8 dwarf that hosts seven transiting planets with orbital periods ranging from 1.5 days to 12.35 days and all with radii of $\approx 1 R_\oplus$ (Gillon et al. 2017). As of this writing, no masses have been measured for the planets, but we can calculate a rough mass based on a mean density of 6 gm s^{-1}. Table 7.1 lists the planets, periods, masses, and *K*-amplitudes for our simulated Trappist-1 system. Note that the planets are listed according to the sequence that they were found in the periodogram analysis. A time series was created using simulated orbits generated from the orbital parameters from Table 7.1 and realistic sampling. As time stamps, we took the sampling of Luyten's star from the CARMENES program (Reiners et al. 2018).

Table 7.1. Simulate Trappist-1 System

Planet	Period (days)	Mass ($M_\oplus$)	K (m s^{-1})
1	1.5	1.6	4.3
2	2.4	1.4	3.4
3	12.4	1.7	2.3
4	9.2	1.2	1.9
5	6.1	0.7	1.2
6	4.1	0.5	1.1
7	18.8	0.5	0.6

[5] *PLATO* (*PLAnetary Transits and Oscillations of stars*) is a space mission of the European Space Agency (ESA). Its goal is to search for transiting planets of up to a million stars using precise light curves.

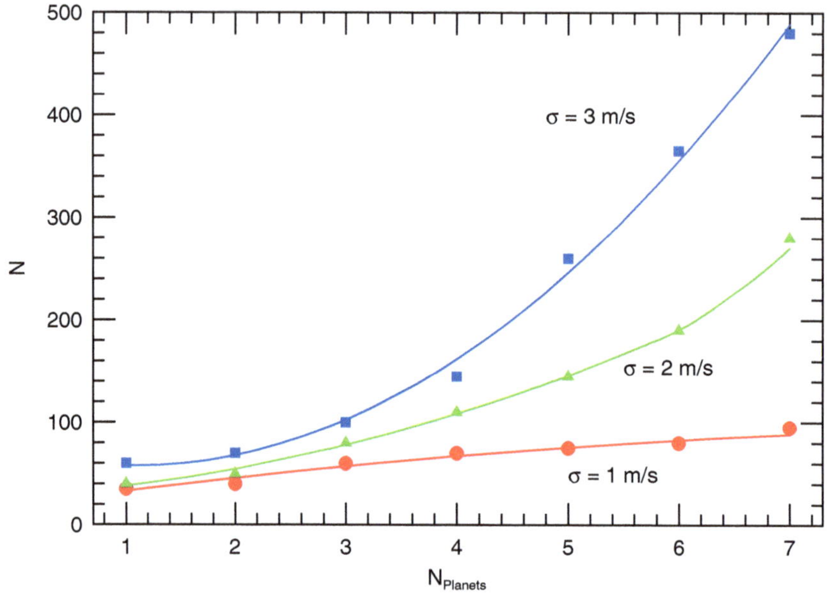

Figure 7.25. Simulations of the detection of a seven-planet Trappist-1-type system (Table 7.1). The abscissa is the number of planets detected in the simulated time series (see text) and the ordinate the number of measurements needed to detect the planet with FAP $\approx$ 0.01. The three curves are for measurement errors of 1 m s^{-1} (red points), 2 m s^{-1} (green triangles), and 3 m s^{-1} (blue squares). The curves represent polynomial fits to the simulations.

Figure 7.25 shows the number of measurements needed to characterize the system for three values of the measurement uncertainty (as random noise added to the time series): 1, 2, and 3 m s^{-1}. N_{Planets} represents the number of planets that can be detected with the last detection having a FAP of 1% or smaller, with a measurement uncertainty of 2 m s^{-1} to detect four planets, the last, with FAP < 0.01, requires at least 100 measurements.

If one has an excellent RV precision of $\sigma = 1$ m s^{-1}, which is equal to or larger than most of the K amplitudes, then one can fully characterize a Trappist-1-like system with approximately 100 measurements. With a measurement uncertainty of 3 m s^{-1}, one would need about 500 measurements. So, for the typical measurements afforded by current state-of-the-art instruments, one should expect to devote several hundred RV measurements to determine the masses of all planets in a multicompact planetary system.

7.13 Frequency versus Period

Before we depart the Fourier world and return to normal space, it is worth making a somewhat personal comment. You will notice that all of the periodograms shown in this chapter have frequency as units for the abscissa. The reason for this is that frequency (inverse of time) is the natural unit for the Fourier transform and thus the periodogram. In recent years, it has become common practice in the literature,

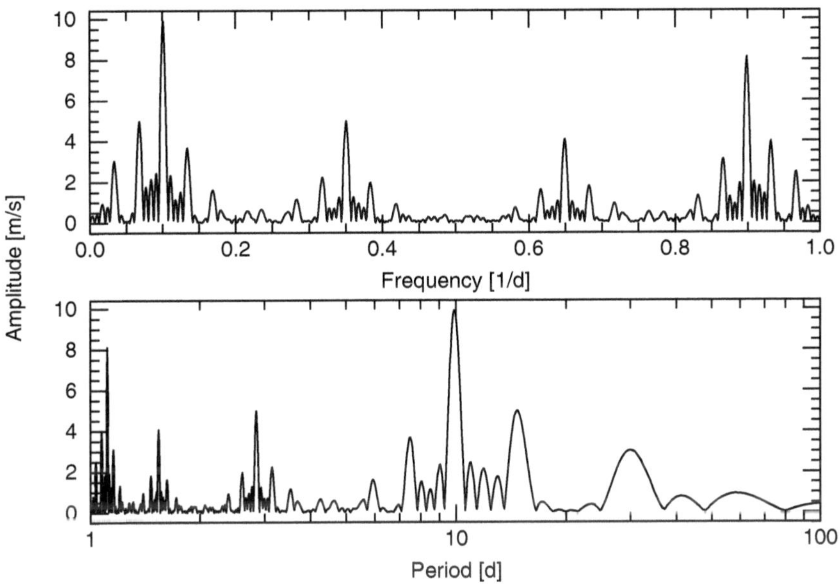

Figure 7.26. The DFT of the two-sine-wave time series (Figure 7.11) plotted versus frequency (top) and period (bottom). Plotting in period distorts the periodogram, making the interpretation of features more difficult.

almost exclusively from the exoplanet community, to use the period (inverse of the frequency) for plotting the periodogram. This is bad practice, but it is understandable why. We are used to thinking in the time domain—planets have orbital periods and not orbital frequencies, although one is as good as the other. This might be tolerable if one is looking for clean, isolated peaks in your periodogram, but this is rarely the case. A periodogram of an observed RV time series has a complex structure consisting of multiple signals, noise, spectral window, etc.

Figure 7.26 demonstrates how plotting the amplitude (or power) versus period can complicate the interpretation of the periodogram. The top panel shows the DFT amplitude spectrum of our simple two-component sine wave (Figure 7.11) plotted in the normal way versus frequency. The lower panel shows the same periodogram plotted on a period scale (and we have to use a logarithmic scale due to the wide range of values in period). Plotting periodograms should be avoided for three important reasons:

1. To re-emphasize, frequency is the natural unit of the periodogram.
2. In the "frequency" domain, the width of the peaks contains information on the length of the time series and thus the frequency resolution. Plotting the power versus period distorts this, with peaks at longer periods having a much larger width than short-period signals. This deceives the reader into thinking there are broad peaks and narrow peaks, when in fact, the individual peaks all have the same width dictated by the length of your time series. When plotted as period, you have no good sense from the periodogram if you have a long or a short time series.

3. One can more readily identify the structure of the spectral window in the periodogram when dealing with frequency. This has roughly the same shape centered on each real signal. The relative separation of alias peaks is fixed in frequency, and this gets lost when converting to period. To the experienced periodogram user, it is much easier to identify true peaks from alias ones.
4. One can more readily identify harmonics due to rotation, resonance of planets, of eccentric orbits. This is obvious by just looking at the periodogram and not having to look at actual frequency values. In "period" space, one is forced to read off the values of the period for each peak and do the mental calculation to realize one is dealing with harmonics. One loses a good deal of power of the "visualization" of the periodogram.

In short, plotting amplitude or power as a function of frequency for periodograms will make it easier for you to start "thinking in Fourier space."

References

Bracewell, R. N. 1978, The Fourier Transform and Its Applications (New York: McGraw-Hill)

Bretthorst, G. L. 2001, in AIP Conf. Proc., 568, Bayesian Inference and Maximum Entropy Methods in Science and Engineering, ed. A. Mohammad-Djafari (College Park, MD: AIP), 241–5

Cochran, W. D., & Hatzes, A. P. 1996, Ap&SS, 241, 43

Cumming, A., Marcy, G. W., & Butler, R. P. 1999, ApJ, 526, 890

Gillon, M., Triaud, A. H. M. J., Demory, B.-O., et al. 2017, Natur, 542, 456

Horne, J. H., & Baliunas, S. L. 1986, ApJ, 302, 757

Kovacs, G. 1981, Ap&SS, 78, 175

Kuerster, M., Schmitt, J. H. M. M., Cutispoto, G., & Dennerl, K. 1997, A&A, 320, 831

Kuschnig, R., Weiss, W. W., Gruber, R., Bely, P. Y., & Jenkner, H. 1997, A&A, 328, 544

Lenz, P., & Breger, M. 2005, CoAst, 146, 53

Lenz, P., & Breger, M. 2014, Period04: Statistical analysis of large astronomical time series, Astrophysics Source Code Library, ascl: 1407.009

Lomb, N. R. 1976, Ap&SS, 39, 447

Luque, R., Pallé, E., Kossakowski, D., et al. 2019, A&A, 628, A39

McArthur, B. E., Endl, M., Cochran, W. D., et al. 2004, ApJ, 614, L81

Mortier, A., Faria, J. P., Correia, C. M., Santerne, A., & Santos, N. C. 2015, A&A, 573, A101

Murdoch, K. A., Hearnshaw, J. B., & Clark, M. 1993, ApJ, 413, 349

Nagel, E., Czesla, S., Schmitt, J. H. M. M., et al. 2019, A&A, 622, A153

Reiners, A., Zechmeister, M., Caballero, J. A., et al. 2018, A&A, 612, 49

Scargle, J. D. 1982, ApJ, 263, 835

Zechmeister, M., & Kürster, M. 2009, A&A, 496, 577

AAS | IOP Astronomy

The Doppler Method for the Detection of Exoplanets

A P Hatzes

Chapter 8

Keplerian Orbits

You have made precise stellar radial velocity (RV) measurements, found a periodic signal in your data, and have decided it might be due to a companion. The next step is to calculate the orbit and determine its mass. In this chapter, we cover Keplerian orbits. An excellent source on this is Michael Perryman's *The Exoplanet Handbook* (Perryman 2018).

8.1 Orbital Parameters

Figure 8.1 shows the basic elements of a Keplerian orbit. These elements include:

reference plane: the plane tangent to the celestial sphere.

line of nodes: the line segment defined by the intersection of the orbital plane with the reference plane.

ascending node: point where the planet crosses the reference plane and moves away from the observer.

descending node: point where the planet crosses the reference plane and moves toward the observer.

Ω = longitude of ascending node. The angle between the vernal equinox and the ascending node. It is the orientation of the orbit in the sky.

Fully parameterizing a Keplerian orbit requires seven parameters: a, e, P, t_{p}, i, Ω, $\tilde{\omega}$:

a: the semimajor axis that defines the long axis of the elliptical orbit.

e: the eccentricity describes the amount of ellipticity in the orbit.

P: orbital period.

T_0: the time of periastron passage.

Ω: longitude of ascending node.

$\tilde{\omega}$, argument of periastron: the angle of the periastron measured from the line of nodes

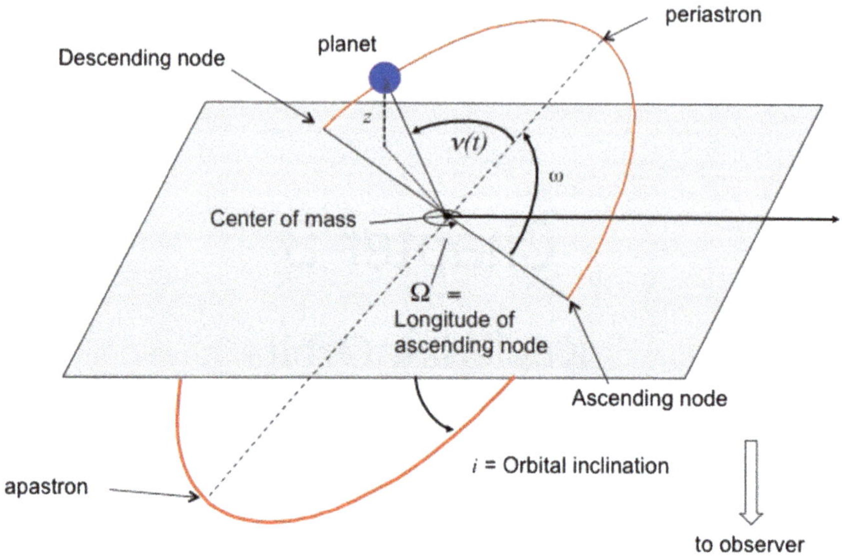

Figure 8.1. The elements of a Keplerian orbit.

i, orbital inclination: the angle of the orbital plane with respect to the line of sight.

In calculating the Keplerian RV measurements, there is an additional parameter and that is the overall velocity of the barycenter of the system due to the space motion of the star. This parameter is often denoted by γ, or the so-called "γ-velocity" of the system. For our purposes, we do not care about the overall space motion of the star, so the γ-velocity is a constant that can be subtracted from the RV variations of the orbit.

RV measurements can determine all of these parameters except for two, Ω and the orbital inclination. The angle Ω is a superfluous parameter that is not needed if one needs to determine the companion mass and the basic shape of the orbit. It is only important in those rare cases where you want to determine the true orientation of the orbit in the sky. The orbital inclination, however, is indeed important. Recall that we measure only one component of the star's motion (the radial one) and the full velocity is needed for the planet mass. We thus can only get a lower limit to the planet mass, i.e., the mass times the sine of the orbital inclination.

For RV orbits, there is one more important parameter that one derives and that is of course the velocity K-amplitude.[1] This is the component along the line of sight of the star's orbital velocity amplitude. We do not measure the mass of the planet directly but through its K-amplitude. The star and planet each have their own semimajor axis, a (with respect to the barycenter), and K-amplitude; the subscript "1" refers to the host star. We are of course solving for the stellar motion, and the semimajor axis of interest is the total, $a = a_1 + a_2$.

[1] Historically, the RV amplitude is denoted by the variable K and is often referred to as the K-amplitude.

To specify the Keplerian RV curve, we only must solve for γ, e, K_1, P, $\tilde{\omega}$, and T_0. Note that P and a are related via Kepler's third law.

8.2 Describing the Orbital Motion

For Keplerian orbits, the star and planet each move about the center of mass, or the barycenter. For elliptical orbits the center of mass is at one focus of the ellipse. The ellipse is described in polar coordinates by (see Figure 8.2):

$$r = \frac{a(1 - e^2)}{1 + e \cos \nu}. \tag{8.1}$$

The eccentricity is related the semimajor (a) and semiminor (b) axes by

$$b^2 = a^2(1 - e^2).$$

The periastron distance, q, and apastron distance, Q are given by

$$q = a(1 - e),$$

$$Q = a(1 + e).$$

In calculating orbital motion, there are several important angles that historically are called *anomalies*:

$\nu(t)$ = true anomaly: the angle between the direction of the periastron and the current position of the planet as measured from the center of mass.

$E(t)$ = eccentric anomaly: this is the corresponding angle referred to the auxiliary circle (Figure 8.2) having the radius of the semimajor axis.

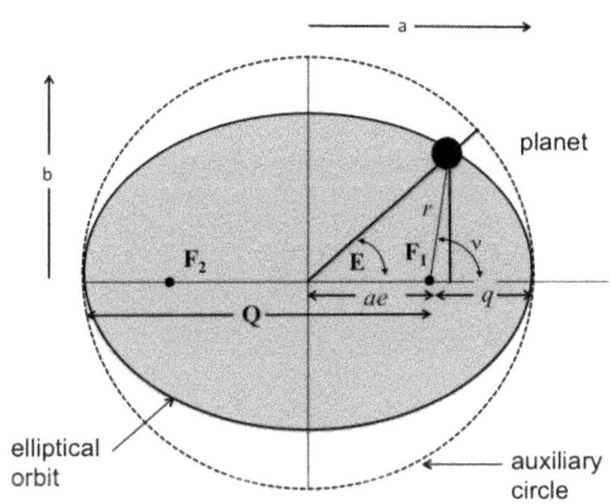

Figure 8.2. Elements of an elliptical orbit. The auxiliary circle has a radius equal to that of the semimajor axis, a. The semiminor axis is b. The true anomaly, ν, describes points on the orbit. Alternatively, one can use the eccentric anomaly, E. The focus, F_1, is the system barycenter.

The true and eccentric anomalies are related geometrically by

$$\cos \nu(t) = \frac{\cos(E(t) - e}{1 - e \cos E(t)}. \tag{8.2}$$

$M(t)$ = mean anomaly: this is an angle relating the fictitious mean motion of the planet that can be used to calculate the true anomaly.

For eccentric orbits, the planet does not move at a constant rate over the orbit (recall Kepler's second law, i.e., equal area in equal time). However, this motion can be specified in terms of an average rate, or mean motion, by

$$n \equiv \frac{2\pi}{P}.$$

The mean anomaly at time $t - T_{rm0}$ after periastron passage is defined as

$$M(t) = \frac{2\pi}{P}(t - T_0) \equiv n(t - T_0). \tag{8.3}$$

The relation between the mean anomaly, $M(t)$, and the eccentric anomaly, $E(t)$, is given by Kepler's equation,

$$M(t) = E(t) - e \sin E(t). \tag{8.4}$$

To compute an orbit, you get the position of the planet from Equation (8.3), solving for E in the transcendental Equation (8.4), and then using Equation (8.2) to obtain the true anomaly, ν.

8.3 The Radial Velocity Curve

We now show how the orbital parameters produce the observed RV curve. Referring to Figure 8.2, a planet moving by a small angle $d\nu$ will sweep out an area $\frac{1}{2}r^2 d\nu$ in a time dt. By Kepler's second law, $r^2\ d\nu\ dt$ = constant. The total area of an ellipse is $\pi a^2(1 - e^2)^{1/2}$, which is covered by the orbiting body in a period, P. Therefore,

$$r^2\frac{d\nu}{dt} = \frac{2\pi a^2(1 - e^2)^{1/2}}{P}. \tag{8.5}$$

The component of r along the line of site is $r\ \sin(\nu + \tilde{\omega})\sin i$. The orbital velocity is just the rate of change in r:

$$V_0 = \sin i\left[r_1 \cos(\nu_1 + \tilde{\omega})\frac{d\nu_1}{dt} + \sin(\nu_1 + \tilde{\omega})\frac{dr_1}{dt}\right] + \gamma. \tag{8.6}$$

The subscripts "1" and "2" refer to the star and planet, respectively. The above equation is thus for the star whose reflex motion we are measuring.[2] Recall that the

[2] For spectroscopic binaries, SB2, where one sees both components in the spectra, we have corresponding expressions for the companion (subscript 2). For exoplanets, we are in the case of an SB1, i.e., single-lined spectroscopic binary.

term γ is the overall radial velocity of the barycenter, which is irrelevant for the mass determination.

We can use Equations (8.1) and (8.5) to eliminate the time derivatives and arrive at

$$V_0 = K_1[\cos(\nu + \tilde{\omega}) + e \cos \tilde{\omega}], \tag{8.7}$$

$$K_1 = \frac{2\pi a_1 \sin i}{P(1 - e^2)^{1/2}}. \tag{8.8}$$

For eccentric orbits there will be a maximum positive RV value of the orbit and a maximum negative value. If we let A_1 and B_1 represent the absolute values of these quantities, then $A_1 \neq B_1$ and

$$A_1 = K_1(1 + e \cos \tilde{\omega}),$$

$$B_1 = K_1(1 - e \cos \tilde{\omega}),$$

and

$$K_1 = \frac{1}{2}(A_1 + B_1).$$

The RV curves from Keplerian orbits can have a variety of shapes depending on the eccentricity and the view angle from Earth. Figure 8.3 shows four examples of Keplerian orbits with $\tilde{\omega}$ fixed at 270°. Circular orbits (top-left panel) show the familiar sine curve. As the eccentricity increases, this turns into a step-like function.

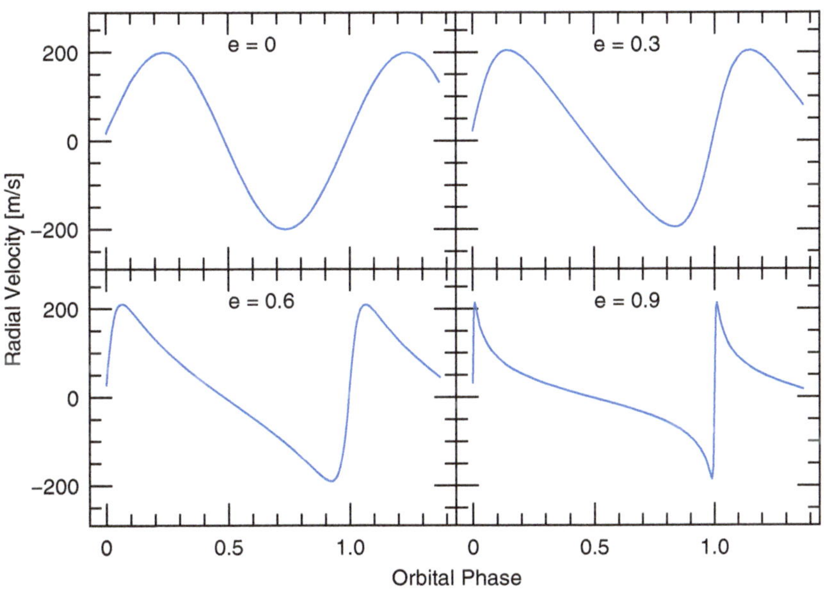

Figure 8.3. Sample RV curves from Keplerian orbits for various eccentricities but fixed argument of periastron ($\tilde{\omega} = 275°$). Clockwise from top left: circular orbit ($e = 0$), $e = 0.3$, $e = 0.6$, and $e = 0.9$.

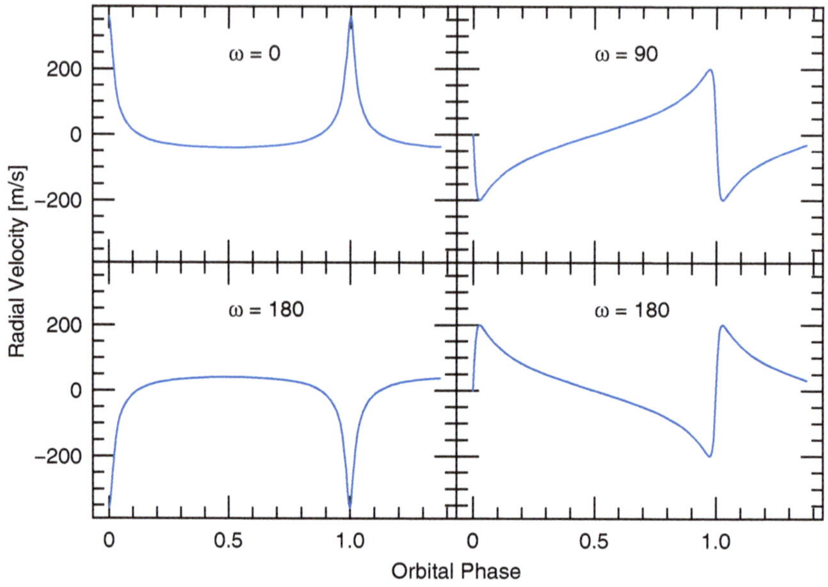

Figure 8.4. Sample RV curves from Keplerian orbits with fixed eccentricity (e = 0.8) but variable omega. Clockwise from top left: $\omega = 0°$, $\omega = 90°$, $\omega = 180°$ and $\omega = 270°$.

It is not just the eccentricity that affects the shape, but also $\tilde{\omega}$. The shape of the RV curve can look markedly different for the same eccentricity, but for different values of $\tilde{\omega}$ (Figure 8.4). Note that for circular orbits, $\tilde{\omega}$ is not defined. A circular orbit has no periastron passage as it keeps a constant distance from the star. In this case, the phase of the orbit is simply defined by the epoch or mean anomaly.

8.4 The Mass Function

Once we have calculated the orbital elements of a planet, we need to derive its mass, and this comes from the mass function. We can write Kepler's law as

$$\frac{G}{4\pi^2}(M_1 + M_2)P^2 = (a_1 + a_2)^3,$$

and where now we have included the components of both the star (1) and planet (2),

$$= a_1^3\left(1 + \frac{a_2}{a_1}\right)^3$$

$$= a_1^3\left(1 + \frac{M_1}{M_2}\right)^3,$$

where we have used $M_1 a_1 = M_2 a_2$,

$$\frac{G}{4\pi^2}(M_1 + M_2)P^2 \sin^3 i = a_1^3 \sin^3 i\left(\frac{M_1 + M_2}{M_2}\right)^3.$$

From Equation (8.8), we can solve for $a_1 \sin i$ in terms of K, P, and e. After substituting and rearranging, we arrive at

$$f(m) = \frac{M_2^3 \sin^3 i}{(M_1 + M_2)^2} = \frac{K_1^3 P(1 - e^2)^{3/2}}{2\pi G} \approx \frac{M_2^3 \sin^3 i}{M_1^2}, \tag{8.9}$$

where for the later expression we have used the fact that $M_1 \gg M_2$ for planetary companions.

Equation (8.9) is known as the mass function, $f(m)$, that can be calculated from the orbital parameters K, P, and e. There are two important things to note about the mass function. First, it depends on the stellar mass, M_1. This means if you want to get a good measurement of the mass of your planet, you have to know the mass of the star. In most cases, this is an educated guess based on the spectral type of the star. Only in cases where you have asteroseismic measurements (e.g., Hatzes et al. 2012) or the host star is a component of an astrometric binary is the stellar mass well known.

Second, you do not derive the true mass of the planet, M_2, only $M_2^2 \sin^3 i$. The Doppler effect only gives you one component of the velocity of the star. The orbital inclination can be measured using astrometric measurements, or for transiting planets. In the latter case, the orbital inclination must be near 90° to see a transit.

It is important to note that the mass of the companion is determined via the mass function. It is a quantity that is constant for a given system, and as such, it should *always* be given when listing orbital parameters derived from RV measurements. The stellar mass may change with refined measurements, but the mass function stays the same. With a published mass function value, it is easy for the reader to calculate a new planet mass given a different mass for the host star.

8.5 Mean Orbital Inclination

RV measurements, on their own, will never give you the true mass of the companion, only the minimum mass, or the product of the mass times the sine of the orbital inclination. Only in cases where you have a transiting planet will RVs alone give you the true mass. A large number of RV measurements coupled with a few astrometric measurements can also give you the companion mass (Benedict et al. 2002, 2006; McArthur et al. 2010).

What if you just so happen to be viewing an orbit nearly perpendicular to the planet? In this case, this might just be a stellar binary companion. Because you do not get the true mass, you may well ask why use the RV method? It is therefore important to ask "what it is the probability that the companion mass is much higher than our measured value?" Also, for a random distribution of orbits, "what is the mean orbital inclination?"

The probability that an orbit has a given orbital inclination is the fraction of celestial sphere that orbit can point to while still maintaining the same orbital orientation i. This gives a probability function of $p(i) = 2\pi \sin i \, di$. The mean inclination is given by

$$\langle \sin i \rangle = \frac{\int_0^{\pi} p(i) \sin i \, di}{\int_0^{\pi} p(i) \, di} = \frac{\pi}{4} = 0.79. \quad (8.10)$$

This has a value of 52°, which gives you, on average, 80% of the true mass.

We have seen that for orbits it is the mass function (Equation (8.9)), $f(m)$, that is important and $f(m) \propto \sin^3 i$. So, for orbits, the mean value of $\sin^3 i$ is what matters:

$$\langle \sin i^3 \rangle = \frac{\int_0^{\pi} p(i) \sin^3 i \, di}{\int_0^{\pi} p(i) \, di} = 0.5 \int_0^{\pi} \sin^4 i \, di = \frac{3\pi}{16} = 0.59. \quad (8.11)$$

How likely is it that we are viewing an orbit perpendicular to the orbital plane ($i \sim 0°$)? The probability that the orbit has an angle i less than a value θ is

$$p(i < \theta) = \frac{2 \int_0^{\theta} p(i) di}{\int_0^{\pi} p(i) di} = (1 - \cos \theta). \quad (8.12)$$

For an orbit to have an inclination less than about 10° the probability is less than 1.5%. So fortunately for RV measurements, viewing orbits perpendicular to the orbital plane is not very likely, but it happens. Vogt et al. (2002) found periodic RV variations of the star HD 33636, which was consistent with a planet having a minimum mass of 10.2 $M_{\rm Jup}$ and an orbital period of 5.8 years. Using astrometric measurements made with the Fine Guidance Sensor of the *Hubble Space Telescope*, Bean et al. (2007) derived an orbital inclination of only 4°, which resulted in a true companion mass of 0.142 $M_{\odot}$. The companion is not a planet but a low-mass star! The probability of this is occurring is small, about 0.2%, but with approximately a thousand planet candidates discovered by the RV method, it sometimes will happen.

The *Gaia*[3] mission is an astrometric space mission that should be able to derive true masses for most of the giant planets discovered by the RV methods. It could be that a few, but not all, of our exoplanets will disappear.

8.6 Eccentric Orbits

One interesting aspect of Keplerian orbits is they can have high eccentricities and thus extreme shapes. The RV curve of a highly eccentric orbit is a shape that is difficult to produce by other phenomena like stellar oscillations or rotational modulation by surface structure. The shape of a highly eccentric RV curve has often been invoked as evidence for a planetary companion (e.g., Frink et al. 2002).

Highly eccentric orbits can also cause problems by introducing biases in surveys and the spurious detection of additional signals in the RV data. They also present challenges in trying to find periods using standard periodogram tools.

[3] *Gaia* is a mission of the European Space Agency that was launched in 2013 and will fly until 2022. Its purpose is to measure the positions, distances, and motions of approximately one billion objects.

8.6.1 Observing Biases Caused by Eccentric Orbits

One somewhat annoying feature of highly eccentric orbits is there can be large stretches of time when nothing seems to happen and other times when things are happening too fast! For instance, if you are on the flat part of the RV curve of an eccentric orbit, you may simply stop observing the star because it appears to be constant. Alternatively, if you hit the part of the orbit when the RV is changing quickly, you might be fooled into thinking you have a stellar companion and again give up on the target. Two examples highlight both cases.

HD 120066 was a star that was followed as part of the McDonald Observatory and California planet search programs. For the first 15 years of observations, the RV measurements showed no significant variations above the RV uncertainties (Figure 8.5). Fortunately, both programs continued to follow this star. After 15 years, the RV variations showed a rapid rise, indicating the presence of a giant planet, $M = 3.2M_{\rm Jup}$ in a highly eccentric orbit, $e = 0.83$ (Blunt et al. 2019). This is the longest period giant planet discovered to date, but it would have been overlooked if not for the patience and perseverance of the research teams, not to mention the support of telescope time allocation committees.

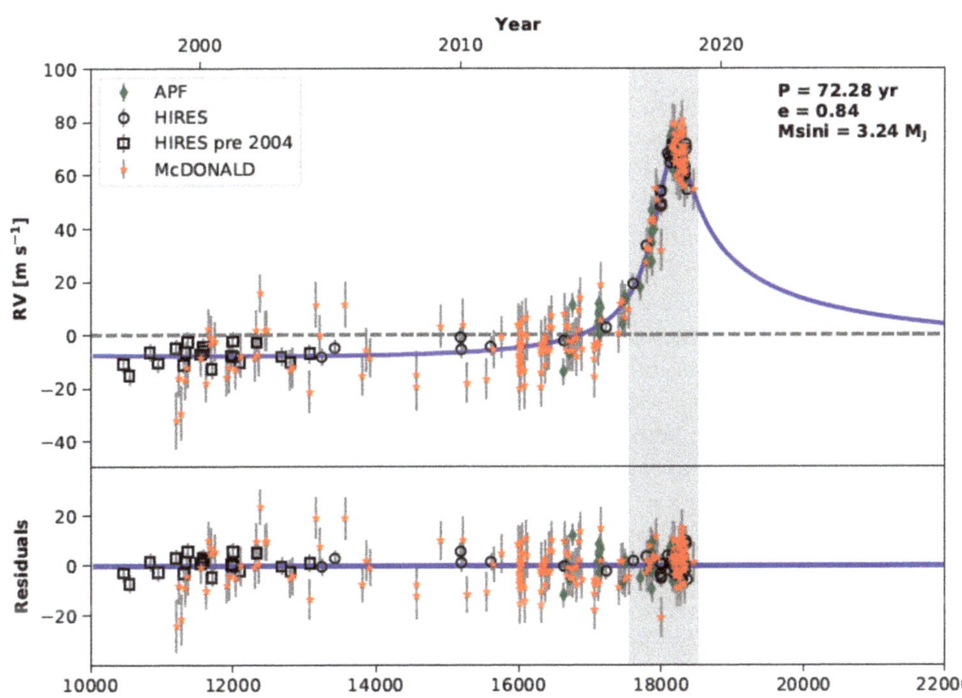

Figure 8.5. RV measurements for HD 120066. Over the first 15 years, the RVs showed no variations. After 15 years, the measurements showed the clear presence of a planetary companion of mass 3.24 $M_{\rm Jup}$ in a 72.28 year orbit with high eccentricity, $e = 0.84$. (Reproduced with permission from Blunt et al. 2019.)

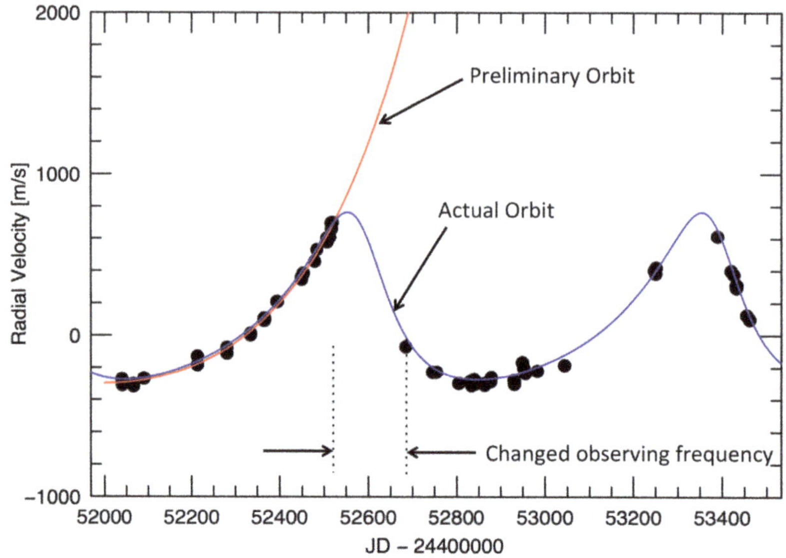

Figure 8.6. RV measurements for the brown-dwarf-hosting star HD 137510. A preliminary orbit of the initial RV measurements indicated a period of ≈3800 day orbit and a companion minimum mass of ≈0.5 $M_{\odot}$. The target was placed on lower priority and observed infrequently (marked by vertical dashed lines) to take into account the long period and the expected "boring" stellar companion. Additional measurements showed the true orbit to be due to a 26 $M_{\rm Jup}$ companion in a 798 day orbit.

HD 137510 b highlights drawing a false conclusion for another reason. Figure 8.6 shows the RV measurements of HD 137510 taken as part of the Tautenburg Observatory Planet Search Program (see Hatzes et al. 2005). Note that there is a large gap in the sampling. The reason for this is that a preliminary orbit (admittedly unconstrained) indicated that the companion was most likely a star with $M \sim 0.5 M_{\odot}$ in a long-period orbit (≈10 years). Because binary stars were not part of the science case of the program, a star was placed on a low- priority list with more infrequent sampling. It was only observed after all other high-priority targets were observed in a given run. When additional measurements were finally taken, these showed that the orbital period was shorter, the eccentricity higher, and the companion mass much lower. Rather than being a "boring" stellar companion, this was a much more "exciting" discovery of a rare short-period (P = 798 day) brown dwarf with a minimum mass of $M \sin i = 26\ M_{\rm Jup}$ (Endl et al. 2004). Eccentric orbits can cause rapid changes in the RV, thus changing the nature of the companion.

Although there are a number of exoplanets that have been discovered in highly eccentric orbits, the number of undetected planets in eccentric orbits can be much higher for two reasons: (1) the difficulty in detecting the eccentric orbit and (2) a "success-driven" bias when searching for exoplanets.

To detect a planet in an eccentric orbit and to get accurate orbital parameters, you need to be looking at the right time. RV measurements must be taken at the maximum (or minimum) RV, and this can occur over just a few days. If the planet has a long-period orbit and you miss the peak of the RV due to bad weather, you may have to wait several months or years for the next opportunity. This factor is simply due to the nature of Keplerian orbits and will result in a planet in an eccentric orbit being undetected.

The second bias is the human factor, as demonstrated in the case of HD 137510 b. Since the discovery of the first exoplanets, the field has been driven by a desire for success. Researchers want to find exciting planets rather than boring constant stars, or stellar binary stars. As a result, there is danger of changing strategies midway in the survey. You drop a star that is constant and add another star, which might show more promise. This results in exoplanets in eccentric orbits not being detected simply because the observer gave up. This also changes your statistics if you wish to draw meaningful conclusions about the frequency of planets in highly eccentric orbits. The proper strategy is to define you sample and stick to the targets, regardless of what the intermediate results tell you.

8.6.2 Eccentric Orbits in the Fourier Domain

Naively, one would assume that periodograms (DFT, LSP) would have difficulty finding the correct period for highly eccentric orbits, which have the shape of a step function. After all, periodograms are designed to find sine functions (i.e., circular orbits). This is not the strictly case. For highly eccentric orbits, the periodogram will find many periodic signals in an eccentric orbit despite only one period being present in the data. Recall that the periodogram is essentially a Fourier transform. This will find the dominant period in the time series, but in order to fit the distorted shape (from a sine wave) of the eccentric orbit, it also requires the harmonics of the orbital period.[4]

Figure 8.7 shows the Lomb–Scargle (LS) periodogram of an RV curve having an eccentricity of 0.75 and a period of 74.5 days ($f = 0.013$ day^{-1}). The LS periodogram shows the correct orbital frequency as the dominant orbit, but also a series of harmonics. All of the visible peaks are integer multiples of the orbital frequency ($\nu = n \times 0.013$ day^{-1}, where n is an integer).

There are two other phenomena that can produce harmonics of the dominant period besides eccentric orbits. The first are planets in resonant orbits. For instance, a planet system in a 1:2:3 resonance would show a periodogram with three peaks, the orbital frequency f_0 as well as $2f_0$ and $3f_0$.

We shall see in Chapter 9 that rotational modulation can also produce harmonics at the dominant period, in this case due to rotation. How can we distinguish between eccentric orbits and the other phenomena? One way is to look at the ratio of amplitudes of the harmonic (side lobe) peaks with respect to the main one as these have fixed ratios.

Figure 8.8 shows the amplitude ratio of the successive "side lobes" with that of the "main lobe" as a function of eccentricity. These are shown for $\tilde{\omega} = 0°$, but the results do not change for nonzero $\tilde{\omega}$. As an example, the first side lobe to main lobe ratio

[4] Ptolemy actually stumbled onto the Fourier series and how it represent a time series. He needed epicycles (harmonics) in order to explain the motion of planets, partly because of their eccentric orbits. Coincidentally, Ptolemy could fit the motion of the planets better than Copernicus, who was unaware that planetary orbits could be eccentric.

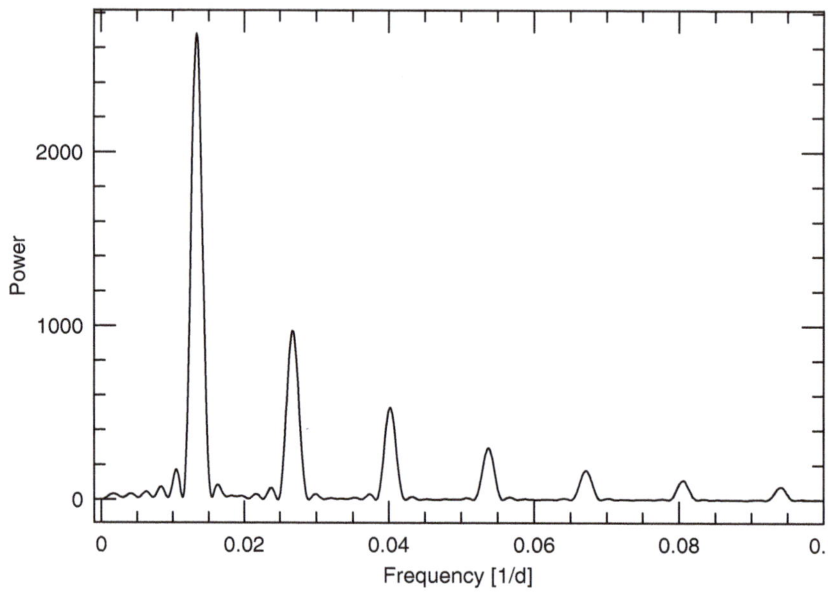

Figure 8.7. Lomb–Scargle periodogram of RV data from an eccentric orbit ($e = 0.75$). The dominant peak is at the orbital frequency $f_{\rm orb} = 0.0133$ day^{-1}. Other peaks occur at harmonics $f = nf_{\rm orb}$, where n is an integer.

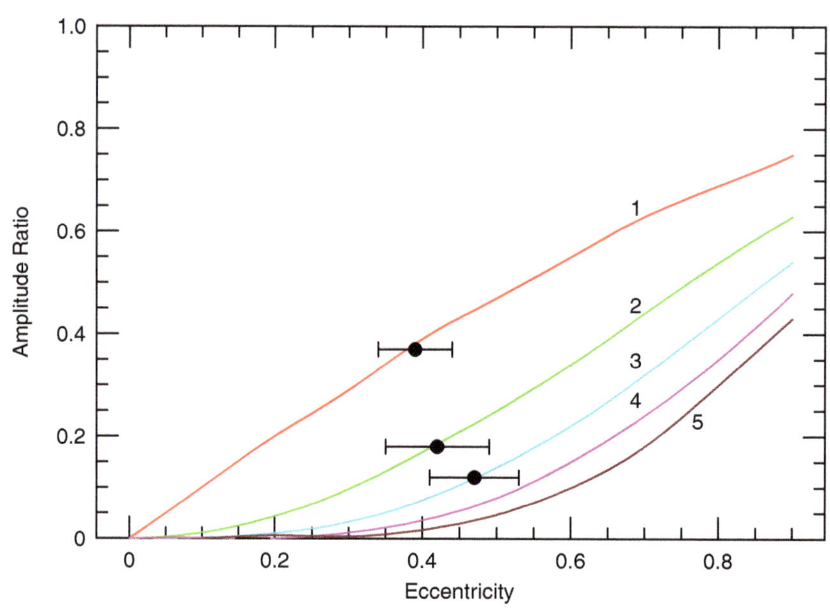

Figure 8.8. Amplitude ratio of the first through fifth (numbered for each curve) side lobes with respect to the main peak in the DFT amplitude spectrum for orbits as a function of eccentricity. The points mark the ratio of the first, second, and third side-lobe amplitudes to the main peak in the periodogram of the RVs for GJ 4276.

is 0.3 for $e = 0.2$ and as high as 0.75 for $e = 0.9$. If one sees a periodogram with harmonics of equal amplitude, then it cannot be explained by eccentric orbits. In the case of rotational modulation, the higher harmonics can even have amplitudes higher than the main lobe.

Let's examine whether the amplitude spectrum in the Fourier domain can differentiate between eccentric orbits versus multiple planets using real data. We will use the example of the planet around GJ 4276. A periodogram analysis of RV measurements for this star showed a dominant signal at 13.35 days (companion mass $M \sin i \sim 16\ M_\oplus$), but a second signal at the first harmonic, $P = 6.67$ days (Nagel et al. 2019). The RV data could be fit either by a single planet on an eccentric orbit ($e = 0.37$) or a two-planet system with a period ratio of 2:1. Understanding the formation and evolution of the system depends critically on whether you have a single planet in an eccentric orbit (and in this case the most eccentric orbit for a planet around an M dwarf at the time) or a two-planet system in resonance. Nagel et al. (2019) slightly preferred the eccentric planet solution, although neither of the two scenarios could be excluded with high confidence. What can a frequency analysis tell us?

A Fourier analysis of the Nagel et al. (2019) RV data easily finds the two dominant frequencies at $f_0 = 0.0749$ day^{-1} and $f_1 = 0.1498$ day^{-1} ($=2f_0$). After removing these, one finds a hint of the third harmonic ($3f_0$) at $f_2 = 0.2247$ day^{-1} Although it is not the highest peak in this frequency range, fitting a sine wave using f_2 and removing this greatly diminishes the amplitude of the surrounding peaks. The amplitude ratios of f_1 and f_2 with respect to f_0 are shown in Figure 8.8. The ratios are consistent with $e \approx 0.4$.

How do the two models compare in the frequency domain? To examine this, we took the eccentric orbit solution, sampled it like the data, and added random noise consistent with the mean measurement error ($\sigma \approx 2$ m s^{-1}). Because the amplitude of the peak at $3f_0$ is low and noise may boost the amplitude of nearby peaks, we created 10 different data sets using different random number generators. We then took the average amplitude spectrum of these, after removing f_0 and f_1. This is, of course, a bit of a cheat since using the mean of several amplitude spectra reduces the Fourier noise, but it does show what a typical amplitude spectrum might look like. The result is shown in the middle panel of Figure 8.9. The peak at the second harmonic $3f_0$ is clearly present.

We then performed the same procedure using the two-planet solution, with the result shown in the lower panel of Figure 8.9. The second harmonic is not present. This exercise does not conclusively favor the eccentric solution; the peak at $3f_0$ in the real data has an unconvincing false alarm probability of $\approx$7%. Furthermore, noise can also make a peak appear at $3f_0$ even if you have a two-planet system. However, with our frequency analysis we can slightly favor the eccentric planet solution and lend support to Nagel et al.'s conclusions.

8.6.3 Keplerian Periodograms

In Chapter 7, we saw how the periodogram finds the best-fitting sine function, in a minimized χ^2 sense, using four parameters: the period, P, amplitude, A, phase and

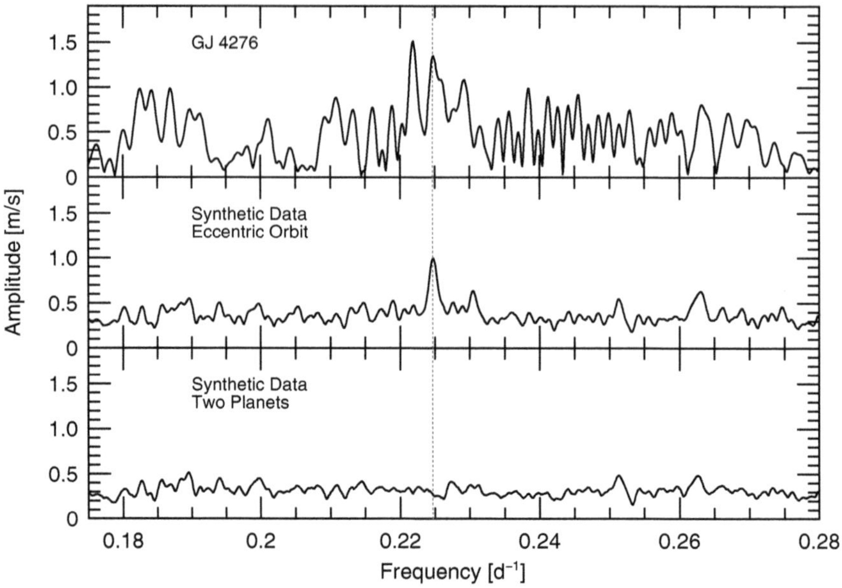

Figure 8.9. (Top) The DFT amplitude spectrum of the RVs for GJ 4276 after removing the dominant frequency of 0.0749 day^{-1} (f_0) and the first harmonic at 0.1498 day^{-1} ($f_1 = 2f_0$). The red dashed line marks the frequency of the second harmonic ($f_2 = 3f_0$). (Middle) The DFT amplitude spectrum of a simulated eccentric orbit (with proper noise and sampling) after removing f_0 and f_1. The frequency $f_2 = 3f_0$ is clearly present. (Bottom) The DFT amplitude spectrum of a two-planet orbit in resonance after removing the contribution of the two planets.

offset, c. The concept of the periodogram can be extended to Keplerian orbits where instead of fitting sine functions, the RV series can be fit with a Keplerian orbit specified by six orbital parameters:[5] γ, K, $\tilde{\omega}$ e, T_0, and P. The so-called Keplerian periodogram was first introduced by Cumming (2004) and elaborated in the context of the GLS periodogram by Zechmeister & Kürster (2009).

The Keplerian periodogram shows how well a trial period (or frequency f) can model the data in terms of χ^2 minimization. Analogous to Equation (7.11), we can write

$$\chi_{\mathrm{Kep}}(\omega) = \frac{\chi_0^2 - \chi_{\mathrm{Kep}}^2}{\chi_0^2}. \tag{8.13}$$

In the Keplerian periodogram, the sine function is replaced by the Keplerian RV curve. Using the substitutions $c = \gamma + Ke\cos\omega$, $a = K\cos\omega$, and $b = -K\sin\omega$, this can be written as

$$RV(t) = c + a\cos\nu(t) + b\cos\nu(t),$$

where $\nu(t)$ is the true anomaly. We can now use the formalism of the periodogram, replacing ωt_i with νt_i. For a fixed period, P, one steps through e and T_0 to yield the best Keplerian fit (i.e., highest power).

[5] Of course, a Keplerian orbit with zero eccentricity reduces to the case of sine fitting.

As an example, we will apply the Keplerian periodogram to an extreme case, the planet HD 20782b. Figure 8.10 shows the RV measurements of HD 20782 from O'Toole et al. (2009). These show evidence for a planet in a highly eccentric orbit, but this high eccentricity is largely determined by one measurement. A standard periodogram analysis, as we shall soon see, fails to find the correct period in the time series.

The Keplerian periodogram is shown in Figure 8.11. The red line shows the best-fit Keplerian orbit to the data. If one normalizes by the best fit at each frequency (see Cumming 2004),

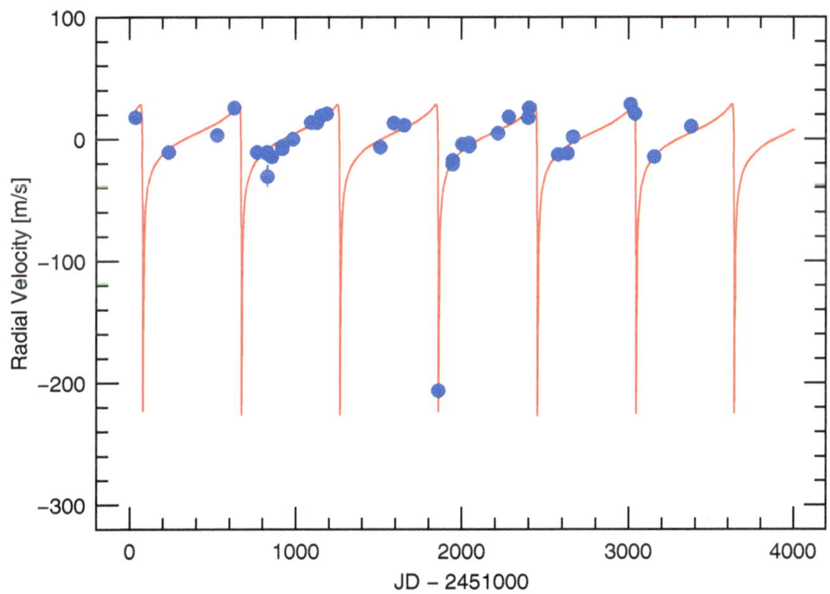

Figure 8.10. RV measurements (points) of HD 20782 (O'Toole et al. 2009) and the Keplerian orbit (red line) after finding the orbital period with the standard periodogram.

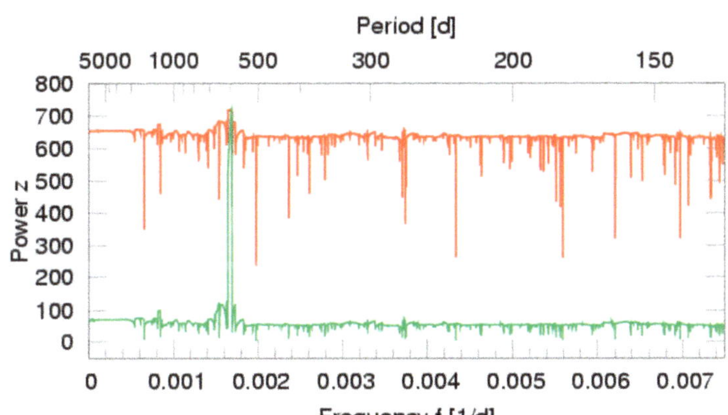

Figure 8.11. Keplerian periodogram for HD 20782 normalized to the best fit (upper red) and the best fit at each frequency (lower green). (From Zechmeister & Kürster 2009, reproduced with permission © ESO.)

$$z_{\text{Kep}}(\omega) = \frac{N-5}{4}\frac{\chi_0^2 - \chi^2(\omega)}{\chi^2(\omega)}, \tag{8.14}$$

the power at the correct frequency, $\omega = 0.00169$ day^{-1} ($P = 591.9$ days), is greatly enhanced (green line in Figure 8.11).

It is of interest to compare the results of the Keplerian periodogram to those of the standard periodogram and to demonstrate that with the proper use of the more standard tool, one can arrive at the same answer as the Keplerian periodogram.

As you gain experience in using the periodogram on a variety of data sets, you will often find that one extreme data point can drastically alter the appearance of the periodogram and thus mask signals that might be in the data. The top panel of Figure 8.12 shows the DFT amplitude spectrum of the RVs for HD 20782. It is essentially flat and featureless, with the highest peak at $f = 0.032$ day^{-1} ($P = 31.2$ day), which is clearly not significant (false alarm probability, FAP = 0.9).

This periodogram can be understood using what we have learned in Chapter 7, especially about Fourier transforms. If one excludes the extreme RV value, one can see clear sinusoidal variations in the RVs of HD 20782. So, we can represent the time series by an extreme value that can be approximated by a δ function superimposed on sine-like variations. In the Fourier (periodogram) domain, this is the sum of the respective Fourier transforms. We have seen that the transform of a δ function is a constant value with frequency in the Fourier domain. Indeed, we see a constant value amplitude spectrum ($\approx$15 m s^{-1}), which completely obliterates the peak due to any periodic variations. Removing this constant value must be done in the time

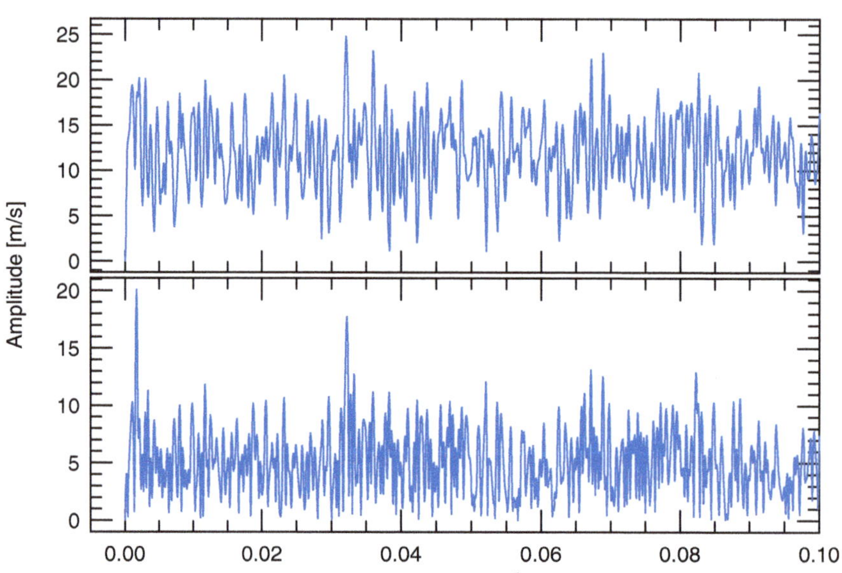

Figure 8.12. (Top) The DFT amplitude spectrum of the RV measurements for HD 20782 (Figure 8.10). (Bottom) The DFT amplitude spectrum after removing the extreme data point due to the high eccentricity of the orbit. The correct orbital frequency is the highest peak.

domain, i.e., remove the outlier. Doing this produces a DFT amplitude spectrum (lower panel of Figure 8.12), which easily shows a strong peak (FAP $\approx 10^{-5}$) at the appropriate frequency of f = 0.00169 day^{-1} (P = 591.7 days). Furthermore, removing this peak and looking at the residuals, one finds a second peak at the harmonic of $2f$ = 0.0034 day^{-1}, a hint that the orbit is eccentric. Armed with the correct period, one can then fit the full data set and recover the correct period and high eccentricity of e = 0.95 (red line in Figure 8.10).

So the lesson learned is that the standard periodogram can still be a powerful tool in finding eccentric orbits you must "think before and after you periodogram."

8.7 Calculating Keplerian Orbits

There are two open source programs that make the computation of Keplerian orbits relatively easy. SYSTEM Console 2 was developed by Stefan Meschari (Meschiari et al. 2009) and can be downloaded at http://www.stefanom.org/console-2/. Exostriker was developed by Trifon Trifonov and can be downloaded at https://github.com/3fon3fonov/exostriker.

Both programs have a graphical interface where you can perform a periodogram analysis to find signals in your data, fit these with a Keplerian orbit, and then look for additional planet signals in the residual data. They also can do analyses of the dynamical stabilty. Exostriker also includes Gaussian processes (see Chapter 11). The reader should consult the respective webpages for the capabilities and use of the programs.[6]

In calculating Keplerian orbits, it is important to have good estimates of the parameters. One may be tempted to solve for all values at once, but there is a risk of getting rubbish results. You can obtain orbits that have unrealistically high eccentricities. In some cases of multiplanet systems, the orbits may cross. In fitting Keplerian orbits, anything is possible as the codes know nothing of gravitation and Newtonian mechanics—that is the responsibility of the user!

The first step is to get a good value of the orbital period, which comes from the periodogram analysis. You then fit the velocity K-amplitude and find an appropriate phase. The eccentricity is the hardest to fit and should be the last parameter to be varied. Fortunately, SYSTEMIC and Exostriker have a nice graphical interfaces where it is easy for the user to vary parameters and see how they fit the data. Once you have a good estimates of P, K, e and $\tilde{\omega}$, you can them optimize all parameters simultaneously. It is best to start with known planet systems that are easy and work your way up the difficulty scale.

8.7.1 Transiting Planets

Transiting planets are a special case in that in calculating the orbit, one specifies the time of midtransit, rather than the time of periastron passage, T_0. (Most transiting planets are in circular orbits so the time of periastron passage is meaningless.) When

[6] These programs could be downloaded as 2019 December. As the readers are aware, the internet is a dynamic and ever-changing platform, so there is no guarantee that these programs will be available in the future.

using the transit time for the T_0, it is important to have the proper phase for the orbit, and in order to do this, one needs to set the argument of periastron to 90°.

Note that the phase of the orbit is a good test that the RV variations are actually due to a transiting companion. At midtransit, the RV should be zero and then go negative. This is sometimes overlooked when claiming to confirm a transiting planet (Dreizler et al. 2003).

In two respects, calculating the Keplerian orbit for a transiting planet is considerably easier than for nontransiting planets. First, it is easier to detect the RV signal of a transiting planet because the orbital period is known. There have been several cases where ground-based RV surveys were not able to detect an RV signal of a planet at a significant level. Later, once a transiting planet was found by the *TESS*[7] space mission, the RV signal could in fact be detected in the earlier RV data (Gandolfi et al. 2018, 2019; Luque et al. 2019). As we saw in Chapter 7, the significance of a peak is greatly increased if a signal is already known to be present at that frequency.

Second, the calculation of the orbit is greatly simplified because there are fewer free parameters. The period and phase of the orbit are known and because a transiting planet is usually close in, the chances are high that it lies on a circular orbit. The eccentricity can generally be fixed to zero as a good first approximation. So, the only parameter that needs to be fit is the RV amplitude.

Because the K-amplitude is the only parameter needed to derive the planet mass, what is the most effective observing strategy to do so? Naively, one would think that making RV measurements at phases 0.25 and 0.75 (using the transit ephemeris), i.e., quadrature, where the RV has maximum positive and negative velocity would be the fastest way to confirm the planet mass. This may be a fast way, but it may not be the most accurate. The RV variations due to a transiting planet are almost never a clean signal. You will most likely have activity-related RV variations, or RV variations due to additional planets in the system. If you take only two measurements at quadrature, then these other signals can have a serious influence on the inferred RV amplitude. You can of course mitigate this by taking multiple measurements, but if the rotation period is near an integral multiple or fraction of the planet orbit (e.g., $2P_{\rm orb}$, $P_{\rm orb}/2$, etc), then even multiple measurements can give you an inaccurate K-amplitude. This also certainly holds for planets in resonant orbits, even if the other planets are nontransiting.

A better strategy is to make RV measurements distributed over the entire orbit. After all, the full shape of the orbit also contains information on the K-amplitude, and it may be less sensitive to stellar variability. These also would reveal whether the orbit has nonzero eccentricity or if other factors (stellar rotation, other planets) are influencing the RV orbit. This was confirmed by Burt et al. (2018), who used simulations to investigate strategies for measuring masses for *TESS*-discovered transiting planets using four different observing strategies for the RV measurements: (1) a uniform sampling of RV measurements along the orbital curve, (2) a random sampling, (3) in-quadrature measurements, and (4) out-of-quadrature measurements.

[7] *Transiting Exoplanet Survey Satellite.*

The simulations showed that when measuring TESS planets at the 1σ, 3σ, and 5σ confidence levels that the uniform sampling performed the best, then random sampling, and followed by in-quadrature measurements. As expected, the out-of-quadrature strategy performed noticeably worse.

We can use *Kepler*-78b to investigate how two different approaches to analyzing RV data can influence the resulting K-amplitude. *Kepler*-78b is an ultrashort-period transiting planet with an orbital period of only 0.355 days that was discovered by the *Kepler* space mission (Sanchis-Ojeda et al. 2013). The transit light curve yields a planet radius of $R = 1 \pm 0.1\ R_{\oplus}$. The *Kepler* light curve shows that the star has a modest level of activity. Two teams obtained precise radial velocity measurements for this star using Keck Hires (Howard et al. 2013) and HARPS-N (Pepe et al. 2013). In particular, the HARPS-N measurements had a modest number ($\approx$20) of measurements taken at quadrature. Using only these measurements results in a K-amplitude of $2.80 \pm 0.93\ \mathrm{m\,s^{-1}}$. On the other hand, using all the data (HIRES + HARPS-N), which span the full orbital curve, results in a significantly smaller K-amplitude of $1.77 \pm 0.26\ \mathrm{m\,s^{-1}}$.

If you want to confirm a transiting planet with RV measurements, what is the minimum number of measurements that are required? The answer is, how precisely do you need to know the planet mass (i.e., K-amplitude)?

Figure 8.13 shows the percent error in the K-amplitude as a function of the number of measurements (N) for three values of the K-amplitude-to-error ratio, $K/\sigma = 0.5$, 1.0, and 2.0. To produce these curves, a synthetic orbit was sampled

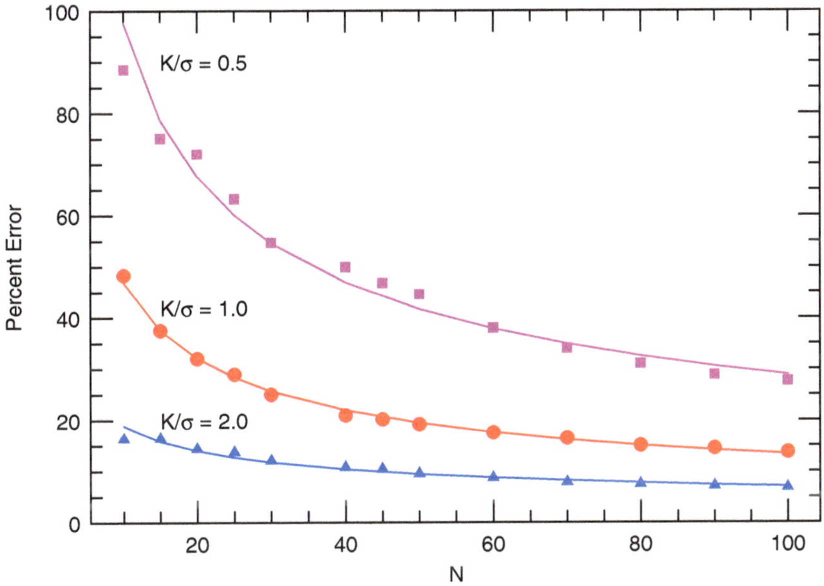

Figure 8.13. Simulations (points) of the percent error in the determination of the K-amplitude as a function of the number of measurements. These are shown for three values of the ratio of K to the measurement uncertainty σ, $K/\sigma = 0.5$ (lilac squares), 1.0 (red circles), and 2.0 (blue triangles). Lines represent fits to the points, and these follow an $N^{-0.5}$ law.

more or less uniformly and random (Gaussian) noise was added. The line represents the fit to the simulations, and as expected the error is proportional to $N^{-0.5}$.

If you have a K-amplitude equal to your typical measurement error, then with 10 measurements you can derive the planet mass only to an error of 50%—not particularly good. Getting a more reasonable error of 20% requires at least four times the measurements. If you want to do serious comparative planetology, which probably requires a precision of about 10% in the planet mass, then you will need at least 100 measurements.

For poor RV precision where your measurement error is twice that of your K-amplitude, then after 100 measurements you will only determine the K-amplitude and thus the planet mass to no better than about 40%. Clearly, if you want to derive accurate planet masses, you should only do this on planet-hosting stars where the expected K-amplitude is at least a factor of 2 greater than your typical measurement error. The reader should keep in mind that these results represent a best case scenario—good sampling of the RV orbit and random noise. In reality your sampling will be poorer and the noise will be anything but Gaussian, particularly if you have a significant amount of activity-related RV variations.

8.8 Dynamical Effects

If planets are in multiple systems, then mutual gravitation can have a large influence on the motion in two ways. First, this can make the system unstable, in which case one or more of your planets are most likely not real. Second, planet interactions may change the orbit, in which case the observed motion cannot be modeled by the simple sum of Keplerian orbits.

8.8.1 Dynamical Stability

When dealing with multiplanet systems it is always wise to check on the dynamical stability. For highly eccentric orbits, it is a simple (but easily overlooked) matter to check that the orbits do not cross. For other systems, dynamical problems may arise for massive planets, or planets with close separations.

The Hill sphere provides a quick check on the stability of a multiple system. The Hill sphere denotes the gravitational influence of a small body (e.g., planet), m, in orbit around a larger body (e.g., star), M. For low-eccentricity orbits (the most favorable for stability), the Hill radius, $r_{\rm H}$, is given by

$$r_{\rm H} \approx a\left(\frac{m}{3M}\right)^{1/3}, \tag{8.15}$$

where a is the semimajor axis of the smaller body.

Consider a two-planet system with mass m_1 and m_2 in orbit around a star of mass M. We can denote the semimajor axis of the outer planet by $a = 1 + \Delta$, where Δ is the fractional separation (Figure 8.14). Let μ_1 and μ_2 be the mass ratios of the two planets with respect to the host star, i.e., $\mu_1 = m_1/M$ and $\mu_2 = m_2/M$. The orbits are most likely stable (Gladman 1993) if

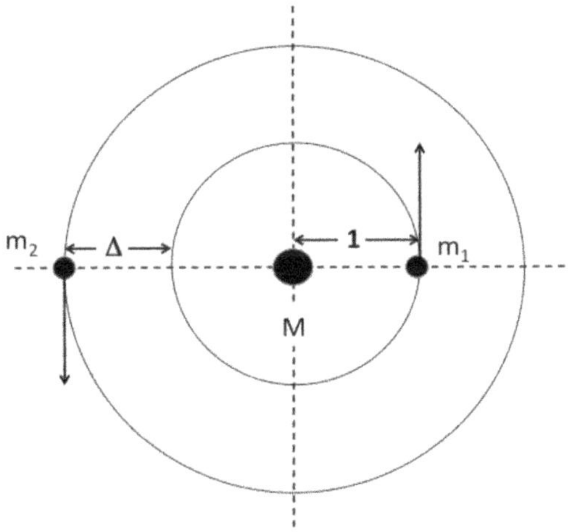

Figure 8.14. A two-planet system with masses m_1 and m_2 in orbit around a host star of mass M. Equation (8.16) defines the minimum value of Δ required for stability.

$$\Delta > 2.4(\mu_1 + \mu_2)^{1/3}. \tag{8.16}$$

Even if the system violates Equation (8.16), the system could still be stable. For example, in the case of γ Draconis where two periodic signals were found, possibly due to planets (Hatzes et al. 2018), one finds $\Delta = 0.56$, larger than $2.4\ (\mu_1 + \mu_2)^{1/3} = 0.12$. However, a dynamical study reveals that there is a 1%–2% chance that the system is actually stable (Hatzes et al. 2018). It always best to do a proper dynamical study. Fortunately, there are programs available to do this, such as Mercury6 (Chambers 2012).

8.8.2 Planet Interactions

When fitting multiple planet orbits, the standard procedure is to find the dominant signal, fit a Keplerian orbit, find the next periodic signal in the residuals, and fit an orbit to those, i.e., a prewhitening procedure. One can be a bit more sophisticated and do a simultaneous fit to all orbits that are present. However, this assumes that the observed motion is a linear combination of the individual orbits. In most cases this is valid, but if the planet masses are large, the orbital separations small, or the planets have some nonzero eccentricity (or all of these), the gravitational interaction of the planets can lead to secular variations which may appear as a fake signal when only taking into account a linear combination of the orbits.

An example of this is shown for the planet-hosting star GJ 876. This M-type star was the first M dwarf found to host a giant with $M \sin i = 1.89\ M_{\rm Jup}$ and an orbital period of 61 days (Marcy et al. 1998, Delfosse et al. 1998). Continued monitoring revealed a second companion with $M = 0.56\ M_{\rm Jup}$ on a 30 day orbit, i.e., a system in a 2:1 mean motion resonance (MMR) resonance (Marcy et al. 2001). Subsequent observations revealed a super-Earth in a 1.96 day orbit (Rivera et al. 2005) and a fourth planet in a 124 day orbit (Rivera et al. 2010).

Let's apply our standard prewhitening to see how many planets we can find in the RV data. We will use the RV data presented in Trifonov et al. (2018). Prewhitening finds seven significant signals, all with false alarm probability less than 0.01 (Left side of Table 8.1). Figure 8.15 shows individual phase diagrams for six of these (we exclude the high RV amplitude planet GJ 876b).

Table 8.1. (Columns 1 and 2) The Prewhitened Periods and Amplitudes Found for the RV Data of GL 876. (Columns 3 and 4) The Prewhitened Periods and Data Found in Synthetic Data Consisting of Four Planets with Dynamical Interactions

P_{data} (days)	K_{data} (m s^{-1})	P_{dyn} (days)	K_{dyn} (m s^{-1})	Comment
61.0	211.0	61.0	211.0	Planet
30.2	88.1	30.2	88.7	Planet
15.0	20.4	15.1	20.2	Interaction
10.0	5.6	10.0	5.5	Interaction
1.94	6.0	1.94	6.0	Planet
124.5	3.5	124.4	3.4	Planet
7.5	1.9	7.5	1.5	Interaction

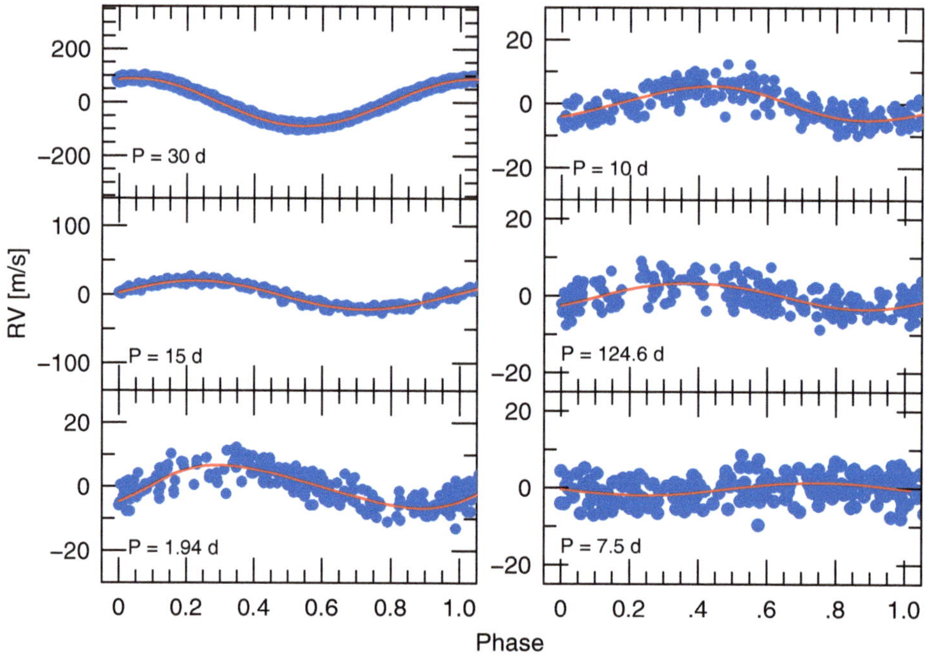

Figure 8.15. Six of the seven periodic signals found by prewhitening of the RVs for GJ 876. The highest amplitude signal at 61 days is not shown. The signals at 7.5, 10, and 15 days are due to dynamical interactions; all others are due to planetary companions (Table 8.1).

Note that two of the signals, $P = 15$ days and $P = 7.5$ days, are the second and third harmonics of the orbital frequency of GJ 876b. A logical conclusion would be that this star hosts seven planets, four of which are in 1:2:3:4 MMR. However, several of these “planets” may be artifacts of dynamical interactions. In fact, Marcy et al. (2001) showed that GJ 876b and GJ 876c interact so strongly that a sum of two Keplerian orbits could not fit the data.

Instead, let’s use actual dynamical fits to the data with just four planets: Planets b ($P = 61$ days), c ($P = 30$ days), d ($P = 1.94$ days), and e ($P = 124$ days). The synthetic model was sampled in the same way as the observations and the appropriate noise added (simulated data were kindly provided by Stefan Dreizler and Trifon Trifonov). The third column of Table 8.1 shows the results of the prewhitening procedure on these synthetic data with only four planets present, but including dynamical interactions. The frequencies that are found are consistent with those of the actual data. Here is a case where the periodogram (i.e., Fourier transform) requires six sine functions to fit the data, yet only four planets are in the system. This emphasizes that one can find real periodic signals in the data. The interpretation of these, however, is left to the investigator.

8.9 Barycentric Corrections

If you take your initial RV measurements and search for a periodic signal and fit a Keplerian orbit, you are guaranteed to find a planetary signal, and from a habitable planet at that! If you have not properly removed the motion of Earth around its barycenter—the so-called barycentric correction—you will find periodic signals in your data at one year (Earth’s orbital motion) or one day (Earth’s rotational motion). The removal of this barycentric motion has to be done with exquisite precision as even a slight uncorrected motion will appear as a low-mass planet.

The case of the first claimed pulsar planet only highlights the pitfalls in an improper removal of Earth’s motion though oversight. Bailes et al. (1991) reported a 10 $M_{\oplus}$ companion orbiting the pulsar PSR 1829–10. The discovery was made using pulsar timing variations for which a barycentric correction had to be applied. The period of the planet was a bit suspicious at exactly six months. The authors, however, were careful to exclude barycentric motion as the cause. Later it was revealed that this was indeed a spurious signal, due to the improper correction for the barycentric motion (Lyne & Bailes 1992). Through an oversight, the astronomers neglected to account for the slight eccentricity of Earth’s orbit. Improper barycentric corrections can fool even the most careful astronomer.

Implicit in our discussion in the calculation of Keplerian orbits (and searching for periodic signals) is the fact that the barycentric motion of Earth has to be removed. The main component of this barycentric motion is the orbital motion of Earth. It orbits the Sun at 1 au in 3.16×10^7 s, which throughout the orbit translates into a velocity of ± 29.85 km s^{-1}. Earth also rotates with a velocity given by

$$V_0 = 464.56 \cos(lat) \cos(\delta),$$

where *lat* is Earth's latitude and δ is the declination of the star. So, the Earth's rotation can cause a velocity shift of up to 464 m s^{-1} (the equatorial velocity of the Earth).

Stumpff (1979) presented one of the first implementations of a barycentric correction for stellar radial velocities. Many planet search programs have used the JPL Ephemeris tables DE200 and DE405, which describe the orbits of the Sun and planets with very high precision over relatively long timescales (Folkner et al. 2009, 2014). These provide an accuracy of a few cm s^{-1}, which is sufficient for most Doppler work.

Wright & Eastman (2014) gave a detailed review of how to calculate barycentric corrections to an accuracy of better than 1 cm s^{-1}. The authors provided a routine in Interactive Data Language (IDL), ZBARYCORR, which has been ported to the Python computing language (Kanodia & Wright 2018). In short, routines for calculating the barycentric of better than a few cm s^{-1} for high-precision Doppler work are readily available.

Most errors in the barycentric correction do not stem from the algorithms used to calculate this, but simple errors in their application. These include the wrong position of stars, a wrong time for calculating the barycentric corrections, or even computer bugs in software preparing your data for the barycentric corrections. One such subtle bug plagued this author. The stellar coordinate of declination was given in the standard three-field form: "degrees minutes seconds." A Fortran computer code parsed these fields to convert to degrees. The code read the sign of declination from the first field, and one star had declination of "−00" degrees (followed by nonzero numbers in the minutes field). Because there is no such thing as a negative zero Fortran interpreted this as +0, i.e., the position of the star was converted to a positive declination resulting in the wrong barycentric correction.

As a basic rule, any time you find a period of exactly 365.25 days in your data, a fraction or even close, you should be very cautious. In Chapter 12, we will discuss in more detail various ways errors in the barycentric correction can creep into your data.

References

Bailes, M., Lyne, A. G., & Shemar, S. L. 1991, Natur, 352, 311

Bean, J. L., McArthur, B. E., Benedict, G. F., et al. 2007, AJ, 134, 749

Benedict, G. F., McArthur, B. E., Forveille, T., et al. 2002, ApJ, 581, L115

Benedict, G. F., McArthur, B. E., Gatewood, G., et al. 2006, AJ, 132, 2206

Blunt, S., Endl, M., Weiss, L. M., et al. 2019, AJ, 158, 181

Burt, J., Holden, B., Wolfgang, A., & Bouma, L. G. 2018, AJ, 156, 255

Chambers, J. E. 2012, Mercury: A software package for orbital dynamics, Astrophysics Source Code Library, ascl:1201.008

Cumming, A. 2004, MNRAS, 354, 1165

Delfosse, X., Forveille, T., Mayor, M., et al. 1998, A&A, 338, L67

Dreizler, S., Hauschildt, P. H., Kley, W., et al. 2003, A&A, 402, 791

Endl, M., Hatzes, A. P., Cochran, W. D., et al. 2004, ApJ, 611, 1121

Folkner, W. M., Williams, J. G., & Boggs, D. H. 2009, IPNPR, 42-178, 1

Folkner, W. M., Williams, J. G., Boggs, D. H., Park, R. S., & Kuchynka, P. 2014, IPNPR, 42-196, 1
Frink, S., Mitchell, D. S., Quirrenbach, A., et al. 2002, ApJ, 576, 478
Gandolfi, D., Barragán, O., Livingston, J. H., et al. 2018, A&A, 619, L10
Gandolfi, D., Fossati, L., Livingston, J. H., et al. 2019, ApJ, 876, L24
Gladman, B. 1993, Icar, 106, 247
Hatzes, A. P., Endl, M., Cochran, W. D., et al. 2018, AJ, 155, 120
Hatzes, A. P., Guenther, E. W., Endl, M., et al. 2005, A&A, 437, 743
Hatzes, A. P., Zechmeister, M., Matthews, J., et al. 2012, A&A, 543, 98
Howard, A. W., Sanchis-Ojeda, R., Marcy, G. W., et al. 2013, Natur, 503, 381
Kanodia, S., & Wright, J. T. 2018, Barycorrpy: Barycentric velocity calculation and leap second management, Astrophysics Source Code Library, ascl:1808.001
Luque, R., Pallé, E., Kossakowski, D., et al. 2019, A&A, 628, A39
Lyne, A. G., & Bailes, M. 1992, Natur, 355, 213
Marcy, G. W., Butler, R. P., Fischer, D., et al. 2001, ApJ, 556, 296
Marcy, G. W., Butler, R. P., Vogt, S. S., Fischer, D., & Lissauer, J. J. 1998, ApJ, 505, L147
McArthur, B. E., Benedict, G. F., Barnes, R., et al. 2010, ApJ, 715, 1203
Meschiari, S., Wolf, A. S., Rivera, E., et al. 2009, PASP, 121, 1016
Nagel, E., Czesla, S., Schmitt, J. H. M. M., et al. 2019, A&A, 622, A153
O'Toole, S. J., Tinney, C. G., Jones, H. R. A., et al. 2009, MNRAS, 392, 641
Pepe, F., Cameron, A. C., Latham, D., et al. 2013, Natur, 503, 377
Perryman, M. 2018, The Exoplanet Handbook (Cambridge: Cambridge Univ. Press)
Rivera, E. J., Laughlin, G., Butler, R. P., et al. 2010, ApJ, 719, 890
Rivera, E. J., Lissauer, J. J., Butler, R. P., et al. 2005, ApJ, 634, 625
Sanchis-Ojeda, R., Rappaport, S., Winn, J. N., et al. 2013, ApJ, 774, 54
Stumpff, P. 1979, A&A, 78, 229
Trifonov, T., Kürster, M., Zechmeister, M., et al. 2018, A&A, 609, A117
Vogt, S. S., Butler, R. P., Marcy, G. W., et al. 2002, ApJ, 568, 352
Wright, J. T., & Eastman, J. D. 2014, PASP, 126, 838
Zechmeister, M., & Kürster, M. 2009, A&A, 496, 577

Chapter 9

Avoiding False Planets: Rotational Modulation

9.1 Introduction

After you have detected a periodic signal in your data and have determined that it is real and not due to noise, the next important step is to determine the nature of the signal. There is a lot of stellar phenomena that can also produce radial velocity (RV) variations, and it is sometimes difficult to assess if you have a true planetary companion. A periodic signal can be found in your data quickly; determining the true nature may take weeks of hard work.

Intrinsic stellar variability is quite often the source of RV signals masquerading as planets, and this is especially true as one tries to extract smaller and smaller RV amplitudes from RV time series. As we shall soon see, there are cases where an RV signal was thought to be due to a planet, and it passed a number of tests, but it turned out that it was actually due to stellar variability.

The main sources of this variability are stellar oscillations and magnetic activity. For late-type main-sequence stars, stellar activity is the dominant form of variability that can influence the RV measurement. Activity-related RV variations are particularly pernicious. They can be periodic, due to the rotation of the star or activity cycles, or stochastic—signals that come and go. In the best case, the activity only adds more RV noise to your data; in the worst case, it can mimic the signal of a planet.

Because of this, there are a number of important steps RV planet hunters use for confirming whether a Doppler signal is indeed due to a planetary companion:

1. A period that is not easily associated with another phenomenon such as stellar rotation or pulsations.
2. An RV signal that is stable, coherent, and long-lived (no amplitude, period or phase variations).
3. An RV amplitude that is constant as a function of wavelength. The Doppler motion due to a planet is independent of wavelength.
4. No photometric variations of the star with the same period as the planet's orbit.

doi:10.1088/2514-3433/ab46a3ch9

5. No variations of activity indicators with the orbital period of the planet.
6. No changes in the spectral line shapes with the orbital period of the planet. A planet causes an overall shift of the line profiles without changing the line shape.

The next two chapters will be devoted to sources of stellar variability that can creep into your RV data. In this chapter, we will focus primarily on rotational modulation as causes for periodic signals appearing in your data that can easily be misinterpreted as arising from the orbital motion of a companion. Chapter 10 will focus on useful activity indicators, sources of stellar "noise," and activity cycles.

Stellar activity introduces RV variations that is often accompanied by surface structure—inhomogeneities that distort the spectral lines and thus introduce a Doppler displacement in your line profile. Figure 9.1 shows some of the surface structure that can be found on the Sun. These include cool dark spots, bright plage, faculae, and granulation due to convection. All of these are associated with magnetic fields.

Periodic RV variations, or rotational modulation, arise because most of these structures rotate in and out of view as the star turns. The characteristic periods are

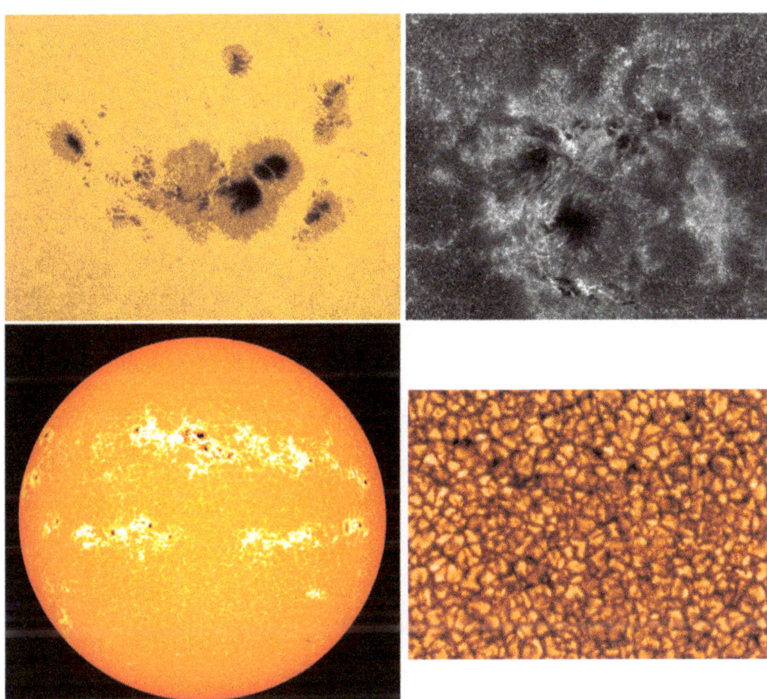

Figure 9.1. Types of surface structure found on the Sun. (Top left) Sunspots (photo credit: NASA/SDO); (top right) plage, which are bright regions near the sunspots (photo credit: Dutch Open Telescope (DOT)); (lower left) the bright faculae (photo credit: Alan Friedman, NASA Goddard Space Flight Center); and the granulation pattern due to convection (photo credit: Astrobite, The Solar Optical Telescope, Hinode. Courtesy of NASA).

thus related to the rotational period of the star. Here we discuss ways which you can recognize rotational modulation in RV data. The study of these surface structures is important for understanding magnetic activity in stars, and certainly RV measurements can play a role in these, but for researchers wanting to find planets with Doppler measurements, these structures present a nuisance.

9.2 Spots

Stellar rotation is the most common nonplanetary periodic signal found in RV data. It arises from stellar surface structure that can create distortions in the stellar line profile, which results in a shift of the line centroid. As the star rotates, these distortions cross the line profile, resulting in a changing RV with the same period as the stellar rotation.

Cool starspots, analogous to sunspots, are a common surface structure found on late-type stars exhibiting magnetic activity. An extreme example of this is HR 1099, an RS CVn-type star. This class of objects are rapidly rotating, often evolved late-type stars that have activity levels several orders of magnitude larger than the Sun. On RS CVn-type stars spots can cover ≈20% of its surface (see review by Strassmeier 2009).

Figure 9.2 shows a sequence of line profiles of Ca I 6439 Å. HR 1099 has a stellar rotational velocity of 40 km s^{-1}, so one can clearly see the distortions due to the cool spot as it moves across the line profile due to stellar rotation. In the first profile, the distortion is in the blue wing of the line, thus causing a shift of the line centroid toward the red—a Doppler redshift. In the last profile, as the spot is receding from the stellar limb, the distortion has moved to the red wing, resulting in a net Doppler blueshift of the line.

For slowly rotating stars, one cannot see the line profile distortions, but they are present and they induce a Doppler shift of the line. The left panel of Figure 9.3 shows a simulated spot distribution of a few percent spot filling factor on a Sun-like star that is rotating at 2 km s^{-1}. This produces RV variations with a peak-to-peak amplitude of 20 m s^{-1} (right panel of Figure 9.3). For a solar-type star with a

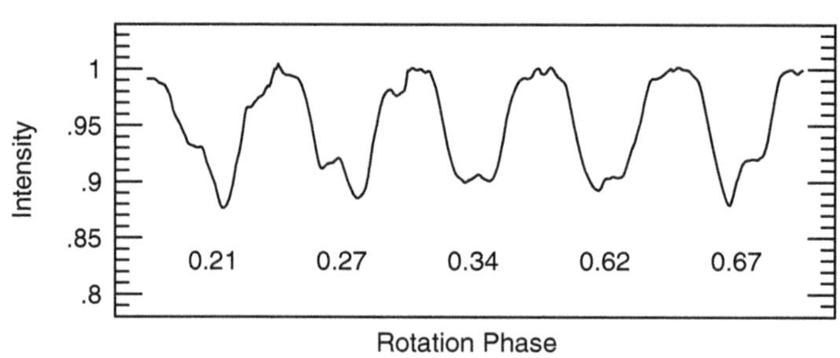

Figure 9.2. Line profile variations in the Ca I 6439 Å line of HR 1099 that are due to starspots. The spectral line is shown at five different rotation phases. This star has rotationally broadened line profiles ($v \sin i = 40$ km s^{-1}) so that one can clearly see the distortions in the line profile due to the starspots. The large spot feature is on the blueshifted limb at rotation phase 0.21, at disk center at rotation phase 0.34, and on the redshifted limb at phase 0.67.

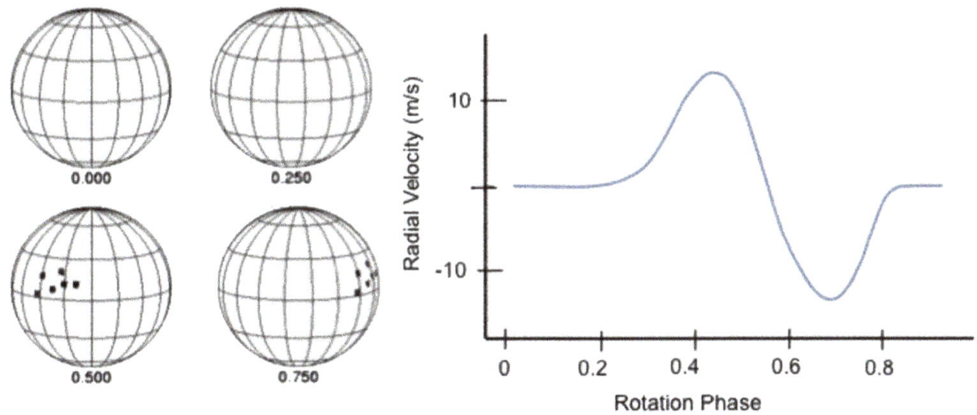

Figure 9.3. (Left) A simulated spot distribution on a solar-like star rotating at 2 km s^{-1}. (Right) The corresponding RV variations due to this spot distribution.

rotation period of approximately 25 days, this would mimic the signal of a 0.3 M_{Jup} companion. Thus, even for a slowly rotating star with modest spot coverage, the resulting RV signal can be substantial and it could seriously hinder the detection of planetary companions or cause a misinterpretation of the results.

Although the rotation period, P_{rot}, is the one most likely to appear in time series RV data, its harmonics can also be present—$P_{rot}/2$, $P_{rot}/3$, etc. (frequency f_{rot}, $2f_{rot}$, $3f_{rot}$ ...).[1] Why do the rotational harmonics appear? A natural, physical explanation comes from the number of spots. If you have only one spot on the surface, you will only see the rotational period. If you have two spots separated by 180° in longitude, you will first see the RV signature of the first spot, but half a rotation later you will see the RV signature of the second one, thus $P_{rot}/2$ will be present in your data. For three equally spaced spots, $P_{rot}/3$, etc., you may even encounter, as we will see below, cases where a harmonic of P_{rot} is actually the dominant rotation signal to be seen in the data.

A better way to consider the shape of the rotation curve is to think in Fourier space. The RV curve results from spots that can be distributed at random on different regions of the star. This will result in a complicated RV curve. However, all variations are tied to the one fundamental frequency (period) and that is the rotation of the star. If you calculate the Fourier series that represents this curve, you should only see harmonics of this fundamental frequency. It is analogous to a periodic square wave function. If you calculate the Fourier sine series, you do not get frequencies at random; all will be integral multiples of the primary frequency of the square wave.

The rapidly oscillating Ap star α Cir nicely demonstrates how one can see many rotation harmonics in the rotationally modulated RV curve for a star. We will later return to Ap stars as they represent some extreme cases for highlighting the effects of

[1] To remain consistent with stellar oscillations, I will refer to $P_{rot}/2$ as the first harmonic, thus P_{rot} is the "fundamental" period.

surface structure and stellar pulsations on the measured RV of the star. Like all Ap stars, this object has surface chemical spots that also cause distortions in the line profile.

Figure 9.4 shows the rotation-modulated RV variations of α Cir. As we shall see later, α Cir is a pulsating star, so for these RV data, binned averages were taken in order to mitigate the variations due to the stellar pulsations. The RV variations can be well fit by the rotational frequency of 0.0805 day^{-1} and its seven harmonics (Table 9.1). The rms scatter about the fit is only 1.7 m s^{-1}.

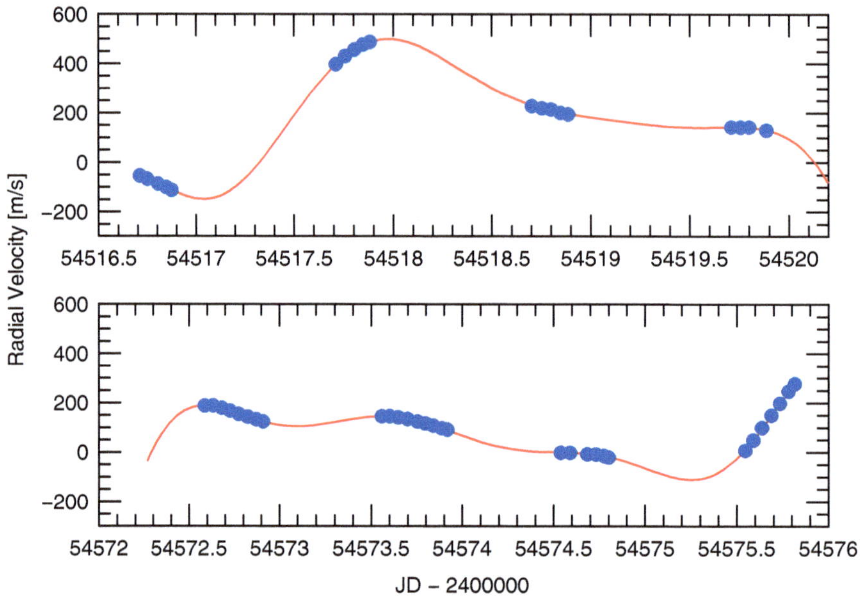

Figure 9.4. The RV variations (points) of the roAp star α Cir due to the rotational modulation of surface features. The curve represents a multicomponent sine fit using the rotational period and its harmonics (Table 9.1).

Table 9.1. Multisine Component Fit to RV Data for α Cir (Figure 9.4)

Frequency (ayd^{-1})	Period (days)	Amplitude (m s^{-1})
0.080537	12.42	1346.33
0.161073	6.22	2698.03
0.240738	4.15	1315.15
0.320983	3.12	1785.66
0.401229	2.49	715.71
0.481475	2.08	961.04
0.56172	1.78	284.18
0.641967	1.56	263.10

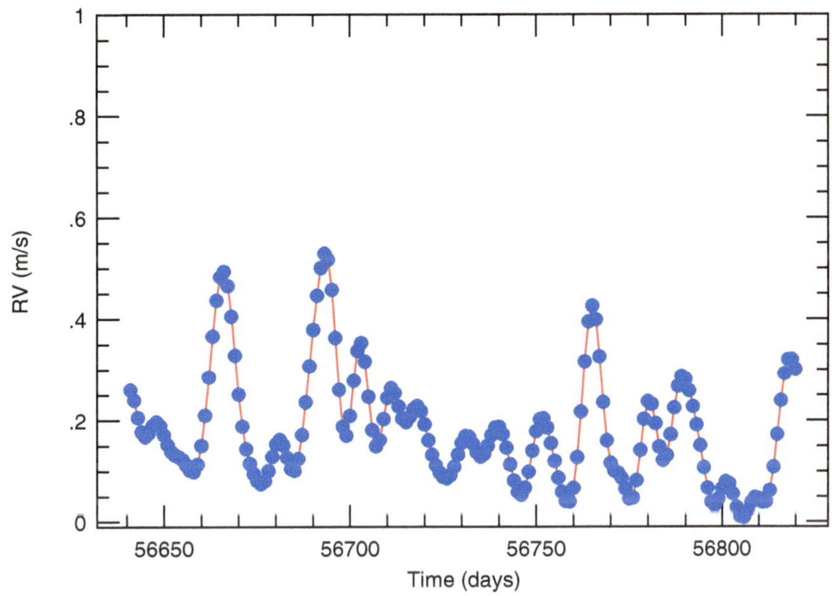

Figure 9.5. The DFT amplitude spectrum of RV variations from sunspots. Data for the daily sunspot area (www2.mps.mpg.deprojectsSun-climatedata.html) were converted to RVs. Points represent actual measurements.

The rather complicated rotational modulation curve for α Cir is most likely due to the fact that different elements can be distributed in different ways on the star. Some may appear only in an isolated spot, others in multiple spots (e.g., Hatzes 1991; Kochukhov et al. 2015; Rusomarov et al. 2016, 2018). Again, the RV curve can be understood in terms of sum of Fourier components related to the rotational frequency of the star.

The Ap stars, however, represent a best case for rotational modulation. The spots on the stars are fixed, and there is no evidence that these have evolved, moved, or changed in any way in the decades that Ap stars have been studied. Unfortunately, nature is not so kind when it comes to starspots on active stars. These can appear, evolve, and disappear at varying timescales, and they can also migrate from their birth location. Differential rotation, the phenomenon where different latitudes on the star rotate at different rates, is surely present as this a key ingredient for driving the magnetic activity.

Figure 9.5 shows the areal coverage of sunspots from Balmaceda et al. (2009) that have been converted into an RV amplitude. (The exact amplitude is not important, merely the functional form of the time series). Table 9.2 shows the Fourier components of these variations. This case is far more complicated than the case for the Ap star and reflects the "dynamic" nature of magnetic spots in terms of their motion and evolution. The dominant signal is roughly the solar rotational period ($P \approx 24$ days) but there are other periods, many not easily assigned to a rotational harmonic. In short, it is a mess due to differential rotation[2] as well as spot evolution

[2] The rotational period of the Sun is about 24 days at the equator and 29 days at the poles. Most of the spots responsible for the variations come from latitudes near the equator.

Table 9.2. Multisine Fit to the RV Variations Due to Spots on the Sun (Figure 9.5).

Frequency (day^{-1})	Period (days)	Amplitude ($m\ s^{-1}$)
0.040597	24.63	0.065
0.004847	206.28	0.053
0.081213	12.31	0.056
0.030385	32.91	0.068
0.014136	70.73	0.0545
0.069609	14.36	0.046
0.057944	17.25	0.0318
0.026488	37.75	0.050
0.112792	8.86	0.02415
0.051832	19.29	0.025
0.103202	9.68	0.0230
0.129373	7.72	0.0221
0.045251	22.10	0.021
0.096368	10.38	0.015
0.142458	7.02	0.0139

and migration. Figure 9.5 should be sobering to the reader. Even in the best case of exquisite sampling and with no large time gaps in the data it is difficult to interpret this frequency spectrum. Ground-based observations taken over weeks to months and with poor sampling and large gaps in time will make the problem far worse. And keep in mind the Sun is not a particularly active star—very active stars will be even more problematic.

The longevity of an RV signal is often used as an argument against spots as a cause for RV variations (e.g., Setiawan et al. 2008). This assumption is reasonable given the relatively rapid evolution of sunspots which typically have lifetimes of days to months. They seldom last for more than one rotation of the Sun. However, this assumption is not valid for other types of active stars. For example, spots can persist for several years on young, active main-sequence stars (Hall & Henry 1994; Strassmeier et al. 1999) and rapidly rotating evolved stars (Vogt et al. 1999; Hatzes & Vogt 1992). These can also have different spot morphologies from the solar case.

Figure 9.6 shows the Doppler image of the spot distribution on the weak T Tauri star V410 Tau in two different years. Briefly, Doppler imaging uses the time series of spectral distortions (Figure 9.2) to map the spot distribution on rapidly rotating stars (Vogt et al. 1987; Rice et al. 1989; Piskunov et al. 1990). In 1994 V410 Tau, had a large, decentered polar spot covering about 20% of the surface (Hatzes 1995).

The Doppler image for V410 Tau in 2004 (right panel of Figure 9.6) shows that after a decade, the spot distribution on this star had largely not changed in shape, size, or location. If one had taken RV measurements of this star over this time, it would have shown variations that were roughly long-lived and coherent with the

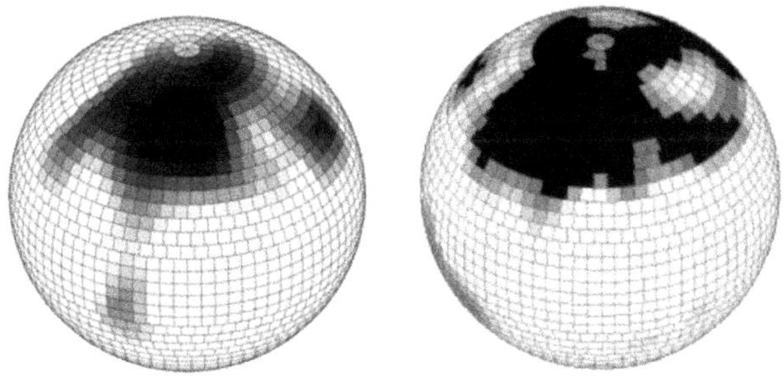

Figure 9.6. (Left) A Doppler image of the spot distribution on WTTS V410 Tau in 1994. (Right) A Doppler image of the spot distribution in 2004 (2004 image courtesy of Tobias Schmidt).

same amplitude, period (the rotation of the star), and phase. The lesson V140 Tau teaches us is that depending on the type of star and its level of activity, it is not a foregone conclusion that a long-lived RV signal cannot be due to spots.

9.3 Plage and Faculae

Cool starspots are not the only surface structure that can influence the measured RV of a star. Plage are bright regions in the chromosphere that are associated with concentrations in the magnetic field of the Sun and other active stars. They are primarily seen in the Balmer Hα or Ca II H&K lines (see Chapter 10). Faculae are bright areas that can be seen in white light most easily near the limb of the Sun. These are also associated with magnetic fields. In the solar case, the bright faculae actually make the Sun appear slightly brighter ($\approx$0.1%) at solar maximum despite the presence of dark sunspots. Both plage and faculae can result in RV variations when observing stars in integrated light. Plage are typically found near cool spots, but because they are bright, their RV signal is anticorrelated with that from the spots (Meunier et al. 2010).

9.4 Granulation and Convective Blueshift

Even in the absence of spots, plage, or faculae, the surface of the Sun would still exhibit surface structure. The outer convection zone of the Sun ensures that there is an inhomogeneous distribution of temperature and velocity as manifested through the granulation pattern. This pattern consists of bright cells having irregular polygonal shapes with sizes 1000–2000 km that are separated by narrow dark lanes.

All late-type stars should show granulation due to convection, and naively, one would think that when observing stars in integrated light that the up and down motion of the bright cells and dark lanes would cancel, but that is not the case. A convective blueshift is seen for spectral lines of the Sun and is expected for all late-type stars.

Figure 9.7 shows the origin of this convective blueshift for the Sun. The polygon-shaped hot cells have more surface area (coverage $\approx$75%), are hot (more flux), and have an upward velocity (blueshift) of ≈ -1.5 km s^{-1}. The dark intergranular lanes

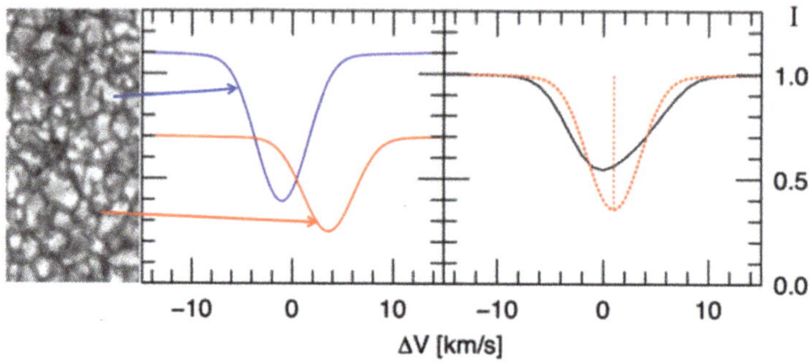

Figure 9.7. Origin of the convective blueshifts caused by convection. (Left) A high-resolution portion of the granulation pattern on the Sun. (Center) The hot rising cell (blue profile and arrow) emits more flux and produces a blueshifted (≈ -1.5 km s^{-1}) profile. The dark lanes (red profile and arrow) have less flux and produce a redshifted ($\approx +3.6$ km^{-1}) profile. (Right) The integrated line profile has an asymmetric shape and is blueshifted compared to the undisturbed profile (red profile), which has a zero-velocity shift and is symmetric (red vertical line).

are cooler (less flux) and have a downflow velocity (redshift) of $\approx +3.6$ km s^{-1}. When the two profiles are integrated, the net result is not only an asymmetric line (see Chapter 10) but also a net blueshift. The magnitude of this blueshift depends on the areal coverage of the hot cells compared to the dark lanes, the upflow and downflow velocities, and the strength of the lines (see below). There are also limb effects (Dravins 1982).

The presence of the magnetic field in spots suppresses the convective blueshift, resulting in a contribution that changes as the spots and plage move in and out of view. This will influence the measured RV. In fact, convection can be the dominant signal with an amplitude of 8–10 m s^{-1} (Meunier et al. 2010).

The stellar granulation pattern can also produce long-term changes in the RV due to activity cycles. Differences in the convective blueshift between active and quiet regions as well as changes in the respective areas covered by the hot cells and cool lanes can lead to long-term changes in the RV due to activity cycles.

In the solar case, this would occur on an 11 year cycle. Coincidentally, this is nearly the same as the 12 year orbital period of the Jupiter. If the solar activity cycle were to induce an RV shift of $\approx$10 m s^{-1}, then if the Sun were seen from afar it would be difficult to disentangle this from the orbital motion induced by Jupiter (Dravins 1982). Clearly, understanding long-term effects of activity on RV measurements are crucial for detecting long orbital period planets.

Meunier et al. (2010) investigated the RV variations due to plage, cool spots (the "flux effect"), and convection using MDI/SOHO magnetograms (Scherrer et al. 1995) to produce simulated RV data from spots, plage, and convection for an entire solar cycle. Figure 9.8 shows the predicted RV variations from the individual components as well as the total contribution to the RV. The predicted amplitude of these are $\approx$ few m s^{-1}, but there are times when these variations are well under 0.5 m s^{-1} The rotational modulation effects due to a single spot can be seen more clearly in Figure 9.9 (Herrero et al. 2016).

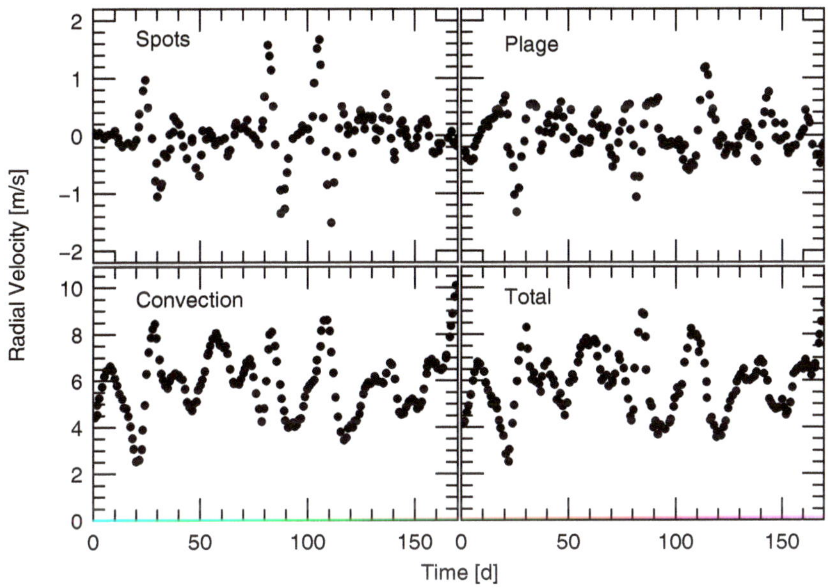

Figure 9.8. The simulated RV variations of solar activity due to spots (top left), plage (top right), convective blueshift (lower left), and the total contribution (lower right). (From Muenier et al. 2010.)

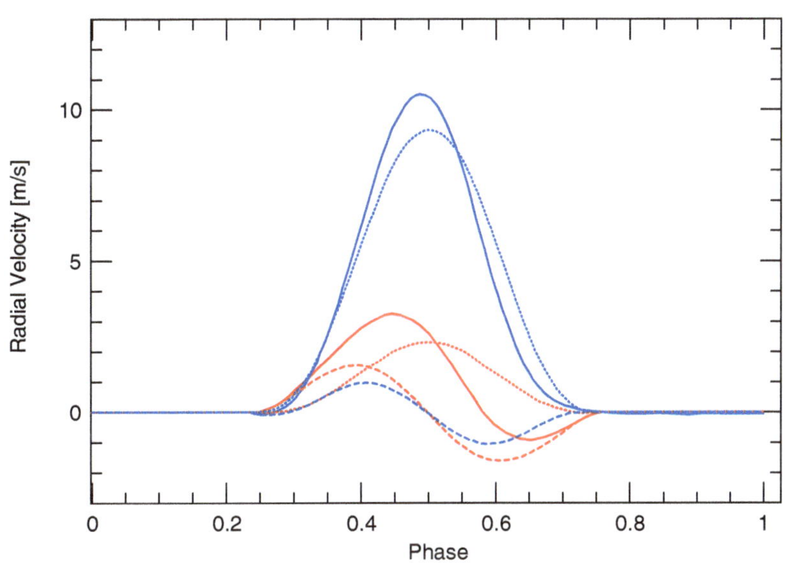

Figure 9.9. The RV shifts produced by a single spot with an area 1.6×10^{-3} the stellar surface. Red lines represent the simulation without plage, and blue lines are with the inclusion of plage. The Sun-like star has a rotational period of 24 days. The dashed line represents the simulation with no convection and is thus due to changes in the flux in the spotted regions. The contribution of convection is plotted with a dotted line. The solid line represents the total contribution of all effects. (From Herrero et al. 2016, reproduced with permission © ESO.)

9.4.1 The Sun Viewed as a Star

Changes in the integrated wavelength shift of integrated sunlight have been studied by a number of investigators. Livingston (1983) found that the Fe 5250 Å line could shift by about 2 mÅ from magnetic to nonmagnetic regions. Deming et al. (1987) measured the RV of integrated sunlight and found that in one day, the changes were less than ≈3 m s^{-1}. Over a three-year period, the data showed an increasing blueshift by about 30 m s^{-1}, which was consistent with a less magnetic inhibition of the convection pattern at solar minimum. Wallace et al. (1988) measured the RVs of integrated sunlight by comparing the wavelengths of lines insensitive to convective blueshifts to those that had a large shift. Over the time span 1976–1986, these showed an upper limit of 5 m s^{-1} in the change of the RV. Measurements of the line shapes indicated that the surface convection caused a change in the RV of 7 m s^{-1} from 1980 to 1986. The important conclusion was that stellar variability comparable to that of the Sun would not be an obstacle to the detection of at least giant planets around other stars.

Observing the Sun "as a star" with RV instruments dedicated to the search for exoplanets started with the work of McMillan et al. (1993), who used a Fabry–Pérot for wavelength calibration. Figure 9.10 shows these data covering five years. The rms scatter of the measurements is ≈6.5 m s^{-1}, and there is a slight trend that possibly stems from the activity cycle. These are consistent with previous measurements of integrated sunlight.

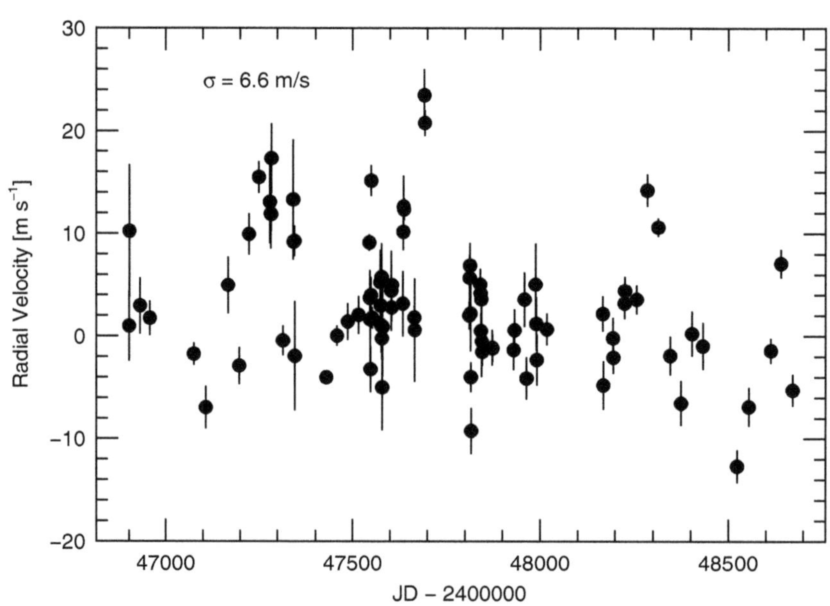

Figure 9.10. Nighttime RV measurements of the Sun from observations of a lunar crater using a Fabry–Pérot as the wavelength calibration The observations span almost five years, or about one-half of a solar cycle. The rms scatter is 6.5 m s^{-1} (McMillan et al. 1993).

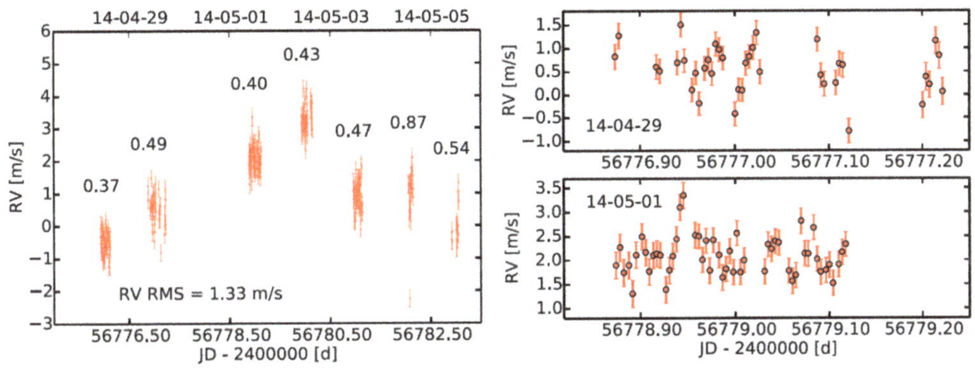

Figure 9.11. (Left) Solar RVs obtained using the HARPS-N solar telescope over one week. The numbers show the rms for the daily observations. (Right) Subset of the daily variations for two days. (Reproduced from Dumusque et al. 2015. The American Astronomical Society. All rights reserved.)

Recently, there has been an interest in monitoring the Sun using high-precision RVs afforded by modern, state-of-the-art spectrographs. Dumusque et al. (2015) have built a solar telescope for feeding sunlight into the HARPS-N spectrograph, an instrument normally employed for nighttime stellar work. Measurements show an RV change of only 50 cm s^{-1} over a few hours. Figure 9.11 shows data taken over a week, which show changes in the RV of ≈4 m s^{-1}, consistent with stellar activity models (Figure 9.8). Changes in the RV of a few m s^{-1} can be seen over several hours. Several of these solar telescopes are planned or have been built for other high-precision RV spectrographs like HARPS and ESPRESSO. As of this writing, the long-term monitoring of integrated sunlight has just begun. Monitoring the Sun over the solar cycle will help understand how stellar activity can influence the measured RVs of stars and how to mitigate these effects.

9.4.2 Velocity Spots

We have seen how spots and plage can suppress the convective blueshift resulting in RV variations. This brings us to the concept of "velocity" spots. Any inhomogeneous surface structure will produce an RV signal—it does not have to be a temperature spot. For instance, the magnetic Ap stars have abundance spots where the atomic species of a certain element has greater or lesser abundance with respect to the surrounding regions, but not large temperature differences. Alternatively, one can think of spots where the velocity distribution within the spot is less than the surrounding "photosphere".

In stars, the granulation pattern and its turbulent velocity field are often parameterized by the so-called "macroturbulent" velocity.[3] This velocity field has a Gaussian distribution, or more appropriately, a "radial–tangential distribution,"

[3] This is to be distinguished from microturbulence, which arises from turbulent elements that are small compared to unit optical depth. Microturbulence is needed to fit the observed equivalent width of lines but is not a large broadening mechanism like macroturbulence.

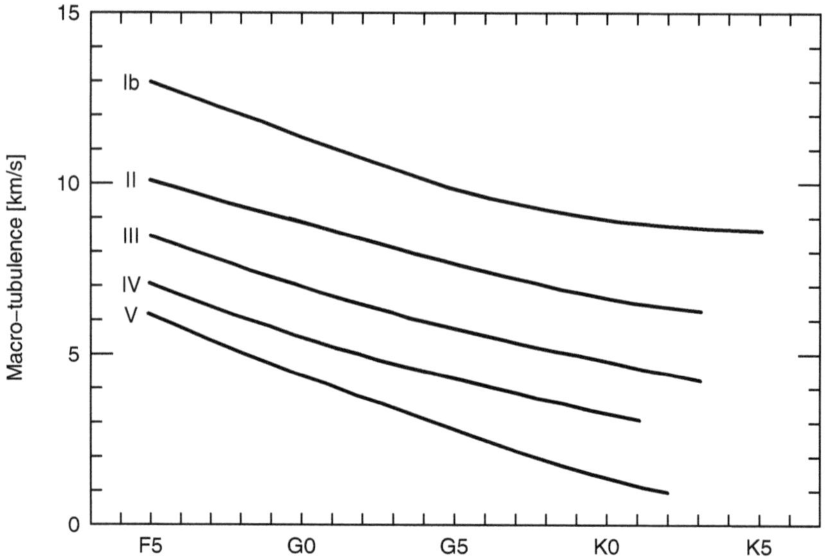

Figure 9.12. The variation of the macroturbulent velocity across the H-R diagram (Gray 2005).

which is more peaked (see *The Observation and Analysis of Stellar Photospheres* by David Gray). The macroturbulence is an additional broadening mechanism needed to fit the observed stellar line profiles. Figure 9.12 shows the macroturbelent velocities for stars across the Hertzsprung–Russell (H-R) diagram (Gray 2005).

Hatzes & Cochran (2000) investigated whether velocity spots could explain the variations in the RV residuals of Polaris. This star is a well-known low-amplitude Cepheid variable, but after removing the variations due to the pulsations, a 40 day signal with an amplitude of 400 m s^{-1} was found. Figure 9.13 shows the RV amplitude as a function of spot radius for two configurations of the spot and inclination of the star. Outside the spot, a macroturbulent velocity of 12.8 km s^{-1} was used, while inside the spot, the macroturbulent velocity was set to zero. The RV amplitude caused by such a "macroturbulent velocity" spot can be as large as 100–150 m s^{-1}, but saturates for large spot areas—large spots now become the "photosphere". Such a velocity spot might arise from large-scale, organized magnetic fields like the types seen in Ap stars. Although this model could not account for the observed RV variations in Polaris (one needs temperature spots as well), it points out that sometimes one needs to "think outside the box" when it comes to interpreting RV signals from stars. Interestingly, if the magnetic field is large enough to suppress the velocity field, but not large enough to alter the spot temperature, one could see RV variations without accompanying photometric variations.

9.5 Testing for Rotational Modulation

The RV signal due to a planetary companion should be long-lived. The signal due to activity and pulsations may come and go, but one due to a planet is like a clock—they are always present in the data, never changing their period, amplitude, or phase

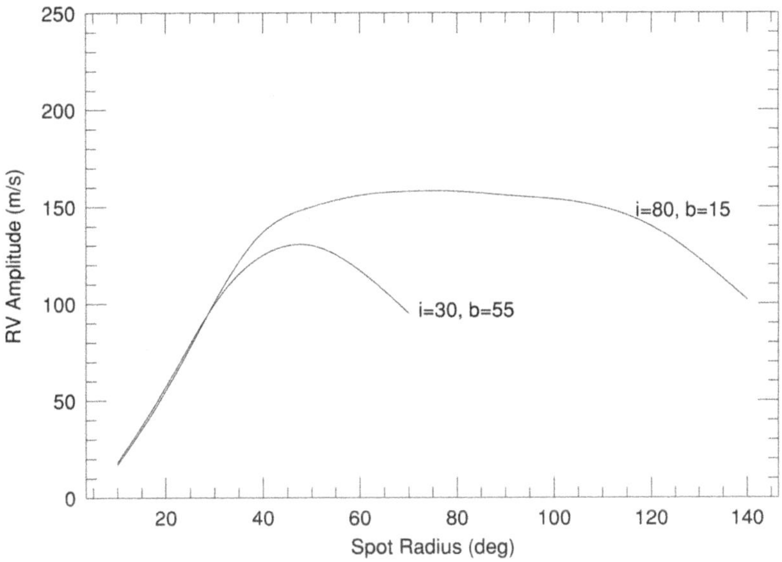

Figure 9.13. RV semiamplitude as a function of spot radius for velocity spots with a macroturbulent velocity of 0 km s^{-1} compared with 12.8 km s^{-1} for the photosphere. Two cases are shown, a spot at latitude 55° on a star with an inclination of 30° and a spot at latitude 15° on a star with an inclination of 80°. (Reproduced from Hatzes & Cochran 2000. The American Astronomical Society. All rights reserved.)

(excluding of course any dynamical evolution of the orbit). Arguably, one of the most convincing cases for support of the planet hypothesis is that it remains present over a long period of time. This can be tested using statistical tests afforded by the Lomb–Scargle (LS) periodogram or by searching for amplitude variations in the periodic signals. You should also make sure that the rotation period of the star is not coincident with your purported planet period.

9.5.1 Determining the Rotation Period of the Star

The simplest test to make sure that the RV period that you find is not due to rotational modulation is to compare it to the rotation period of the star. How can you measure the rotation period of the star? The best way is to perform a periodogram analysis on ancillary data not used for RV work, in particular photometric measurements, but also activity indicators (to be discussed in the next chapter). These will show variations only due to activity (rotation and activity cycles) and not due to the orbital motion of the star. If your star was a target of the *Kepler*, *CoRoT*, or *TESS* space mission, then high-quality light curves with good temporal resolution can be used to measure the rotational period of the star.

It will often be the case that data from activity indicators will not be available, and you will have to ascertain the rotational period of the star solely from the RV data. Both the rotational and orbital periods will be seen in the RV time series; you have to decide which is which. One can estimate the rotational period of the star from the stellar radius and rotational velocity,

$$P_{\rm rot} = \frac{2\pi R_{\rm star}}{V_{\rm rot}}. \tag{9.1}$$

So first, you need the stellar radius. Estimates of the stellar radius can come often from the spectral type of the star and evolutionary tracks. Fortunately, the *Gaia* mission will have accurate distances for a large number of stars. From the stellar distance, brightness, and effective temperature of the star, one can derive a fairly accurate stellar radius. Asteroseismic data from the *PLATO* mission will also deliver accurate stellar radii. All in all, one should be able to determine the stellar radius and rotation period to about 10% in each quantity.

The final ingredient is the rotational velocity, and unfortunately, one can only measure the projected rotational velocity of the star, $v \sin i$, and not the true rotational velocity. Because $V_{\rm rot} \geqslant V \sin i$, this means Equation (9.1) only gives you the maximum rotational period. So, you can only be sure that an RV period is not due to rotation if it is larger than that of the estimated rotational period. It is usually sufficient to measure the $v \sin i$ from fitting the spectral line profiles, although in Chapter 13 we will discuss more accurate methods of doing this.

A characteristic feature of rotation is that in the periodogram, one not only sees the dominant frequency due to rotation, $f_{\rm rot}$, but also its harmonics, $2f_{\rm rot}$, $3f_{\rm rot}$, etc., at roughly equally spaced frequency intervals. The rotational period can be measured by taking the mean spacing of the harmonic frequencies.

Of course, planetary systems in resonant orbits will also produce peaks at equally spaced frequency intervals. In this case, the first harmonic is just another planet in a 2:1 resonance. A dynamical study is required to shed light on the true nature of the signal.

The Autocorrelation Function

Aside from the standard periodogram, the autocorrelation function (ACF) is often employed to search for periodic signals due to stellar rotation (McQuillan et al. 2013). The ACF has the form

$$r_k = \frac{\sum_{i=1}^{N-k}(x_i - \bar{x})(x_{i+k} - \bar{x})}{\sum_{i=1}^{N}(x_i - \bar{x})^2}, \tag{9.2}$$

where r_k is the autocorrelation coefficient at lag k for the time series x_i ($i = 1,\ldots N$) and $\bar{x}$ is the mean value. Each lag k corresponds to $\tau_k = k\Delta t$ for a cadence Δt (see Shumway & Stoffer 2010).

Figure 9.14 compares the ACF method to a traditional periodogram for a synthetic signal (from McQuillan et al. 2013). The standard periodogram shows a single peak at the input period whereas the ACF shows an oscillatory behavior with equispaced peaks at multiples of the input period. Also shown are the effects of introducing correlated noise. See McQuillan et al. (2013) for a more detailed description of measuring periods using the ACF.

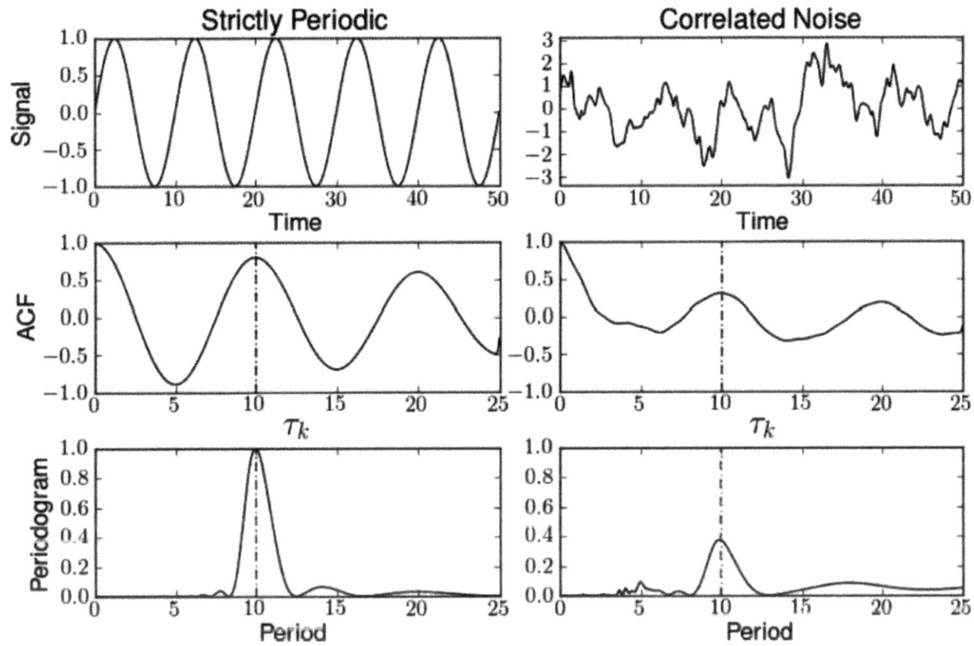

Figure 9.14. A synthetic periodic signal with a period of 10 days (top-left panel). The corresponding ACF (middle-left) and periodogram (lower-left). The right panels show the effect of introducing correlated (top-right) noise and the corresponding ACF (middle-right) and periodogram (lower-right). The dashed vertical lines mark the input period. (Reproduced from McQuillan et al. 2013. Copyright of OUP Copyright '2013'.)

9.5.2 Evolution of Statistical Significance

For a long-lived signal, the statistical significance should increase with the number of measurements. A signal from stellar activity may come and go, but the RV signal due to a planet is a reliable clock that is always present. As one acquires more data, one beats down the Fourier noise and the signal becomes more significant. A signal whose statistical significance increases with the number of measurements in a way that is expected for the quality of data is a necessary condition for the planet hypothesis, but not always sufficient.

The star ε Eridani was one of the first long-period giant planets discovered by RV surveys. Hatzes et al. (2000) reported the presence of a 0.86 M_{Jup} planet in a 6.9 year orbit that was highly eccentric. ε Eri is also a young active star whose RV jitter dominates the K amplitude of the star. For this reason, it took a large number of measurements spanning several decades to detect the planet. However, because of the high level of activity, subsequent studies challenged the planet hypothesis. Either it was not detected with new data (Zechmeister et al. 2013) or the planetary signal was attributed to an activity cycle (Anglada-Escudé & Butler 2012). Thirty years of RV monitoring has established that ε Eri b is indeed present (Mawet et al. 2019).

The planet hypothesis is supported by the increasing significance of the signal. The right panel of Figure 9.15 shows the statistical significance (false alarm probability, FAP) of the 6.9 year period in the RV of ε Eri as a function of the

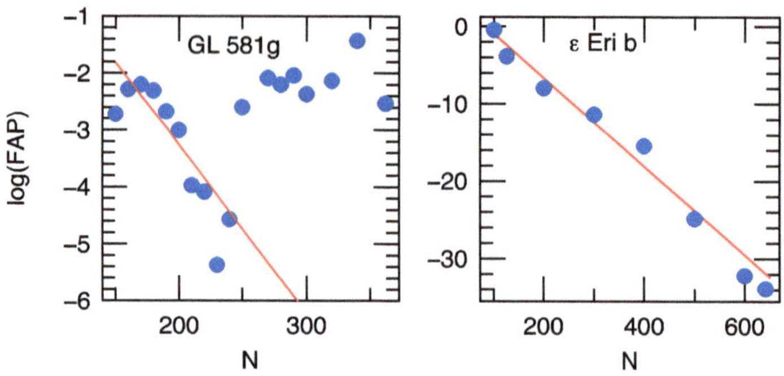

Figure 9.15. (Left) The logarithmic value of the false alarm probability (FAP) for RV variations due to GL 581g (other planet signals removed) as a function of the number of measurements (N). (Right) The same for RV variations due to ε Eri b. The red lines show the expected behavior from simulations with a coherent signal present.

number of measurements. The red line shows a simulation taking the orbital solution, sampling it in the same way as the data and with random noise consistent with the measurement error and the presence of the so-called activity RV jitter (see Chapter 10). The fact that the actual FAP follows the simulated data provides strong evidence in support of the planet hypothesis—or at least that the signal is indeed long-lived and coherent.

The situation is much different for a purported fifth planet around the M-dwarf star GL 581. Initially, four planets were reported around this star (Bonfils et al. 2005; Mayor et al. 2009). Using 11 years of RV measurements made with the Keck HIRES spectrograph and combined with the HARPS measurements, the presence of a fifth planet, GL 581g (Vogt et al. 2010), was reported. The period of 36 days placed this planet firmly in the habitable zone of the star. However, Forveille et al. (2011) presented 121 additional HARPS measurements and, after analyzing all data, showed that GJ 581g was not present.

The test of statistical significance can immediately show the problems with this planet. The FAP indeed decreases with the number of measurements (left panel of Figure 9.15) and nearly approaches FAP $\sim 10^{-5}$. However, with the addition of more measurements, the FAP jumps to a much lower significance of FAP $\sim 10^{-2}$. Furthermore, the actual behavior of the FAP does not follow the simulated data.

The case of GL 581g presents a sobering case. After over 200 RV measurements, the 36 day signal was present at a highly significant level (FAP $\sim 10^{-5}$). This is strong evidence for a real signal. Furthermore, the behavior of the FAP with the number of measurements followed that of simulated data for the first 200 measurements. By most researchers' reckoning (including this author), this is a bona fide planet. Vogt et al. (2010) indeed reached a reasonable conclusion given the data in hand. It was only with considerably more measurements, taken over a longer time, that the true nature of the signal was revealed.

More sophisticated versions of the periodogram have been developed to help researchers determine the nature of periodic signals. Mortier & Collier Cameron (2017) presented the Bayesian Generalized Lomb–Scargle (BGLS) periodogram as a way of identifying the periodicities caused by stellar activity. It is based on the concept that the power or probability for stable signals increase with time (or equivalently, the number of data points). We have already seen this for GL 581 and ε Eri (Figure 9.15). The BGLS is computed for the first n data points. The subsequent data points are added, and the new periodogram is computed. All periodograms are then normalized to their maximum values, and each one calculated with incremental n is then stacked on each other. The x-axis represents the period or frequency while the y-axis represents time (or equivalently the number of measurements). By color-coding the power, one has a visual representation of the time evolution of the periodogram. Also, unlike the case where we examine just one frequency, one can see the temporal evolution of the full periodogram more easily. Recall that with BGLS one has a relative probability between peaks in the periodogram.

Figure 9.16 shows the process applied to Gl 581, where the periodogram of the 5.37 day period is shown on the left, the 12.9 day period in the center, and the 66.9 day signal to the right. The 5.27 day and 12.9 day periods are stable and show a temporal evolution of increasing power with time. The 66.9 day period, on the other hand, shows a much more unstable evolution, with a sharp rise in power followed by a sharp drop, and then again a rise. This is more indicative of activity despite an overall increase in power from $n = 50$ to $n = 250$.

The reader should note, as pointed out by Mortier & Collier Cameron (2017), that a signal which shows significance is not proof of the existence of a planet. More analysis is required. Likewise, a signal whose significance fluctuates is not unequivocal proof of an activity signal. This is especially true if the rotational period of the star is close to that of the planet's orbital period such that the activity signal can interfere constuctively and destructively with the RV variations of the planet. This is

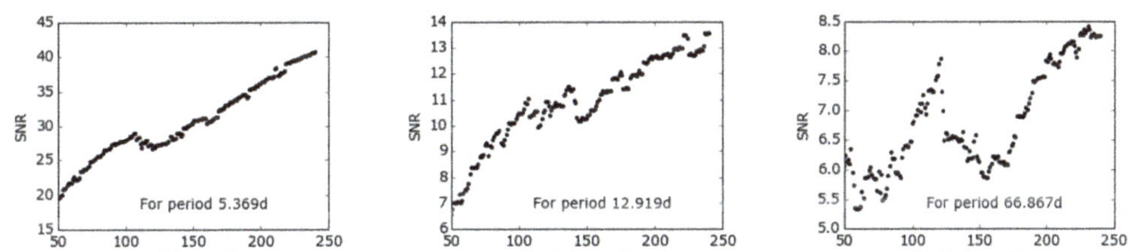

Figure 9.16. The stacked BGLS periodograms of Gl 581. (Bottom) The significance of the detection (signal-to-noise ration = SNR) versus the number of observations for the most significant period. (Left to right) The full data set with a dominant period at 5.369 days, the subtracted best fit of the 5.37 day period showing a dominant period of 12.92 days, and the subtracted best fit, which leaves a dominant period of 66.9 days, which is suspected to be due to activity. (From Mortier & Collier Cameron 2017, reproduced with permission © ESO.)

demonstrated by a simple simulation. We took a planet signal with orbital period P = 8.9 days and an amplitude K = 10 m s^{-1} We added a rotational modulation (RM) signal with the same amplitude but with P = 9.2 days. We then generated a time series where the first one-third of the time series had only the planet signal and for the second one-third the planet plus RM with a phase such that it constructively interfered with the planet signal. For the final one-third of the data set, the RM signal was still present but with a different phase. Random noise (σ = 3 m s^{-1}) was also added.

Figure 9.17 shows the evolution (points) of the LS power and thus the significance as a function of the number of data points for the signal at P = 8.9 days (planet). Also shown is the expected behavior (red line) of a pure planet signal with the appropriate noise added. The LS power follows the theoretical curve (as expected because there is no RM signal at the start), but sharply drops once the RM signal contributes to the RV. It then slowly rises before turning over again. If this were real data, the logical conclusion would be that the $P \approx 9$ day signal is unstable and thus due to activity even though a stable and coherent signal is actually in the data. The evolution of the LS power is only suggestive. An assessment of the true nature of a signal requires detailed simulations including all possible signals that can be in the data. Even then, the statistical test may "confirm" a planet, later refute it, only to have the planet return with more data!

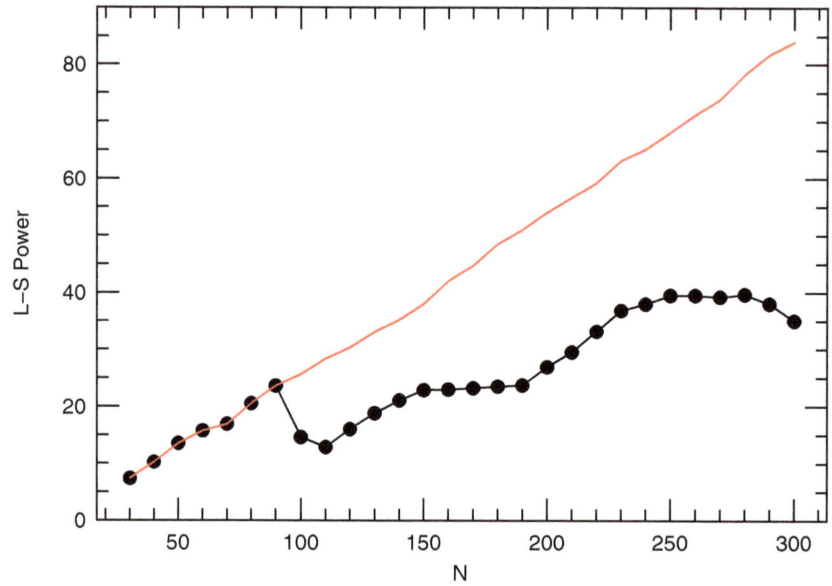

Figure 9.17. The growth of the LS power as a function of data points for a planetary signal with a period of 8.9 days and an amplitude of 10 m s^{-1} along with a rotational modulation (RM) signal with P = 9.2 days and K = 10 m s^{-1} due to a spot. For the first 90 measurements, only the planet signal is present. At N = 90, the RM is present, but out of phase with the planetary signal. At $N \approx 180$, the RM signal changes phase. In this simulation, the growth of the LS power looks unstable in spite of the presence of a stable planetary signal.

9.5.3 Amplitude Variations

A search for amplitude and phase variations in the RV signal may also give hints as to the nature of the signal. The Doppler motion due to a planet should have the same amplitude and phase with time. This is not necessarily true for a signal due to activity. Spots and plage are evolving features. They form, grow, and eventually fade away. You will thus see a change in the amplitude of the RV signal. They also migrate, or new features are born on another part of the star. This would result in a change in phase.

The case of GJ 15Ab represents a subtle case where a harmonic of the rotation period was mimicking a planet signal, but this only became apparent after a closer look at the evolution of the significance and amplitude variations of the signal after taking more data. Howard et al. (2014) first reported several periodic signals in the M-dwarf star GJ 15A using 117 radial velocity measurements spanning 1997–2011 that were taken with the HIRES spectrograph on the 10 m Keck Telescope. The strongest peak in the periodogram was at 11.44 days followed by one at ~44 days (Figure 9.18). The FAP of the signal was convincingly low at about 10^{-5}.

To confirm the nature of the RV signals, the authors used the standard tools (see Chapter 10). First, they examined the Ca II S_{HK} index and found a period of 44.8 days (frequency = 0.022 day^{-1}), which coincided with the second highest peak in the periodogram. Photometric measurements also revealed power in the periodogram at 43.8 days. The logical conclusion was that the rotation period of the star was

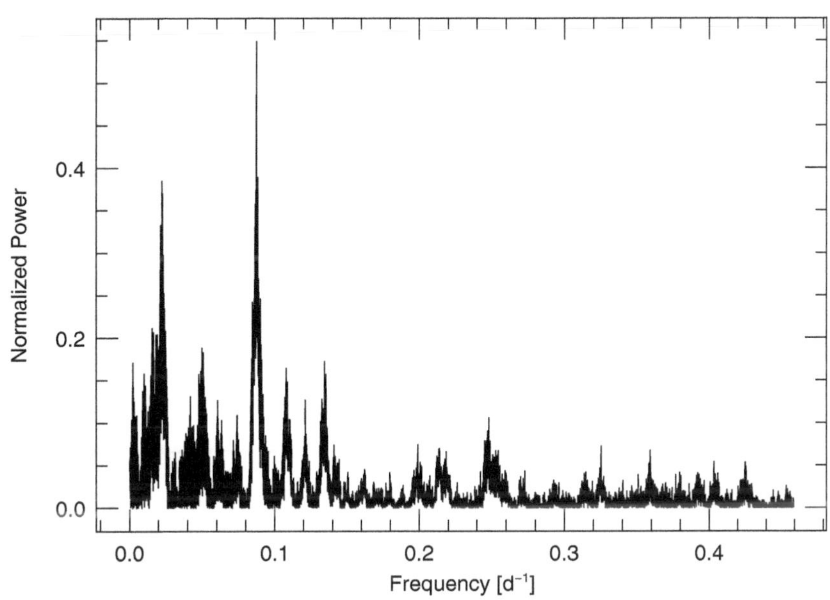

Figure 9.18. The GLS periodogram of the RVs of GJ 15A from Howard et al. (2014). The secondary peak at frequency = 0.022 day^{-1} (P_{rot} = 43.8 days) is presumed to be due to rotational modulation. The primary peak at frequency = 0.087 day^{-1} (P = 11.44 days) was believed to due to the orbital motion of a planet. However, this peak is close to the third harmonic, $P_{rot}/4$.

approximately 44 days, and the 11.44 day period was due to a planet, although this is uncomfortably close to the third harmonic of the rotation period ($P_{rot}/4$).

GJ 15A was included in the CARMENES RV survey of M-dwarf stars. Ninety-two RV measurements spanning 2016–2017 showed no strong evidence for the 11.44 day planet signal (Trifonov et al. 2018). The authors then combined the CARMENES data with the earlier HIRES measurements, as well 358 additional HIRES measurements (Butler et al. 2017). Only with this extensive data set, now spanning 20 years, were they able to determine that not only did the significance of the 11.44 day period decline with time, but also the *K* amplitude.

The top panel of Figure 9.19 shows that after 100 measurements the GLS power, a measure of statistical significance, steadily decreases as one adds measurements. At the same time, the *K* amplitude determined by fitting subsets of the data also decreases with the number of measurements. The most likely explanation is that the 11.44 day period arises from stellar rotation, in this case the harmonic of the rotational period. This only highlights that a simple examination of activity

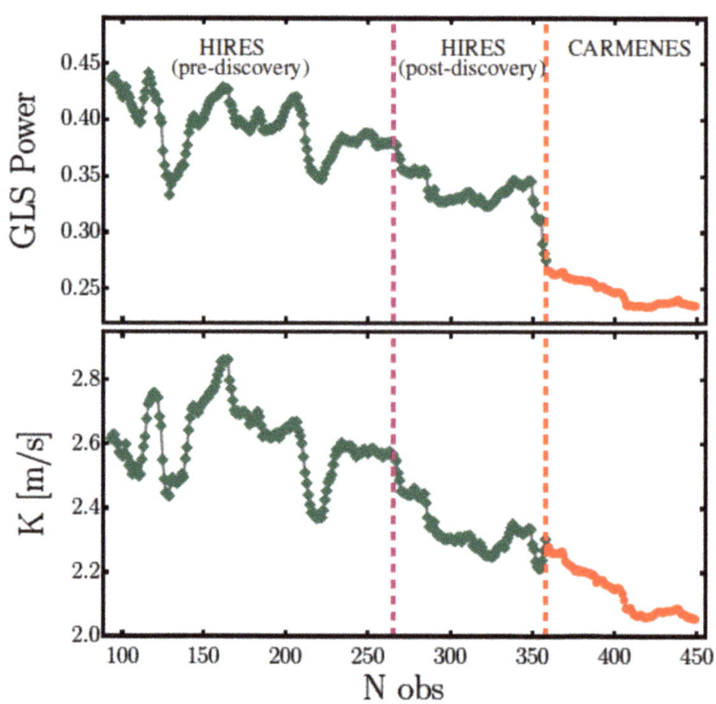

Figure 9.19. The GLS power and fitted *K* amplitude of the 11.4 day period as a function of the number of data using all available data. The vertical magenta line separates pre- and postdiscovery HIRES RV data for GJ 15A, while the vertical red line indicates the beginning of the CARMENES measurements. The curves shows a fading trend in both the statistical GLS power (statistical significance) and *K* amplitude for the signal. Note that the downward trend is already evident in the HIRES-only data. (From Trifonov et al. 2018, reproduced with permission © ESO.)

indicators may still fail to reveal that an RV signal is due to activity. This may only become evident after many more years of observations.

So, if the 11.44 day period is due to rotation, why did it not appear in the activity indicators? This may possibly arise from the different epochs where the data sets were taken. The RV measurements spanned 1997–2011, with the median occurring in 2010. The CA II S_{HK} data showed the rotation period of 44 days primarily in data taken in 2011. The photometric data were taken in 2008–2011, but again, the rotation period was only seen in the high-cadence measurements from 2011.

One plausible explanation is that the 11.4 day RV period was $P_{rot}/4$, which was dominant in the first half of the measurements, whereas the activity indicators sampled a time when the P_{rot} had become the dominant signal. The case of GL 15A demonstrates that even when using activity indicators, one can still be misled. If one is using photometry to confirm the planetary nature, these should be contemporaneous with the RV measurements.

Luyten's star (=GJ 273) is another example where a look at the amplitude and phase variations in the signal sheds light on the nature of the RV variations. Two planets were reported around this M3.5 dwarf as part of the HARPS M-dwarf survey (Astudillo-Defru et al. 2017b). The two planets had periods of P_1 = 18.7 days and P_2 = 4.72 days, and amplitudes of K_1 = 1.61 m s^{-1} and K_2 = 1.06 m s^{-1}, respectively. A third signal with a period of ≈99 days was also detected, but this was attributed to stellar activity as the estimated rotational period from photometry was also 99 days (Astudillo-Defru et al. 2017a).

This star was also observed as part of the CARMENES M-dwarf survey (Reiners et al. 2018). Combined with HARPS and CARMENES data, over 500 measurements have been made for this star. Let's see if the amplitude variations could identify which signal is a planet and which is activity, focusing only on the 4.72 day and the ≈99 day periods.

An analysis of the full data set easily finds the two planets reported by Astudillo-Defru et al. (2017b) as well as the suspected activity signal with a period of 98 days. Figure 9.20 shows the amplitude and phase variations for the 98 day and 4.74 day periods found in Luyten's star. In each case, the contributions from all other signals was removed in order to isolate the period of interest. The horizonal lines mark the nominal K-amplitudes and errors using the entire data set. The data were then subdivided into epochs, and a fit was made keeping the period fixed to the nominal value calculated with the full data, but allowing the amplitude and phase to vary.

The signal at 4.72 days shows an amplitude and phase that are more or less constant and within the errors on the nominal values from the full data set. The variations of the 98 day period tell a different story. These show large variations in both amplitude and phase well above the error and well beyond the nominal values. Even without photometric data, or ancillary measurements from activity indicators, the logical conclusion would be that the 98 day period is due to rotational modulation and the 4.72 day period is a real planet. A similar analysis also suggests that the 18.7 day period is also planetary in nature.

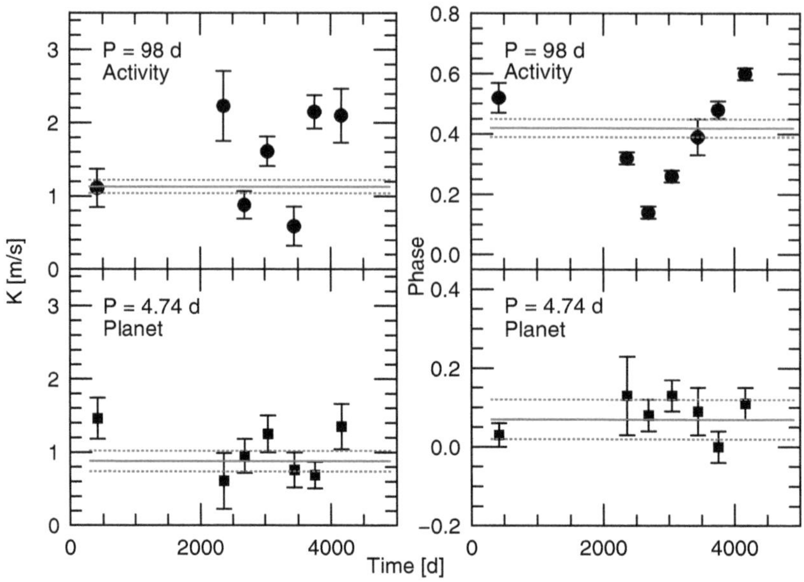

Figure 9.20. The amplitude (left panels) and phase (right panels) as a function of time for two periodic RV signals found in Luyten's star. The horizontal solid red lines represent the values using the full data set and the dashed lines mark the $\pm 1\sigma$ values. The 4.74 day period is believed to be due to a planet, and its amplitude (lower left) and phase (lower right) are roughly constant with time. The 98 day period shows strong amplitude (top left) and phase (top right) phase variations more than the measurement error and outside of the nominal values using the full data set. This period is due to activity and is consistent with the ~100 day period found in photometry.

One characteristic feature of amplitude variations is that in periodograms (i.e., the Fourier domain), these will manifest themselves as two peaks closely spaced in frequency. This is easy to understand in terms of the beat phenomenon between the two frequencies. Over time, these will interfere constructively and destructively, and this will look like amplitude variations of a single, much longer period.

This is demonstrated in the RV variations of γ Draconis. Over 14 years, these showed a single period of 702 days ($f = 0.0014$ day^{-1}), but with an amplitude that changed with time (Hatzes et al. 2018). Fitting the RV data over short time spans with a fixed-period 702 day sine function but variable amplitude showed clear variations with a 10.6 year period (top panel Figure 9.21). In the Fourier domain the peak at 0.0014 d^{-1} is clearly asymmetric. Prewhitening reveals that it comprises two peaks with $f_1 = 0.00125$ day^{-1} ($P_1 = 666.7$ days and $f_2 = 0.00125$ day^{-1} ($P_2 = 800$ days). The beat frequency is the difference of these or $f_b = 0.00025$ day^{-1}, or a period of $\approx$4000 days, which is consistent with the period of the amplitude variations.

How can one distinguish between the two closely spaced signals of a single frequency with amplitude variations? One must rely on other arguments. For instance, in γ Draconis the RV variations can be explained by a two-planet system

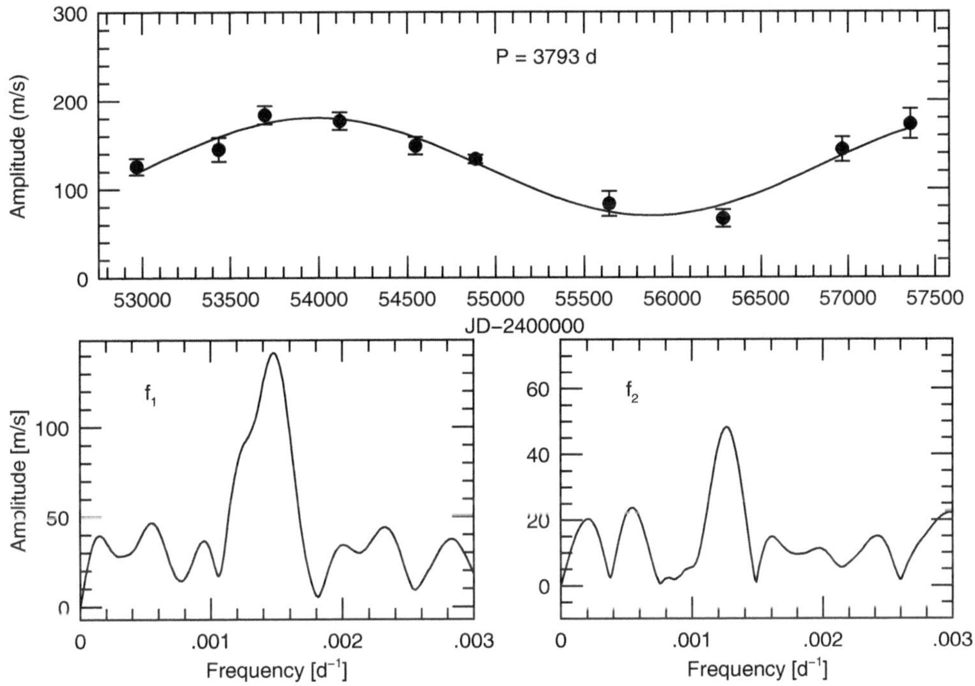

Figure 9.21. (Top) The amplitude of the 702 day period detected in the RV variations of γ Draconis as a function of time (Hatzes et al. 2018). (Lower left) The amplitude spectrum of the RV measurements for γ Draconis. The peak is at $f_1 = 0.00150$ day^{-1} ($P = 666.7$ days). (Lower right) Removing f_1 reveals a second peak at $f_1 = 0.00125$ day^{-1} ($P = 800$ days).

of giant planets with periods of 667 and 800 days. However, such a system is most likely unstable as a dynamical study shows that there is only a few percent chance that such a system could be stable over the lifetime of the star (Hatzes et al. 2018). Alternatively, the two frequencies may point to a new type of oscillations in K giant stars (Hatzes et al. 2018).

References

Anglada-Escudé, G., & Butler, R. P. 2012, ApJS, 200, 15

Astudillo-Defru, N., Delfosse, X., Bonfils, X., et al. 2017a, A&A, 600, A13

Astudillo-Defru, N., Forveille, T., Bonfils, X., et al. 2017b, A&A, 602, A88

Balmaceda, L. A., Solanki, S. K., Krivova, N. A., & Foster, S. 2009, JGRA, 114, A07104

Bonfils, X., Forveille, T., Delfosse, X., et al. 2005, A&A, 443, L15

Butler, R. P., Vogt, S. S., Laughlin, G., et al. 2017, AJ, 153, 208

Deming, D., Espenak, F., Jennings, D. E., Brault, J. W., & Wagner, J. 1987, ApJ, 316, 771

Dravins, D. 1982, ARA&A, 20, 61

Dumusque, X., Glenday, A., Phillips, D. F., et al. 2015, ApJ, 814, L21

Forveille, T., Bonfils, X., Delfosse, X., et al. 2011, arXiv:1109.2505

Gray, D. F. 2005, The Observation and Analysis of Stellar Photospheres (Cambridge: Cambridge Univ. Press)

Hall, D. S., & Henry, G. W. 1994, IAPPP, 55, 51
Hatzes, A. P. 1991, MNRAS, 248, 487
Hatzes, A. P. 1995, ApJ, 451, 784
Hatzes, A. P., & Cochran, W. D. 2000, AJ, 120, 979
Hatzes, A. P., Cochran, W. D., McArthur, B., et al. 2000, ApJ, 544, L145
Hatzes, A. P., Endl, M., Cochran, W. D., et al. 2018, AJ, 155, 120
Hatzes, A. P., & Vogt, S. S. 1992, MNRAS, 258, 387
Herrero, E., Ribas, I., Jordi, C., et al. 2016, A&A, 586, A131
Howard, A. W., Marcy, G. W., Fischer, D. A., et al. 2014, ApJ, 794, 51
Kochukhov, O., Rusomarov, N., Valenti, J. A., et al. 2015, A&A, 574, A79
Livingston, W. C. 1983, in IAU Symp. 102, Solar and Stellar Magnetic Fields: Origins and Coronal Effects, ed. J. O. Stenflo (Dordrecht: Reidel), 149–52
Mawet, D., Hirsch, L., Lee, E. J., et al. 2019, AJ, 157, 33
Mayor, M., Bonfils, X., Forveille, T., et al. 2009, A&A, 507, 487
McMillan, R. S., Moore, T. L., Perry, M. L., & Smith, P. H. 1993, ApJ, 403, 801
McQuillan, A., Aigrain, S., & Mazeh, T. 2013, MNRAS, 432, 1203
Meunier, N., Desort, M., & Lagrange, A.-M. 2010, A&A, 512, A39
Mortier, A., & Collier Cameron, A. 2017, A&A, 601, A110
Piskunov, N. E., Tuominen, I., & Vilhu, O. 1990, A&A, 230, 363
Reiners, A., Zechmeister, M., Caballero, J. A., et al. 2018, A&A, 612, A49
Rice, J. B., Wehlau, W. H., & Khokhlova, V. L. 1989, A&A, 208, 179
Rusomarov, N., Kochukhov, O., & Lundin, A. 2018, A&A, 609, A88
Rusomarov, N., Kochukhov, O., Ryabchikova, T., & Ilyin, I. 2016, A&A, 588, A138
Scherrer, P. H., Bogart, R. S., Bush, R. I., et al. 1995, SoPh, 162, 129
Setiawan, J., Henning, T., Launhardt, R., et al. 2008, Natur, 451, 38
Shumway, R., & Stoffer, D. 2010, Time Series Analysis and its Applications (Berlin: Springer)
Strassmeier, K. G. 2009, A&ARv, 17, 251
Strassmeier, K. G., Stępień, K., Henry, G. W., & Hall, D. S. 1999, A&A, 343, 175
Trifonov, T., Kürster, M., Zechmeister, M., et al. 2018, A&A, 609, A117
Vogt, S. S., Butler, R. P., Rivera, E. J., et al. 2010, ApJ, 723, 954
Vogt, S. S., Hatzes, A. P., Misch, A. A., & Kürster, M. 1999, ApJS, 121, 547
Vogt, S. S., Penrod, G. D., & Hatzes, A. P. 1987, ApJ, 321, 496
Wallace, L., Huang, Y. R., & Livingston, W. 1988, ApJ, 327, 399
Zechmeister, M., Kürster, M., Endl, M., et al. 2013, A&A, 552, A78

AAS | IOP Astronomy

The Doppler Method for the Detection of Exoplanets

A P Hatzes

Chapter 10

Avoiding False Planets: Indicators of Stellar Activity

In this chapter, we continue our investigation into intrinsic stellar variability and how it can produce a fake planet signal. It will first focus on several useful activity diagnostics that can be used to ascertain the nature of the periodic radial velocity (RV) signals that you find. We then discuss sources of the so-called RV "jitter." This is an additional error source that stems from intrinsic stellar variability. Finally, the chapter ends with a short discussion on activity cycles, which can also introduce a periodic signal in the RV data but with periods much longer than the rotation period of the star.

10.1 Activity Indicators

RV variations due to stellar magnetic activity almost always manifest themselves in variations of activity indicators. By comparing the periodic signals found in these indicators to those found in the RV data, one can determine which signals are actually due to the Doppler reflex motion of the star. Here we discuss a number of activity indicators that are often used to test whether a periodic signal in RV data is in fact due to stellar activity.

Of course, finding a periodic signal in these activity indicators may not provide irrefutable proof against the planet hypothesis. You could have some form of star–planet interaction, or the orbital period of the planet may just happen to coincide with the period of the stellar rotation or that of an activity cycle. Our own solar system is an example of this. If extraterrestrial planet hunters were to monitor our Sun from afar with RV variations, they would detect the orbital period of Jupiter at 11.86 years. Measuring the activity indicators of our Sun, they would see the solar cycle of 11 years and conclude, wrongly, that the RV variations were due to activity. So, if you find variability in other quantities near the period of the "planet," this may

doi:10.1088/2514-3433/ab46a3ch10

not disprove the planet hypothesis—you just have to take the extra steps to confirm this.

One of the most important activity indicators is photometric variability. Surface structure like cool spots influences the light level of the star, and photometric measurements are the best way to unambiguously determine the rotation period of the star. Knowing a rotation period makes the interpretation of the RV measurements easier. However, it is difficult to obtain photometry simultaneously with your RV measurements. Here we focus on activity indicators that can be measured with the same spectra used for computing the Doppler shifts.

10.1.1 Ca II H & K

The Ca II H and K lines at 3933.7 Å and 3968.5 Å are spectral features that are the "classic" indicators of stellar magnetic activity (Figure 10.1). In solar-type stars, these resonance lines are among the strongest features found in the spectrum. Recall from basic stellar atmospheric physics that strong lines are formed higher up in the stellar atmosphere, and this is particularly true for the H & K lines. For a single line, the core is formed higher up in the atmosphere than the wings. In the case of the Ca II lines, the cores of the lines actually come from the chromosphere, where the source function starts to increase with height (i.e., the temperature increases from the photosphere to the chromosphere). For active stars, this results in a core reversal, or emission peak in the core of the line which is indicative of the presence of a chromosphere.

There are two ways of measuring the strength of this core reversal and thus the level of activity of the star. The first is the so-called Mt. Wilson *S* index, developed

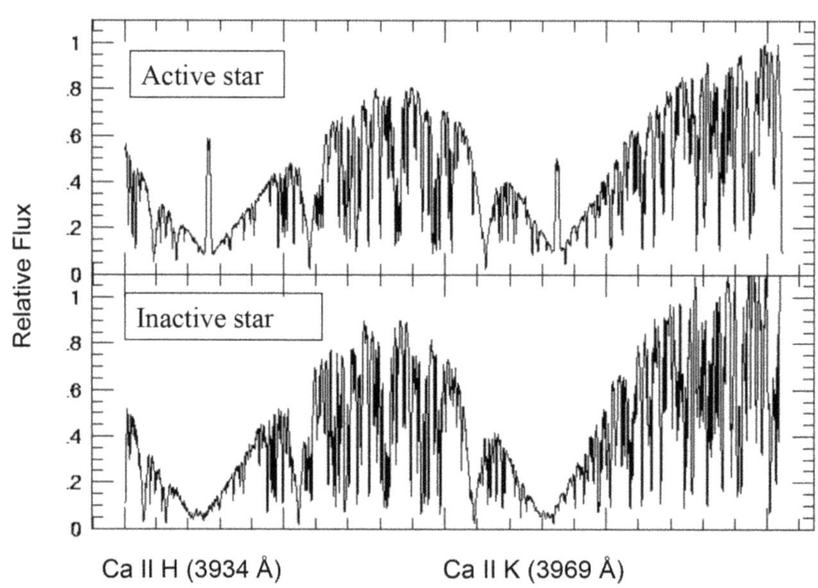

Figure 10.1. The Ca II H & K lines in an active star (top) and an inactive star (bottom).

for the long-term monitoring of active stars from the Mt. Wilson Observatory (e.g., Baliunas et al. 1995). This index is defined as

$$S = \alpha \frac{F_{\rm H} + F_{\rm K}}{F_{\rm B} + F_{\rm R}}, \tag{10.1}$$

where $F_{\rm H}$ and $F_{\rm K}$ are the fluxes in a triangular band pass 1 Å wide centered on the H and K lines and $F_{\rm B}$ and $F_{\rm R}$ are the fluxes in 20 Å band passes in the continuum that are centered on 3901 Å and 4001 Å, respectively (Soderblom et al. 1991; Baliunas et al. 1995). The term α is a calibration constant, which is needed if you want to put the S index on an absolute scale so as to compare to other stars. Active stars can have values of the S index as high as 0.3.

Alternatively, you can use the $R'_{\rm HK}$ index as a measure of chromospheric activity:

$$R'_{\rm HK} = \frac{F'_{\rm HK}}{F_{\rm bol}}, \tag{10.2}$$

where $R'_{\rm HK}$ is the total chromospheric Ca II H&K surface flux corrected for the photospheric component and $F_{\rm bol}$ is the bolometric flux of the star. Thus $R'_{\rm HK}$ is the normalized purely chromospheric component of the H&K flux. Very active stars can have $\log(R'_{\rm HK})$ as high as −3.

For the confirmation of exoplanets, it is not the absolute values of the S index or $R'_{\rm HK}$ that are important, but whether the relative changes of the values correlate with the measured RV. The RV measurements for HD 166435 (top panel Figure 10.2)

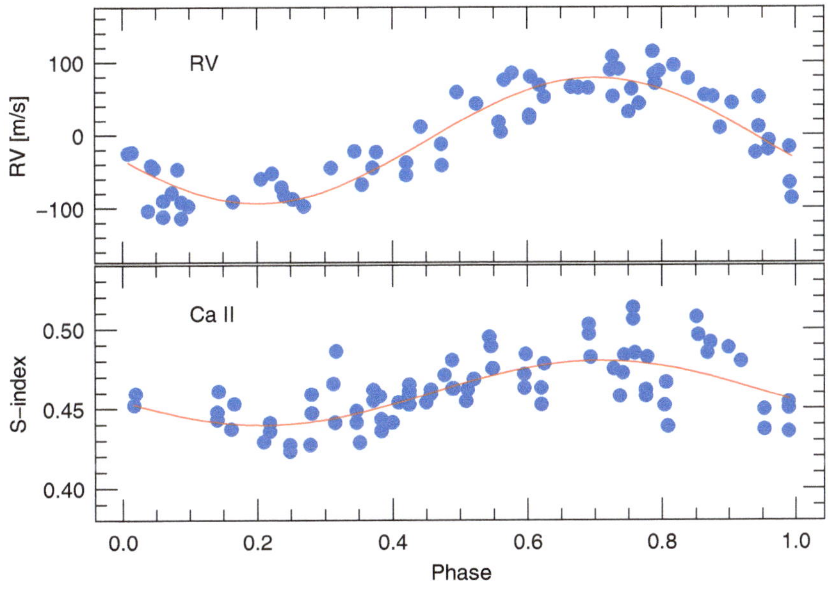

Figure 10.2. (Top) The RV variations of the active star HD 166435 phased to a period of four days. (Bottom) The variations of the S index for this star phased to the same period. This confirms that the RV variations stem from stellar activity and that four days is the rotational period of the star. (Data from Queloz et al. 2001.)

show variations with a four-day period that are consistent with the presence of a hot Jupiter (Queloz et al. 2001). The fact that the *S*-index measurements vary with the exact same period (lower panel Figure 10.2) confirms that these RV variations are due to the rotational modulation of stellar activity features.

Many spectrographs do not cover the Ca II H&K lines, particularly those designed for RV measurements in the near-infrared. As an alternative, the Ca II infrared triplet found at 8498 Å, 8552 Å, and 8662 Å can be used as a measure of stellar activity (see Larson et al. 1993).

10.1.2 Hα

The Balmer Hα line at 6563 Å is another feature that can appear in emission in very active stars (Figure 10.3). In less active stars like the Sun, there may not be an obvious reversal in the core of Hα, but emission may partially fill the core of the line. Measurements of the equivalent width of the hydrogen line may reveal variations due to stellar activity.

The Hα line has emerged as a powerful diagnostic for excluding the RV variations due to stellar activity in M-dwarf stars. Kürster et al. (2003) first used this to show that RV variations due to Barnard's star correlated with changes in the equivalent width of Hα. A planet with an orbital period of 67 days was reported around the M-dwarf star Gl 581 (Mayor et al. 2009; Vogt et al. 2010), Gl 581d. Robertson et al. (2014) challenged this planet by showing evidence that the Hα index (a measure of the equivalent width) correlated with the 66 day RV variations due to the planet. Although Anglada-Escudé & Tuomi (2015) argued that this result stemmed from an

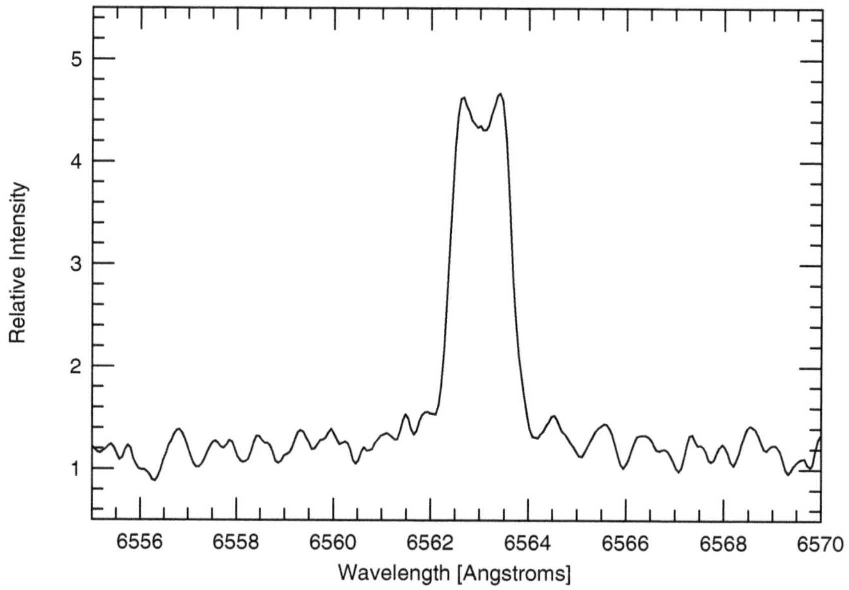

Figure 10.3. The Hα line in the M-dwarf AD Leo. The line appears in emission because of the high level of activity in the star.

improper use of periodograms on residual data, as we will show, a Fourier analysis revealed that there are indeed periodic variations in Hα with the purported orbital period of Gl 581g (Hatzes 2016).

The top panel of Figure 10.4 shows the DFT amplitude spectrum of the Hα index measurements of Robertson et al. (2014). This shows three peaks: a dominant one at 139 days (ν = 0.0072c day^{-1}), and two equal-amplitude peaks at 125.4 days (ν = 0.0080c day^{-1}) and 66 days (ν = 0.0152c day^{-1}). The latter is very close to the orbital period of Gl 581g. We are not interested so much in the nature of the two longer periods, so we can use prewhitening to remove the contribution of these and isolate the variations at the 66 day period (lower panel Figure 10.4). These show clear periodic variations out of phase with the orbital RV variations of the purported planet (Figure 10.5). The RV variations of Gl 581g are most likely an activity signal. This once again highlights how prewhitening can be a useful tool for investigating multiperiodic signals.

Interestingly, the out-of-phase variations of Hα with RV was also seen evident in the anticorrelation of these quantities in Barnard's star. Kürster et al. (2003) attributed these to changes in the convective blueshift from the star.

10.1.3 Na D

The sodium Na I D features at 5895.92 Å and 5889.5 Å are another set of resonance lines that have been shown to be useful indicators of stellar activity in cool stars (e.g., Gomes da Silva et al. 2011). Robertson et al. (2013) found that the sodium lines can be more sensitive to RV shifts due to activity. The sodium index I_D is the ratio of the flux in a window centered on the NA D lines to the flux in the nearby continuum (Díaz et al. 2007). Using 0.5 Å windows produces an I_D that correlates more closely with

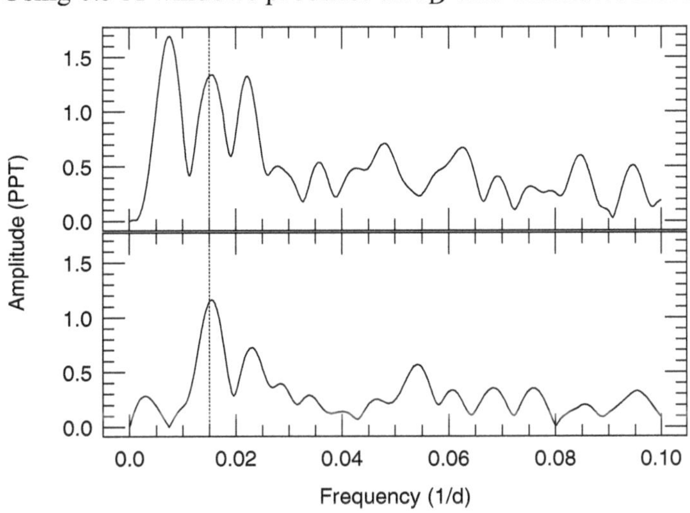

Figure 10.4. The prewhitening procedure performed on the Hα-index measurements of Gl 581 by Robertson et al. (2014). (Top) DFT of the original measurements. The amplitude is in parts per thousand (PPT). The dashed vertical line marks the orbital frequency of GL 581d. (Bottom) DFT of the residuals of the Hα-index measurements after removing the contribution of both peaks to either side of the orbital frequency.

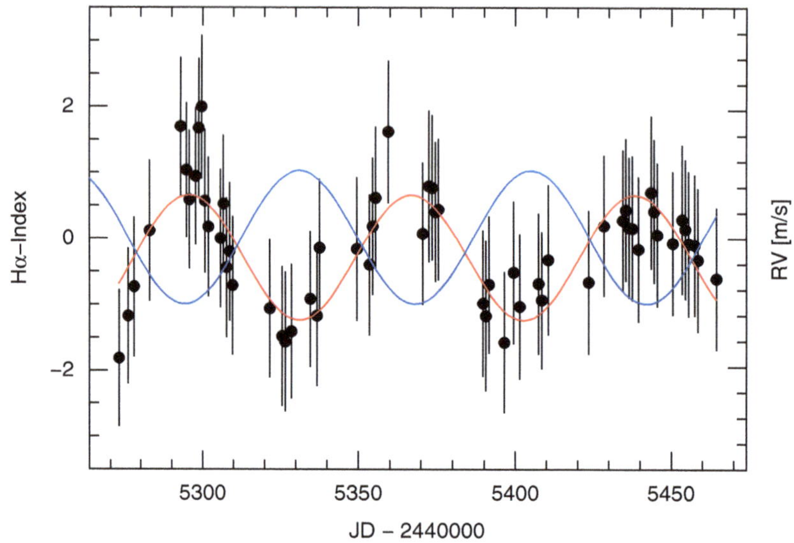

Figure 10.5. The time series of the Hα-index measurements (points) of Gl 581 from Robertson et al. (2014). The red line represents a sine fit to the data. The blue line shows the orbital fit to the RV variations from the purported planet GL 581d. The Hα variations have the same period as the RV data, but are out of phase. The RV variations are thus due to stellar activity.

the *S* index (Gomes da Silva et al. 2011). Robertson et al. (2015) found possible variations in NA D with the period of the planet GJ 176d, which suggested the RV signal was intrinsic to the star (Figure 10.6).

Compared to other activity indicators, the NA D lines are more problematic for a variety of reasons:

1. In general, Na D is not as strong an activity indicator as more traditional ones like Ca II H & K.
2. Na D can be contaminated by telluric water vapor lines.
3. Many astronomical sites can have contamination of NA D emission lines from street lighting.
4. For distant stars, contamination by interstellar NA D absorption will be a problem.

10.1.4 TiO Bands

Searching for variability due to starspots is facilitated by looking for spectral features that are only found, or at least strengthened, in the cool regions. The molecular bands of TiO at 7055 Å and 8860 Å are such features because their strength increases with decreasing temperature. The idea of using the TiO bands to measure starspot temperatures and filling factors was first proposed by Ramsey & Nations (1980) and Vogt (1979, 1981). The method was used by Neff et al. (1995) and O'Neal et al. (1996) to measure spot filling factors in RS CVn-type stars.

The top panels of Figure 10.7 show the molecular bands of TiO at 7055 Å and 8860 Å in a cool M giant star $T_{\text{eff}} = 3550$ K. The lower panel compares the 7050 Å

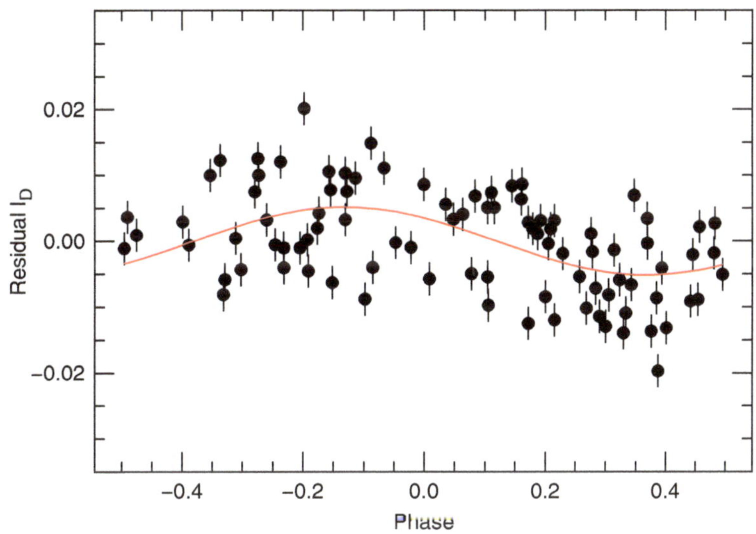

Figure 10.6. The sodium D index, I_D, for GJ 176 phased to the 73 day orbital period of the presumed planet GJ 176d (reproduced from Robertson et al. 2015. The American Astronomical Society. All rights reserved).

region of an inactive star to the active star EI Eri. The increased absorption of TiO due to spots in the active star is readily apparent. Measurements of the strength of spectral features around the TiO bands can provide additional diagnostics to help distinguish between RV variations due to planets from those due to activity.

10.1.5 Hydroxyl 1.563 μm Absorption

At infrared wavelengths, one can use the OH 1.563 μm feature to detect the presence of cool spots on stars. Figure 10.8 shows that the equivalent width for OH increases for decreasing effective temperature for giants and subdwarfs (O'Neal et al. 2001). Excess OH absorption due to starspots has been detected on several active stars of the RS CVn and BY Dra classes. The OH feature shows a similar behavior for dwarfs and giants, but it is much weaker in M dwarfs than in M giants for the same effective temperature. Therefore, it may be of limited use as a spot diagnostic for M-dwarf stars.

10.2 Line Depth Ratios

Line depth ratios (LDRs) using spectral line pairs can provide sensitive measurements of changes in the effective temperatures, $T_{\rm eff}$, of solar-type stars (Gray & Johanson 1991; Gray 1994). You take the LDR of two spectral lines, one that is sensitive to changes in $T_{\rm eff}$ and the other not. Pairs closely spaced in wavelength should be used in order to minimize systematic errors in the continuum fitting and changes in the instrumental profile with wavelength.

One such pair of spectral lines good for LDR measurements is V I 6251.83 Å and Fe I 6252.57 Å. The V I line shows large changes in depth with $T_{\rm eff}$ while Fe I does

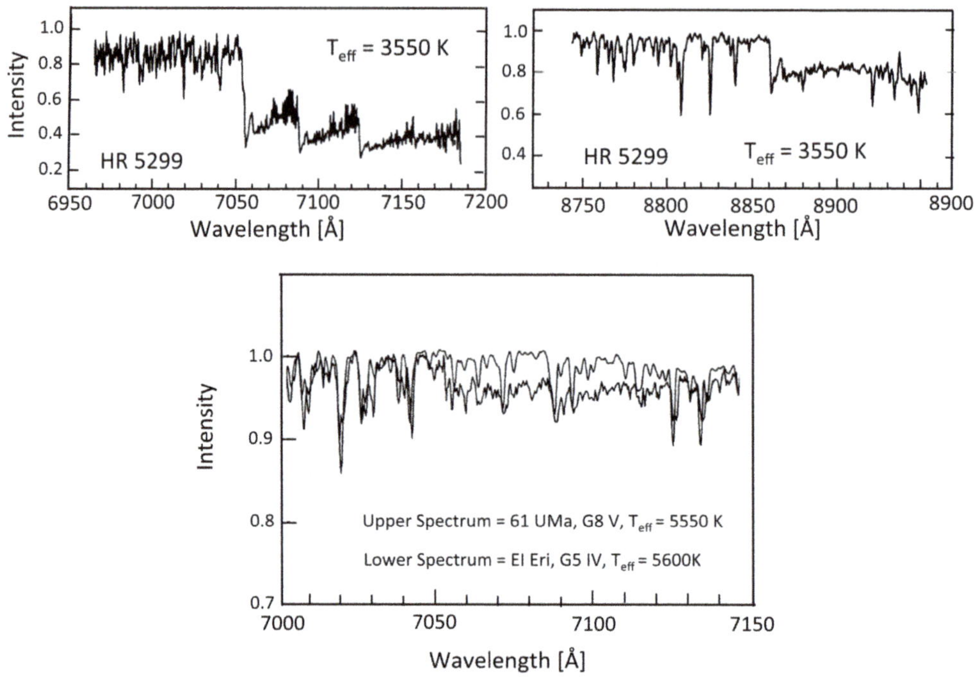

Figure 10.7. (Top panels) The TiO band at 7050 Å (left) and 8850 Å (right) in the M-giant star HD 5299 (T_{eff} = 3550 K). (Reproduced from Neff et al. 1995. The American Astronomical Society. All rights reserved.) (Lower panel) The spectrum at the TiO band at 7050 Å of the active G3V star EI Eri (lower spectrum) compare to an artificially broadened spectrum of the inactive star 61 UMa. The excess absorption in the EI Eri spectrum is due to spots. (Reproduced from O'Neal et al. 1996. The American Astronomical Society. All rights reserved)

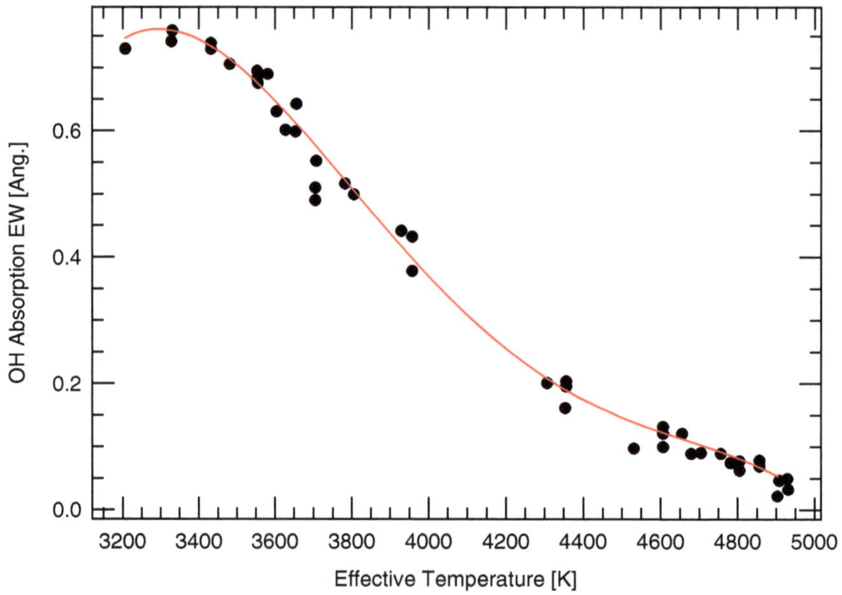

Figure 10.8. The equivalent width of the OH 1.5627 μm feature as a function of effective temperature T_{eff} for giant and subgiant stars (from O'Neal et al. 2001). The red curve is a polynomial fit to the data.

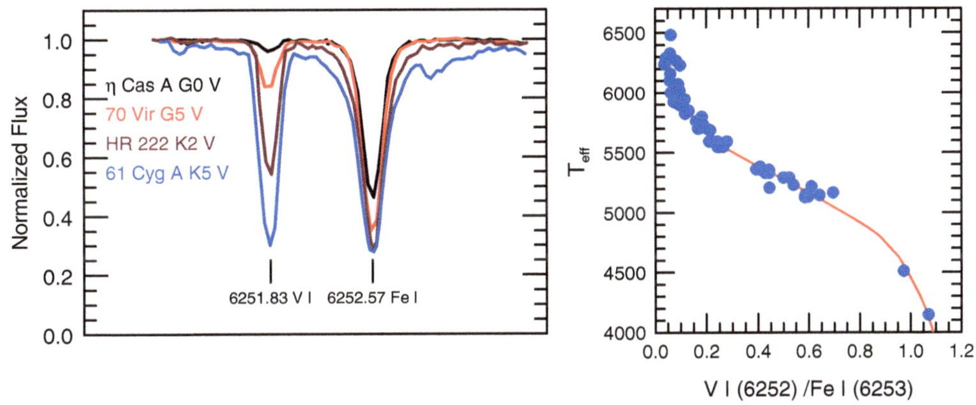

Figure 10.9. (Left) The change in line depth with spectral type (temperature) in the line pair VI 6251.83 Å and Fe I 6252.57 Å. The depth of the Fe I is relatively insensitive to changes in the temperature compared to V I. (Right) The calibration curve for $T_{\rm eff}$ versus the ratio V I/Fe I. (Reproduced from Gray & Johanson 1991. The American Astronomical Society. All rights reserved.)

not (left panel Figure 10.9). By using stars of known effective temperatures, one can calibrate the LDR as a function of $T_{\rm eff}$.

For the V I/Fe I ratio, Gray & Johanson (1991) determined a calibration of

$$T_{\rm eff} = 6660.5 - 9941.7R + 35297.7R^2 - 67336.1R^3 + 61565R^4 - 21767R^5, \quad (10.3)$$

where R is the depth ratio.

Figure 10.10 shows the LDR measurements of V I/Fe I as a function of orbital phase for 51 Peg. These show an rms of 0.00149, which translates into a $\Delta T_{\rm eff} = 1.7$ K (Hatzes et al. 1998a). Therefore, these LDRs support the planet hypothesis for the RV variations of 51 Peg.

Table 10.1 lists line pairs that are suitable for LDR measurements. Also listed are the excitation potential (E.P.), ionization potential (I.P.), and whether the lines are sensitive to temperature changes.

10.3 Spectral Line Shapes

Arguably, the most stringent tests for confirming exoplanets found via the Doppler method is the examination of spectral line shapes. The principle is simple: the Doppler reflex motion of the star due to a planetary companion produces an overall shift of the spectral lines. Stellar surface inhomogeneities in the form of spots, plage, faculae, etc. results in the distortion of the stellar lines, which changes as the star rotates (Figure 9.2). This distortion shifts the measured centroid of the line, resulting in a Doppler shift that can mimic a planetary companion. Stellar oscillations, particularly nonradial ones, can also alter the shape of the spectral lines (e.g., Hatzes 1996). In short, any changes in the spectral line shapes that correlate with the measured RV of the star excludes the planetary hypothesis as a cause of the variations.

There are two methods commonly used by the exoplanet community to measure the changes in the spectral line shapes. The first is the measurement of the so-called

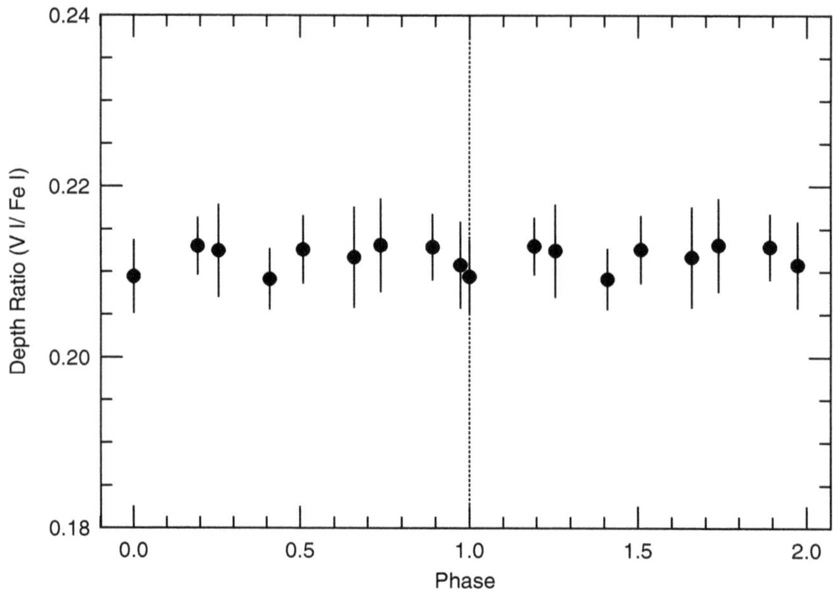

Figure 10.10. The line depth ratio of V I/Fe I for 51 Peg as a function of the orbital phase of the planet (Hatzes et al. 1998a). The variations indicate a temperature change of no more than 1.7 K. Measurements are repeated after the vertical dashed line.

Table 10.1. Absorption Lines for Line Depth Ratios

Species	Wavelength (Å)	E.P. (eV)	I.P. (eV)	T-sensitive
Ni I	6223.99	4.10	7.63	N
V I	6224.51	0.29	6.75	Y
Fe I	6226.74	3.88	7.90	N
Fe I	6229.23	2.84	7.90	N
V I	6233.20	0.28	6.75	Y
V I	6242.84	0.26	6.75	Y
Si I	6243.83	5.61	8.15	N
Fe II	6247.56	3.89	16.18	N
V I	6251.83	0.29	6.75	Y
Fe I	6252.57	2.40	7.90	N
Fe I	6253.83	4.73	7.90	N

spectral line bisector and the second is the width of the spectral lines as measured by their FWHM.

10.3.1 Line Bisectors

The spectral line bisector has become a standard tool for the measurement of asymmetries in spectral line profiles. The bisector is the locus of points marking the

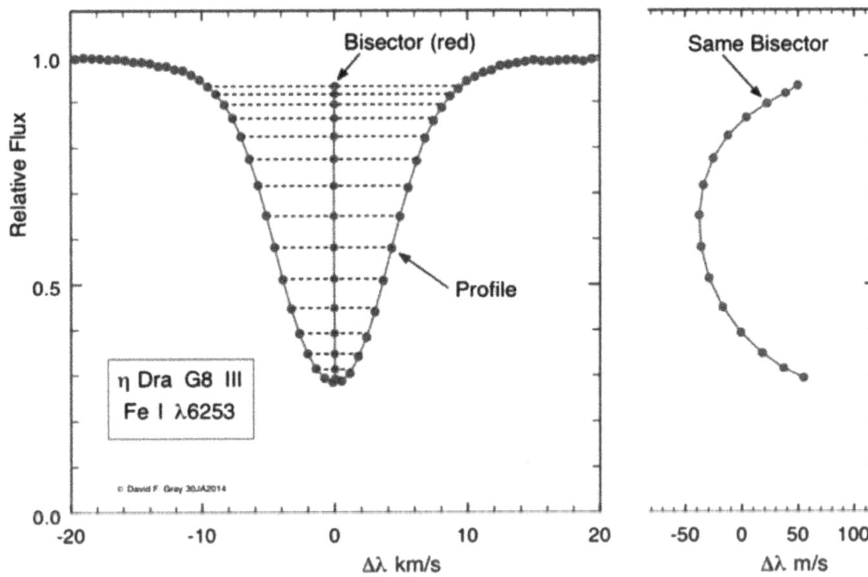

Figure 10.11. (Left) The bisector is defined as the locus of midpoints between equal flux levels of the spectral line profile. The bisector is shown on the same velocity (wavelength) scale as the profile. (Right) The bisector shown on a velocity scale that has been expanded 100 times. Here one can see the characteristic "C-shape" found in late-type stars. (Figure courtesy of D. Gray from http://astro.uwo.ca/~dfgray/Home.html.)

midpoint of a horizontal line segment spanning the line profile (left panel Figure 10.11). For a symmetric line profile, the bisector is a straight vertical line. For late-type stars, it takes on a characteristic "C-shape" (right panel of Figure 10.11) due to convection. The line bisector is a powerful tool for studying the convection pattern on stars as a function of stellar types (e.g., Dravins 1987a, 1987b; Gray & Nagel 1989).

The errors in bisector measurements are easily calculated from the shape of the line profile. Let us define the slope of the line profile as $dF/d\lambda$, in flux units, F, in the fraction of the continuum, and the photometric error (given by the signal-to-noise ratio) as δF. The bisector error, $\delta\lambda_b$, from pure photon statistics is (Gray 1983)

$$\delta\lambda_b = \frac{1}{\sqrt{2}}\delta F\left(\frac{dF}{\delta\lambda}\right)^{-1}. \qquad (10.4)$$

We see that $\delta\lambda_b$ becomes very large in the wings and in the core, where dF/λ are small. Thus, in calculating bisectors, it is best to avoid both of these regions of the line. Typically, the bisector is computed up to depths of about 0.85–0.95 of the continuum.

The typical bisector quantities that are measured are the bisector span and curvature. The span is the velocity difference of two points (near the ends) on the bisector, which is a measure of the slope or first derivative. The curvature is the difference in span between the upper and lower halves of the bisector and is a measure of the second derivative. It is best to measure both the velocity span and curvature because some configurations of the surface structure can produce stronger variations in one as opposed to the other quantity.

Bisector measurements require data with high signal-to-noise ratios, S/N > 200. Even in this case, the bisector for a single line will still be quite noisy; therefore, it is best to average as many strong lines, with roughly the same shape. To increase the S/N of the bisector measurement, it is common practice to compute the bisector span and curvature measurements using the cross-correlation function (CCF). This exploits the fact that the CCF essentially represents the mean line profile shape. This works, because in the case of planet confirmation we are not interested in the absolute shape of the bisector, but rather in the changes with respect to a mean shape. If you are interested in doing a detailed analysis of the bisector shape, say for investigations of the convection pattern on the star, it is not recommended to use the bisector of the CCF.

Because you are investigating subtle changes in the shapes of the line bisector, it is best to have data taken at high spectral resolution. For slowly rotating stars ($v \sin i < 3$ km s^{-1}), data should have $R > 100{,}000$. Lower resolution data translates into lower amplitude bisector variations. Figure 10.12 shows the amplitude of the bisector span variations caused by the same spot distribution on a star and measured with simulated data having resolving powers of $R = 50{,}000$ and 100,000. The amplitude of the bisector span variations is 60% higher in this case for the higher resolution measurements.

Rapid rotation in a star also gives you more "leverage" in the bisector measurements so that high-quality measurements can be made at a much lower resolution. This is demonstrated in Figure 10.13, which shows the results from bisector span

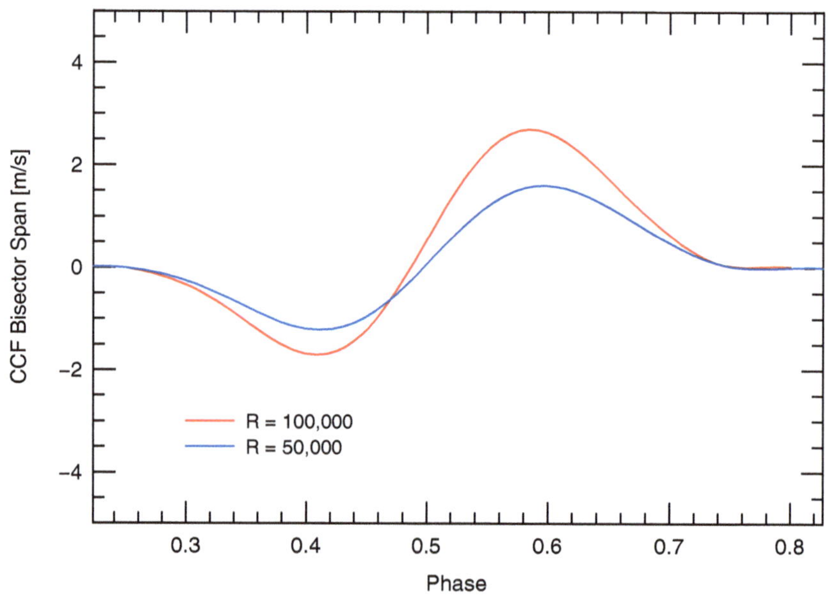

Figure 10.12. The simulated bisector span variations caused by a spotted star using spectral data with resolving power $R = 50{,}000$ (blue) and $R = 100{,}000$ (red).

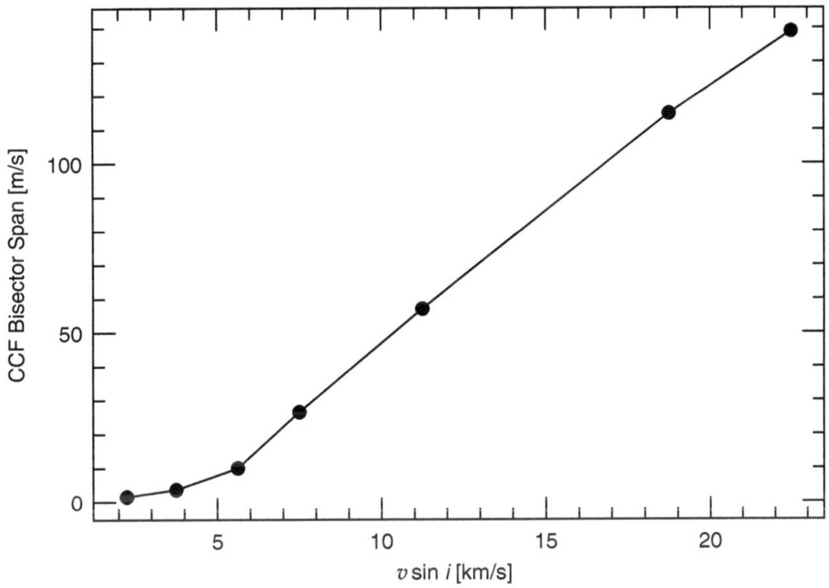

Figure 10.13. The bisector span of the CCF as a function of the $v \sin i$ from a single spot using simulated data with resolving power R = 50,000.

variations for the same spot distribution on a star as a function of $v \sin i$ for spectral data with R = 50,000. These increase roughly linearly with projected rotational velocity.

As an exoplanet confirmation tool, bisectors were first used for confirming the planet 51 Peg b (Hatzes et al. 1997, 1998a, 1998b). Figure 10.14 shows the bisector span and curvature measurements for 51 Peg (Hatzes et al. 1998a) from spectral data having a resolving power of R = 200,000 and S/N = 200. A total of eight spectral lines were used to compute the bisector quantities. The span and curvature measurements have an rms of 1.3 m s^{-1} and 4.4 m s^{-1}, respectively. These demonstrate the power of having high spectral resolution and high S/N ratios when computing bisector quantities.

The first use of line bisector variations to refute the planet hypothesis was done for HD 166435. Radial velocity measurements of this star taken with the ELODIE fiber-fed spectrograph showed RV variations with a period of 3.8 days (Queloz et al. 2001), suggesting the presence of a close-in giant planet. Photometric observations made to search for transits actually revealed that there were variations with the same period as the planet. Queloz et al. (2001) showed that changes in the bisector correlated with the RV variations (Figure 10.15). The RV variations were not due to a planet, but rather to starspots.

The Limitations of Bisector Measurements

Although high-quality bisector measurements can be a powerful tool for the confirmation of exoplanets discovered with the Doppler method, it does have its limitations. It has become a popular tool for interpreting the nature of RV

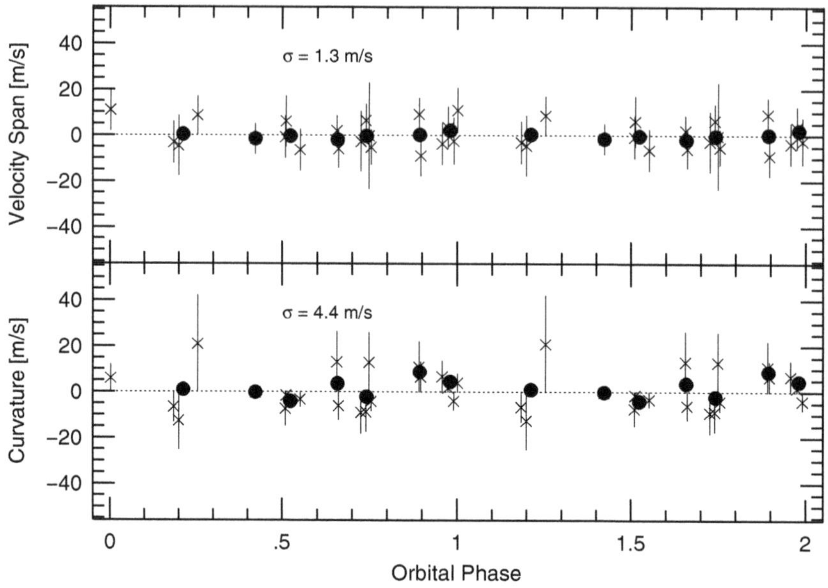

Figure 10.14. (Top) The mean bisector velocity span measurements as a function of phase (crosses) for 51 Peg using eight spectral lines. (Bottom) The mean bisector curvature measurements (crosses) averaged over all spectral lines as a function of phase. The solid points in both panels represent phase-binned values. (From Hatzes et al. 1998a.)

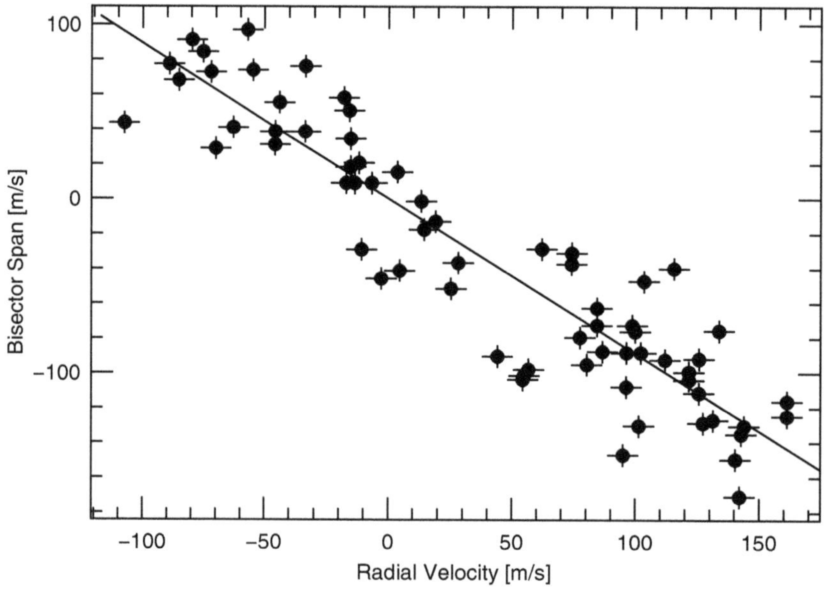

Figure 10.15. The bisector span versus the RV for HD 166435 (Queloz et al. 2001). This anticorrelation is typical for cool starspots.

variations, but researchers often put too much faith in bisector measurements. The danger is that apparent bisector variations might refute a planet that is real, or alternatively, confirm a planet that is not actually present. There are examples in the literature of both cases.

The star 51 Peg offers an example of apparent bisector variations used to refute a planet that was in fact real. Gray (1997) reported bisector variations in 51 Peg with the same period as the planet's orbit (4.23 days). The false alarm probability (FAP) was about 0.2%, a value that would nominally qualify as a real signal. The bisector data were of high quality, coming from spectral data with $R = 100{,}000$ and S/N $\approx$ 400. Furthermore, the shape of the curvature measurements as a function of phase could be well fit using a model of stellar nonradial pulsations (Gray & Hatzes 1997). The amplitude of the phased bisector curvature measurements was 43 m s^{-1}, a value clearly refuted later by measurements made at much higher spectral resolutions (Hatzes et al. 1998a) as seen in Figure 10.14. Additional measurements taken with the same instrumental setup as the data used by Gray & Hatzes (1997) also did not show evidence for bisector variations (Gray 1998).

The bisector case of 51 Peg offers the reader two important lessons. First, even with high-quality bisector measurements, you may see variations that are not real. Second, signals with an FAP as low as 0.2% can still turn out to be spurious.

The planet around TW Hya offers us a case of how a lack of bisector variations cannot be used to confirm an exoplanet discovery. Setiawan et al. (2008) reported RV variations with a period of 3.56 days and K amplitude of $\approx$200 m s^{-1} in this star. This was interpreted as the signal of a giant planet with a minimum mass of 9.8 M_{Jup}. This discovery, which was of particular interest since TW Hya is a T Tauri star and was thus quite young, with an age of $\approx$10 Myr. It would be the first hot Jupiter found around a star with its protoplanetary disk still present.

T Tauri stars are known to be active and have spots, so there is naturally a concern that the RV signal stems from a surface spot. The planet hypothesis was favored, based on (1) a relatively long-lived RV signal and (2) a lack of correlation between the bisector span (right panel of Figure 10.16) and curvature measurements with the RV. However, the longevity of the spot was based on observations separated only by several months, and V410 Tau has already shown us that spots on young stars can be stable for many years. But what about the bisector results? Let's take a more careful look at these.

Before we investigate this we need to know a bit about T Tauri stars. These objects are young, solar-like stars that typically have fast rotational rates and high levels of stellar activity (an early warning sign). Doppler images of their slightly more evolved counterparts, weak T Tauri stars, show large, decentered polar spots like V410 Tau (see Figure 9.6). So, for the case of TW Hya, it is reasonable to expect that it has a long-lived decentered polar spot. Based on the Doppler images of V410 Tau we know that such a spot can indeed last for a long time.

It is also clear that TW Hya is a star that is viewed nearly pole on, with an inclination angle between the rotation axis and the line of sight of only 7° (Setiawan et al. 2008). The key question is: what do the RV and bisector variations look like from such a spot configuration?

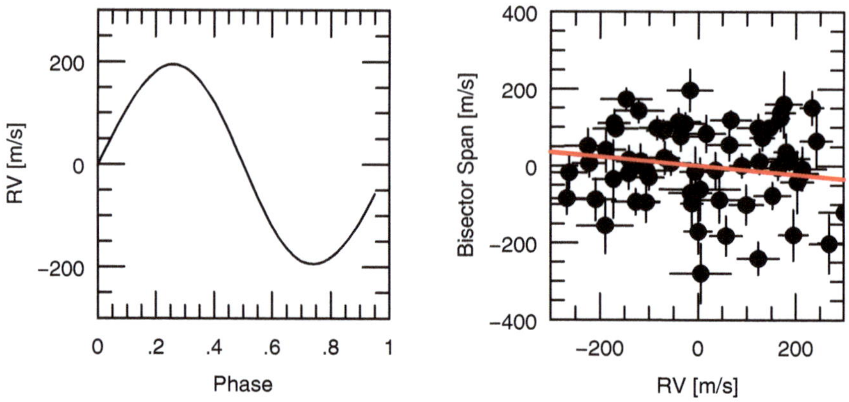

Figure 10.16. Simulations of the RV and bisector span variations from a polar spot on TW Hya viewed nearly pole on. (Left) The predicted RV variations. (Right) The predicted RV variations (red line) compared to the actual bisector span measurements from Setiawan et al. (2008).

To answer this, we assumed that the period of the RV variations represents the rotational period of the star. We placed a large spot with a coverage of about 10% of the visible sphere placed at a stellar latitude of 75°. The stellar inclination was taken to be 7°. We then used SOAP 2.0 (Dumusque et al. 2014) to produce data at a spectral resolving power of R = 50,000, the same as for the FEROS spectrograph used to measure the RVs for the star.

The left panel of Figure 10.16 shows the expected RV variations for this spot distribution as a function of stellar rotation phase. It shows a nice sinusoidal variation with an amplitude of $K \approx 200$ m s^{-1}, the same as for TW Hya. The right panel of the figure shows the bisector velocity span measurements of Setiawan et al. (2008). The red line shows the predicted velocity span measurements for our spot distribution of TW Hya. These have an amplitude of $\approx$20 m s^{-1}. The typical bisector error of $\approx$40 m s^{-1} for the actual measurements is twice as large. The scatter of the measured velocity span is even larger at about 100 m s^{-1}. Clearly, if a spot were responsible for the RV variations of TW Hya, it would produce undetectable bisector span variations given the quality of the data. These bisector measurements do not have the precision needed to confirm the planetary nature of the signal.

TW Hya offers the reader another important lesson regarding bisector measurements for planet confirmation: a lack of bisector variatios is a necessary but not a sufficient condition for confirming the planetary nature of an RV signal. If you find bisector variations with the planet RV signal, then it is most definitely not a planet. However, as we saw for 51 Peg, these measurements must be of high quality as noisy bisector measurements can sometimes give a false signal. If you find no correlation between the bisector and RV then you have proved nothing—the signal can still be intrinsic to the star. In short, other criteria in addition to a lack of bisector variations should be used to confirm the planetary nature of RV detections. If you find no variations, it is always good practice to compare this to what you should have seen given realistic assumptions.

Although bisector measurements may be of limited use in confirming planet detections, they are quite useful in excluding a blend scenario for transit detections. One of the most common false positives in the detection of transiting planets is a blend due to a spectroscopic binary. If the stellar companion is weak or blended with the lines of the primary, such tools like the CCF method may not detect the presence of the stellar companion. Blends due to a stellar companion should produce large line shape changes that are easily measured with line bisectors.

10.3.2 Line Widths

Rotational modulation by stellar surface structure is often accompanied by variations in the FWHM of spectral lines. Consider a cool spot moving across the line of sight to the star (Figure 9.2). When the spot is on the approaching limb of the star, it produces a distortion in the blue wing of the line profile. This shifts the centroid of the line toward the red, resulting in a maximum positive Doppler shift. The line shape, however, is narrow, resulting in a minimum in the FWHM. When the spot is at disk center, the distortion appears in the core of the line, so it produces no shift of the line centroid (zero Doppler shift). The distortion causes the line profile to appear shallower and fatter, resulting in a maximum in the measured FWHM. Because the FWHM is correlated with the rotational modulation due to spots it can serve as a useful proxy of photometric variations.

Finally, when the spot appears on the receding limb, the distortion shifts the line centroid to the blue and one measures a maximum negative Doppler shift for the star. The line profile, however, is again narrow, i.e., a minimum in the FWHM. Because the maximum in the FWHM is when a spot is at disk center, yet the RV has an extrema when the spot is near the stellar limb, there is approximately a 0.25 rotational phase shift between the two variations.

Figure 10.17 compares the FWHM and RV variations in CoRoT-7. The top panel shows the residual RV measurements phased to the 22 day rotation period of the star. The RV contributions due to planets have been removed so as to isolate the rotational modulation signal. The lower panel shows the FWHM measurements phased to the same period. One can see a ≈0.2 phase shift between the two variations, indicating that these are caused by cool spots on the surface.

10.4 Chromatic RV Variations

The contrast between cool spots and the hot photosphere decreases as one goes to longer wavelengths, and this ratio can be estimated using the blackbody law:

$$F_{\rm p}/F_{\rm s} = \frac{e^{hc/k\lambda T_{\rm s}} - 1}{e^{hc/k\lambda T_{\rm p}} - 1}, \tag{10.5}$$

where λ is the wavelength of light, and h and c the Planck and speed of light constants, respectively. $T_{\rm p}$ is the photospheric temperature, which is about 5800 K for a solar-like star, and $T_{\rm s}$ is the sunspot temperature, about 3000 K for sunspots. So, at 5000 Å, the contrast ratio is ≈103, whereas at 1 μm it is only ≈7. The amplitude of RV variations due to a cool spot should be less at infrared wavelengths

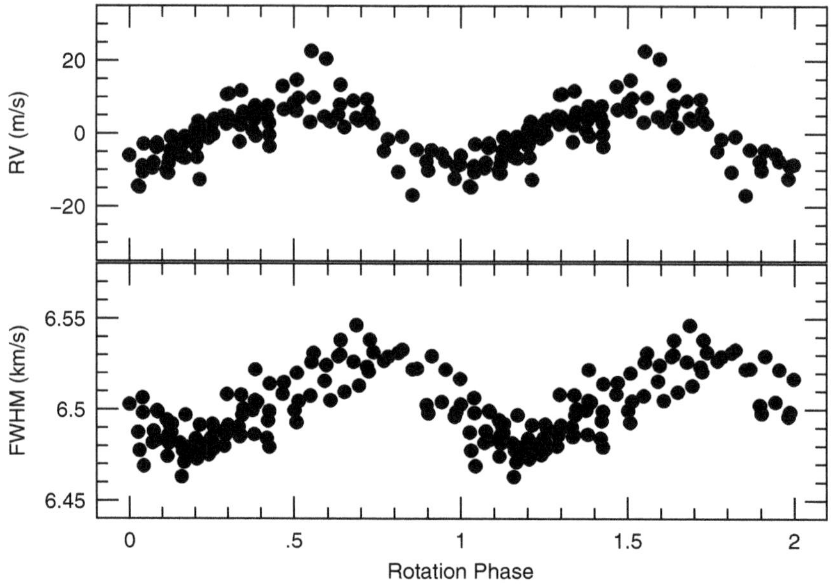

Figure 10.17. (Top) The residual RV variations (the signal of all planets removed) of CoRoT-7 phased to the rotational period of approximately 22 days. The RV contribution of the planets have been removed. (Bottom) The variations of the FWHM of the cross-correlation function phased to the rotational period.

compared to optical regions. On the other hand, the velocity reflex of a star due to a companion should be the same regardless of the wavelength you use for the Doppler measurement. In short, if you measure the same RV amplitude in the optical and infrared, then it is most likely due to a companion.

The "planet" around TW Hya provides a textbook case of using RV measurements in the infrared to determine the true nature of the RV variations for this star. As we discussed earlier, in the optical regime, the star shows a K amplitude of 200 m s^{-1} (top panel of Figure 10.18), which was originally attributed to a close-in Jupiter (Setiawan et al. 2008). We just showed that the lack of bisector variations in this star could not confirm the planet with a strong degree of confidence. RV measurements in the IR also do not support the planet hypothesis.

Huélamo et al. (2008) performed RV measurements at 1.588 μm using the CRyogenic high-resolution infraRed Echelle Spectrograph (CRIRES; Käufl et al. 2008). These revealed a K amplitude that was about one-third the one found at optical wavelengths (lower panel of Figure 10.18). They could model the RV variations using a cold spot covering 7% of the stellar surface and located at a stellar latitude of 54°, parameters comparable to our earlier simulation. The observed RV variations of TW Hya are most likely due to a spot on the stellar surface, a spot configuration similar to the one we used to simulate the bisector variations.

Of course, these IR measurements cannot completely rule out that part of the RV signal is in fact due to a planetary companion. However, it would be extremely difficult to disentangle the contribution of the two. When claiming an important

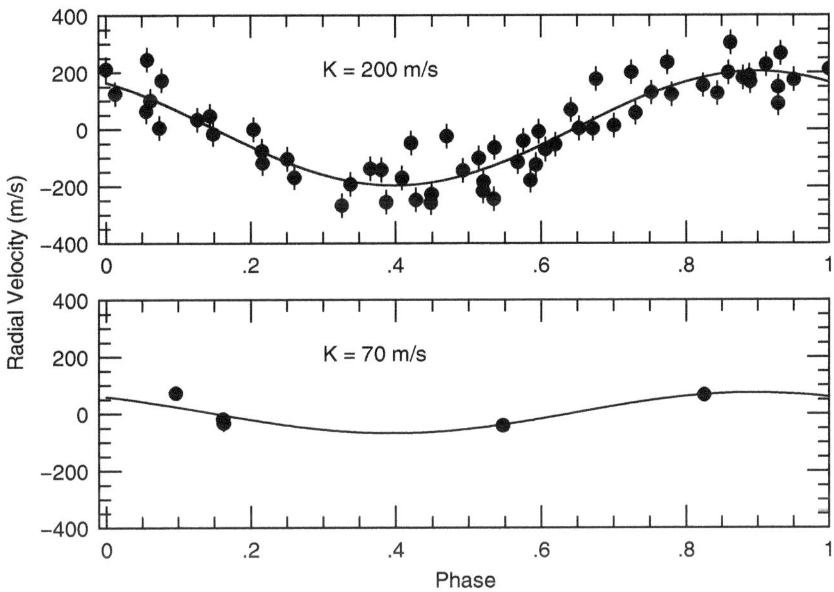

Figure 10.18. RV measurements for TW Hya at optical (top) and near-infrared (bottom) wavelengths. Both are phased to the orbital period of the hypothetical planet. The amplitude in the infrared (70 m^{-1}) is almost a factor of 3 smaller than the amplitude (200 m^{-1}) at visual wavelengths.

discovery that can inspire theoretical work, one must be absolutely certain of the result: "extraordinary claims require extraordinary proof".

If one does not have access to a high-resolution IR instrument, one can still use the broad wavelength coverage offered by modern echelle spectrographs. The central wavelength of an echelle spectral order increases with increasing order number. By measuring the RV in individual spectral orders, one can get information on the wavelength dependence of the RV.

Zechmeister et al. (2018) defined a so-called "chromatic RV index," β, which is a measure of the RV as a function of the wavelength of each order. If λ_v is the weighted RV using all spectral orders and λ_m is a representative wavelength of spectral order m, the velocity of each order can be defined as

$$v(m) = \beta \ln \frac{\lambda_m}{\lambda_v}. \quad (10.6)$$

Figure 10.19 shows the chromatic index for two observations of the active M-dwarf star YZ CMi using 42 spectral orders of the CARMENES spectrograph. In one observation, the slope, β, is positive, whereas for another observation, it is negative. The slope for an individual observation is not important. It can arise from cool spots, hot spots, or some unknown active surface structure. What is important is the changes in the slope, which can be an indication that activity is responsible for the RV variations.

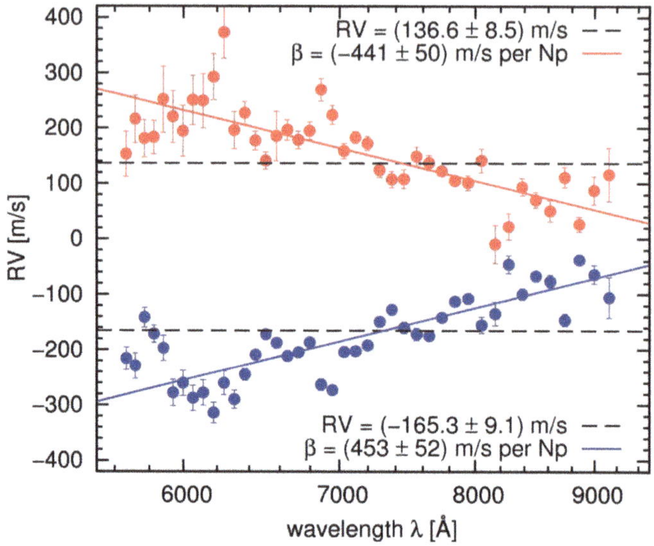

Figure 10.19. The RVs measured in 42 orders of the visual channel of CARMENES for the M-dwarf YZ ZMi for two observations (red and blue). The weighted average is represented by the horizontal dashed line. The chromatic index for each observation is shown by the solid red and blue lines (from Zechmeister et al. 2018, reproduced with permission © ESO).

Figure 10.20 summarizes how many of the various activity indicators just presented can be used to confirm the nature of the RV variations for an M-dwarf star in the CARMENES sample of M dwarfs (Reiners et al. 2018). Also shown are the RV, Hα, Na D, and Ca I infrared triplet variations. All measurements were from spectra taken with the optical arm of the CARMENES spectrograph. This nicely shows how the period (frequency) of the RV variations can be seen in all activity indicators. Most cases will not be as nice as you will see variations in some indicators, but not the others. This is due to the nature of the activity features causing the variations as well as the quality of the data.

10.5 Use of Individual Lines

In order to achieve the highest RV precision, one normally calculates the Doppler shifts using as broad a wavelength as possible, incorporating hundreds of stellar lines. It is these "integral" RV measurements that enable us to achieve an RV precision of $\approx$1 m s^{-1}. Measuring the Doppler shift of a single spectral line will give a measurement error of 10–50 m s^{-1}, which is not ideal for exoplanet detection. However, these kinds of Doppler measurements can be a powerful tool to help us discern the true nature of the RV variations. Also, by using fewer but carefully selected lines that have good "RV behavior," one can actually arive at a better precision due to the reduced activity noise.

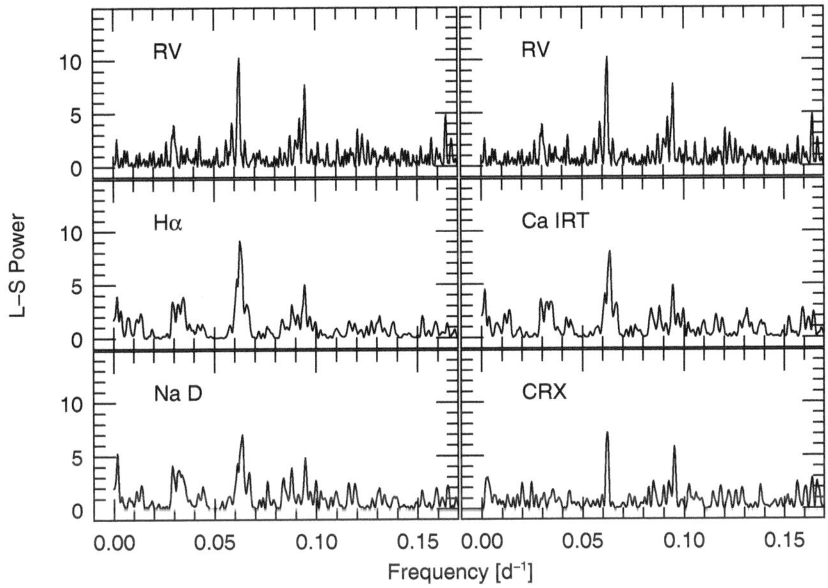

Figure 10.20. Lomb–Scargle periodograms of RV and activity indicators for the active M dwarf. (Top panels) Periodograms of the RV measurements (repeated for easier comparison to activity indicators). (Middle panels) Periodograms of the Hα (left) and Ca II infrared triplet (right). (Bottom panels) Periodograms of the Na D (left) and chromatic RV (CRX) measurements (right).

10.5.1 Radial Velocities

The conditions forming individual spectral lines can be different, with each line having a different sensitivity to temperature and magnetic field strength. Thus, a line formed in a spot or plage on an active star will have a different shape. Figure 10.21 shows the line bisector of the Fe 5250.6 Å line in the Sun observed in and out of magnetic regions (Livingston 1983). The bisector in the magnetic region has a different span and stronger curvature, plus it is Doppler-shifted by $\approx$100 m s^{-1}. If there is a strong contribution from the spotted regions, or the mean magnetic field on star changes with the activity cycle, one would see periodic changes caused by the changing contribution of magnetic regions, which produce a different line shape. This would be more apparent in lines that are more sensitive to magnetic fields.

Spectral lines are also formed at different depths in the stellar atmosphere, with weak lines formed, on average, deeper in the atmosphere compared to stronger lines. At these different depths, the velocity and temperature distribution in the atmosphere may be different. It is reasonable to expect that for RV variations from stellar variability, individual lines will show different Doppler shifts. The reflex motion of the star, however, will produce the same Doppler shift for all spectral lines. This can be exploited to confirm planet discoveries. Thus a "line-by-line" Doppler measurement may provide insights into the nature of the RV variations.

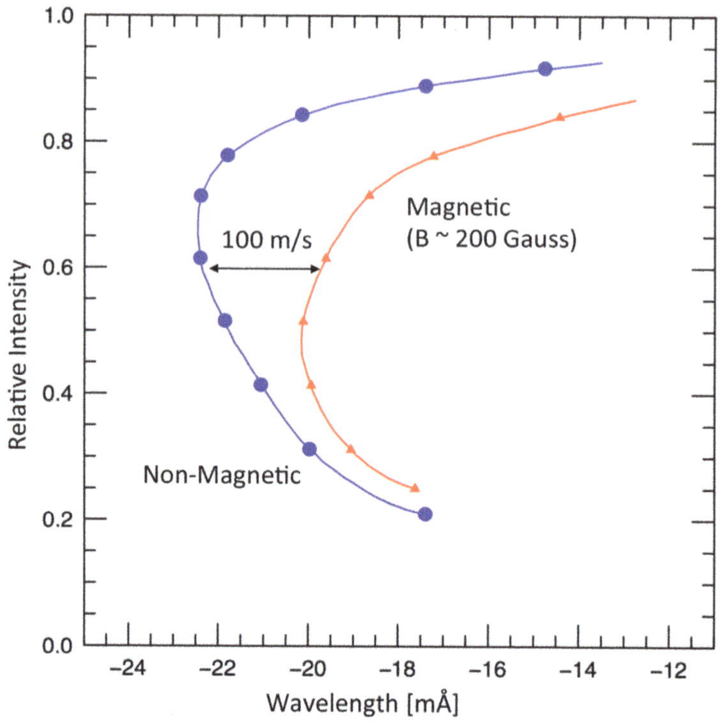

Figure 10.21. Line bisectors of Fe 5250.6 Å in and out of magnetic regions. (Reproduced from Livingston 1983.)

The rapidly oscillating Ap stars provide an extreme and obvious case as to how line-by-line Doppler measurements produce different results. As noted before, these chemically peculiar stars oscillate with periods of 6–15 minutes. An amazing property of these stars is that individual lines can show drastically different RV amplitudes (Kanaan & Hatzes 1998; Mkrtichian et al. 2003; Hatzes & Mkrtichian 2005; Mkrtichian et al. 2008). HD 101065 (Przybykski's Star[1]) exhibits stellar oscillations with a dominant period of 12.1 minutes. Figure 10.22 shows the RV variations of two spectral lines, Pr II 5999 Å and Eu II 6445 Å, both phased to a pulsation period of 12.1 min. The Pr II line has an amplitude of $K = 623$ m s^{-1}, more than five times the amplitude $K = 121$ m s^{-1} of the Eu II line. The reason is that the chemical distribution in this star is stratified, and the two lines are formed in different heights of the atmosphere where the pulsational characteristics can be different. There can also be large phase variations. In the case of 33 Lib, lines of the same element, Nd, but with different ionization stages, pulsate 180 degrees out of phase with each other (Mkrtichian et al. 2003).

In active, solar-type stars, spectral lines that are temperature senstive can have different Doppler shifts due to the inhomogeneous temperature across the star

[1] Przybykski's is an F8-type star that is the most chemically peculiar star. It has a rich spectrum of absorption lines with over half of these yet to be identified.

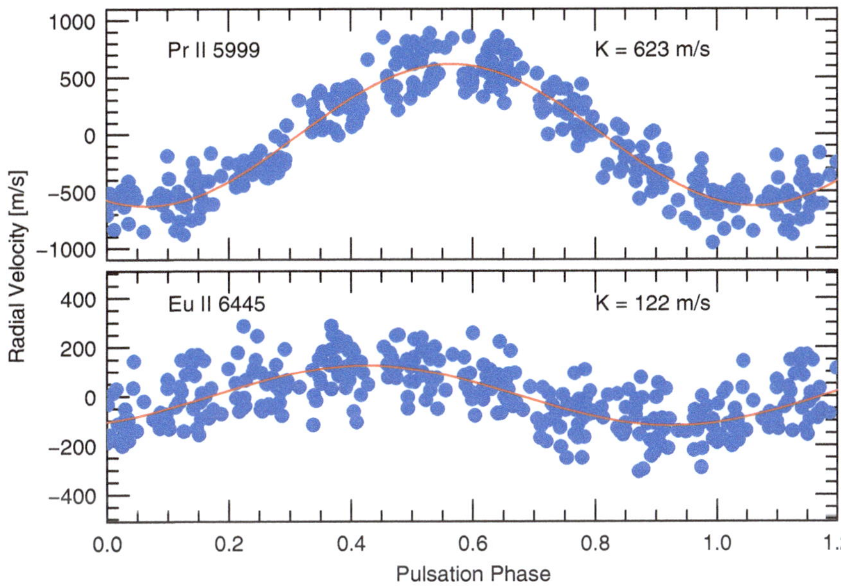

Figure 10.22. RV variations for individual lines in the roAp HD 101065 phased to the pulsational period of 12.14 minutes. The Pr II 5999 Å line (top) has an amplitude more than five times that of the Eu II 6445 Å line.

(spots). Figure 10.23 shows the spectral line profiles of Fe I 6430 Å and Ca I 6439 Å observed in the rapidly rotating active star HR 1099, which is heavily spotted. The Fe I 6430 feature has a strength that is the roughly the same in and out of the spotted regions The distortion that is seen in the line profile results mostly in differences in the flux between the spot and photospheric regions (see Vogt & Penrod 1983 and Chapter 13). The line strength of Ca I 6439 Å, on the other hand, increases with decreasing temperature. This partially compensates for the flux effect which produces a distortion that is less pronounced than for Fe I 6430 Å. If one were to measure Doppler shifts of the spectral lines, Fe I would produce a larger amplitude by virtue of the stronger distortions. Calculating the RV for temperature-sensitive versus temperature-insensitive lines would show amplitude differences.

In order to increase the RV precision, it is best to look at an ensemble of spectral lines that are expected to show the same activity signal. For example, one can calculate the RV using only those lines with high-temperature sensitivity and compare these to the results of temperature-insensitive lines. If one is using the CCF method, then one simply creates a mask that isolates temperature-sensitive lines. The trick is to find which lines are the most sensitive to activity.

Dumusque (2018) explored this in a study of α Cen. He first calculated the RVs using all of the spectral lines and then those of the individual lines. By looking at the correlation between the RV of individual lines to those from the "integral" RV measurements, he was able to separate activity-sensitive spectral lines from the nonsensitive ones.

Figure 10.24 shows the RVs of all spectral lines compared to the activity-sensitive and nonsensitive lines. The RV amplitude of the variations of the activity-sensitive

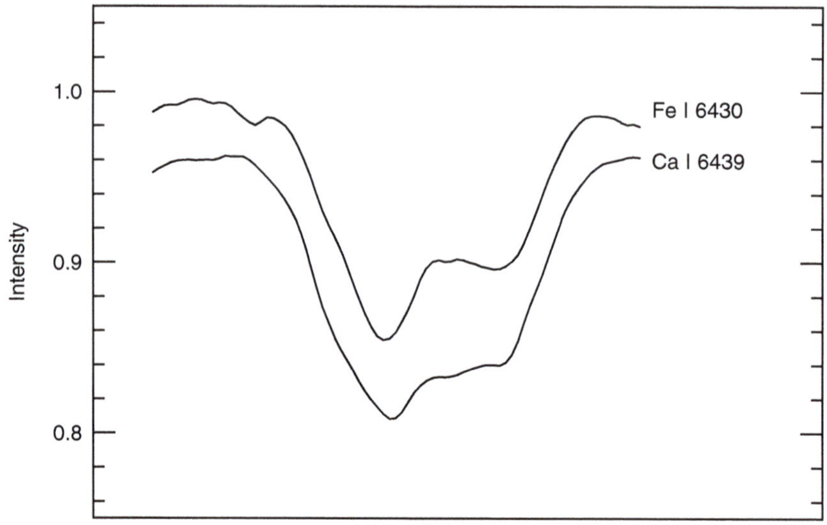

Figure 10.23. The line profile of Fe I 6430 Å (top) and Ca I 6439 Å (bottom) of the rapidly rotating active star HR 1099. The distortion in the profile is due to a large, cool starspot. Ca I 6439 Å is a temperature-sensitive line whose strength increases with decreasing temperature. This makes the distortion less pronounced than in the Fe I 6430 Å line.

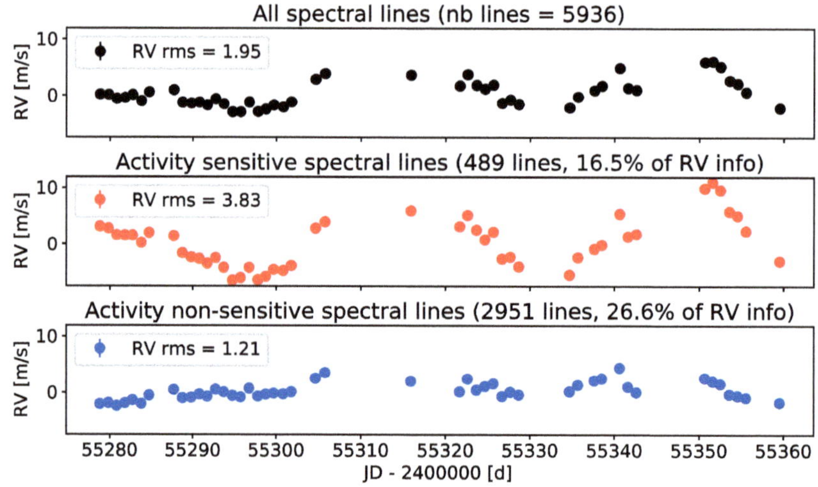

Figure 10.24. RVs for α Cen B showing variability due to activity. (Top) RVs calculated using all spectral lines. (Middle) The RVs calculated using only those lines that are sensitive to activity. (Bottom) RVs calculated using lines that are not senstive to activity. These show a much lower amplitude. (From Dumusque 2018, reproduced with permission © ESO.)

lines is slightly higher than for the integral values, whereas the nonsensitive lines clearly show a much lower amplitude. By comparing the RVs of activity-sensitive and nonsensitive lines, one can establish if the variations are due to a companion. Such investigations may also hold the key to overcoming the obstacle stellar activity presents when trying to extract the RV signals of small planets.

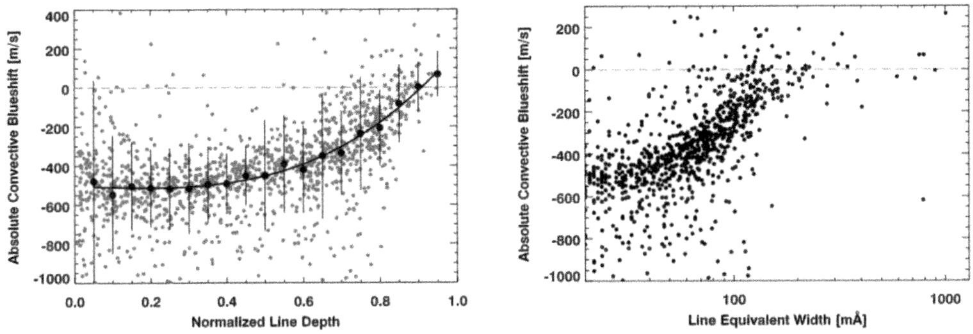

Figure 10.25. (Right) Absolute convective blueshift of Fe I lines. (Left) Blueshift measurements as a function of line depth. The median blueshift (large points) is shown in bin sizes of 0.05. (Right) Blueshift measurements as a function of the equivalent width of the line from Allende Prieto & Garcia Lopez (1998), reproduced with permission © ESO.

10.5.2 Convective Blueshifts versus Line Strength

The convective blueshift (CBS) also depends on the line strength (Figure 10.25). Thus, different spectral lines can show different CBS velocities. Reiners et al. (2016) provided an empirical fit to the CBS velocity:

$$\Delta \nu = -504.891 - 43.7963x - 145.560x^2 + 884.308x^3 \text{ [m s}^{-1}\text{]}, \tag{10.7}$$

where x is the relative line depth between 0.05 and 0.95.

Again, the reason for this change in the CBS velocity is the fact that weaker lines are formed deeper in the stellar atmosphere where the velocity field may be different from that higher up in the atmosphere. Intuitively, one can understand why weaker lines should have a higher blueshift. Imagine a hot convective cell rising from deep in the atmosphere and moving to higher layers. As it rises, the cell first has a higher velocity in the deep layers (where weak lines form), but as it rises, it slows down. By the time it reaches the outermost layers during its ascent, it will eventually stop, move horizontally, cool, and then become part of the sinking (redshifted) cool lanes.

The problem is that for most active stars, the mean magnetic field may change during the activity cycle, and this may affect the velocity flow deep in the interior. For instance, suppose you have a strong line where the CBS is zero and a nearby weak line (normalized depth = 0.2) with a CBS of -500 m s^{-1}. If, during the magnetic cycle the weak line changes its CBS velocity by 50 m s^{-1}, then this will result in a Doppler shift (using both lines) of $\approx$10 m s^{-1}. Monitoring the Doppler shift of weak lines as an ensemble over suspected activity cycles may help identify RV variations due to the changing convection pattern and thus convective blueshift.

10.6 Radial Velocity Jitter

The chances are high that if you make precise RV measurements of a star using a high-resolution spectrograph, you will find that the observed RV scatter is much higher than that expected from the internal errors. The term "RV jitter" is used to describe the extra "error" that must be added in quadrature to the estimated error in order to make these consistent with what is observed. This RV jitter has two sources: (1) there may be unknown systematic errors creeping into your measurements, and these are often difficult to assess; (2) the star shows real intrinsic RV variability.

RV jitter is a real contribution to the error budget, but it can be misinterpreted. It is the opinion of this author that the estimation of the true RV uncertainty due to your instrument is difficult to ascertain, and for many spectrographs, these are often underestimated. It is too convenient for researchers to make the performance of their RV machine look good by "blaming it on the star" and bundling the instrumental error into the jitter term. That said, it is clear that stellar variability is a real and unavoidable source of noise for Doppler measurements.

RV jitter is important for two reasons:

1. It gives you a better estimate of the true error when calculating an orbit and determining the uncertainty on the orbital parameters.
2. It can give you an estimate of how much intrinsic noise will be present in your measurements. If the expected RV amplitude is small, this can tell you how many observations you will need, or whether it is even worth trying to detect the RV signal of the planet.

10.6.1 RV Jitter and Orbit Fitting

The inclusion of RV jitter is important for deriving realistic parameters with errors when fitting orbits. A good way to estimate the RV jitter is to first calculate the orbit using the best estimate of your errors, look at the rms scatter about your solution, and add a term in quadrature to the standard error to make the rms scatter consistent, for example. If you have done things properly, your reduced χ^2 should be near unity.

In most cases, the inclusion of a jitter term will not affect the values of your parameters very much, but it will give you a more realistic estimate of their errors. Table 10.2 gives the orbital solutions for the planet candidate HD 13189b calculated with and without jitter. In this case, the nominal errors are approximately 5 m s^{-1}, but the jitter correction is at least five times larger (27 m s^{-1}) due to stellar oscillations in the K-giant host star.

10.6.2 Sources of Jitter

For planet searches with the Doppler method, there are three dominant sources of intrinic stellar jitter: (1) stellar granulation, (2) stellar oscillations, and (3) stellar magnetic activity (see Table 10.3). We have already seen how spots and other surface structure can cause periodic RV variations due to rotational modulation. We do not formally include this as jitter if one (in the best of cases) can actually model this rotational modulation and remove it. Rather, RV jitter refers to stochastic processes (stellar oscillations) or RV due to surface structure for which you have inadequate sampling to fit the variations. For example, you observe a star with a spot, but the next time you observe it, the feature is gone. In essence, it is an additional source of noise.

We now will look into some sources of jitter in more detail.

Table 10.2. Orbital Solutions for HD 13189b

Parameter	No Jitter	With Jitter
Period (days)	446.74 ± 0.07	448.80 ± 0.166
K (m s^{-1})	144.1 ± 18.4	154.2 ± 18.4
T1	2,450,586.44 ± 12.1	2,450,575.3301 ± 27.2
e	0.372 ± 0.07	0.410 ± 0.07
ω	171.27 ± 24.2	169.6 ± 24.2
χ_{reduced}	44.15	1.07

Table 10.3. Sources of RV Jitter

Phenomenon	Timescales	Amplitude
Oscillations	5–15 minutes	~0.5–1 m s^{-1}
Spots and surface structure	~days to months	~1–100 m s^{-1}
Magnetic cycles	~3–30 years	1–20 m s^{-1}
Short-term granulation	~days	~1 m s^{-1}
Long-term granulation	years to decades	1–20 m s^{-1}

10.6.3 Stellar Oscillations

The types of stellar oscillations most likely encountered by RV planet surveys are *p*-mode oscillations. These are acoustic oscillations where pressure is the restoring force. Stellar oscillations are ubiquitous in the Hertzsprung–Russell (H-R) diagram, and a classic example is the Cepheid class of stars which oscillate in low-order radial modes with periods of days and amplitudes of ~km s^{-1}.

Stellar oscillations can be described mathematically by spherical harmonics with "quantum" numbers n, ℓ, and m, where n is the number of nodes in the radial direction; ℓ is the degree of the node and specifies the number of surface nodes that are present; and m is the azimuthal order of the mode.

In the asymptotic limit (Tassoul 1980, 1990) where $n \gg \ell$, *p*-modes follow a spacing

$$\nu_{n,\ell} = \Delta\nu_0\left(n + \frac{\ell}{2} + \alpha\right) + \varepsilon_{n,\ell}, \tag{10.8}$$

where α is a constant of order unity and $\varepsilon_{n,\ell}$ is a small correction factor. The large spacing, $\Delta\nu_0$, is the inverse travel time of a sound wave from the surface to the core and back. So, the frequency of *p*-mode oscillations can give us information about the internal structure of the star, hence the term asteroseismology.

Solar-like *p*-mode oscillations show a power spectrum of modes that are equally spaced in frequency and representing a specific nonradial mode (Figure 10.26). The large spacing is given by twice the observed frequency spacing because adjacent ℓ

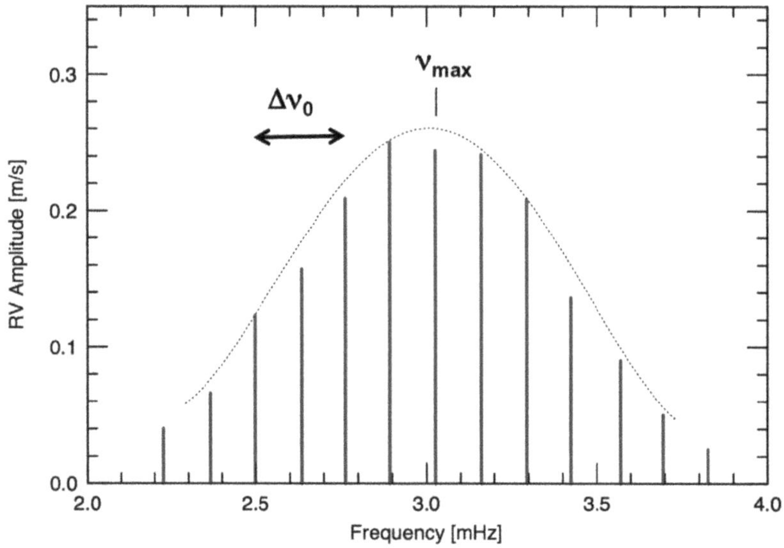

Figure 10.26. Schematic of the RV amplitude spectrum of p-mode oscillations in the Sun. These are characterized by a large spacing $\Delta\nu_0$ which is twice the observed spacing of the modes. The large spacing is 135 μHz for the Sun. The large spacing is related to the mean density of the star. The modes amplitudes have an envelope whose maximum is at ν_{max}, which is about 5 minutes (3.05 mHz) for solar *p*-mode oscillations.

odd and even modes appear at one-half the large spacing. In the case of the Sun, the large spacing is 134.9 μHz. These peaks are modulated by a broad envelope centered at a frequency ν_{max}. For the Sun, ν_{max} is about 3.05 millihertz (mHz), corresponding to a period of about 5 minutes, hence the famed "5 minute oscillations". The peak RV amplitude for these modes is about 0.23 m s^{-1}.

These stellar oscillations will increase the scatter ("noise") in your Doppler measurements, although this is not strictly noise. In fact, precise RV measurements have been employed to study these oscillations (e.g., Bedding et al. 2006, 2007), so "one person's noise is another person's signal". For our work, we are only interested in the RV amplitude of these oscillations and the timescales involved.

Kjeldsen & Bedding (1995) derived relationships for the expected RV amplitudes and frequencies for stellar oscillations that are scaled to solar values. The expected RV amplitude, v_{osc} in m s^{-1}, is given in Equation (10.11) in terms of the stellar luminosity, L, and mass, M. Equation (10.10) gives ν_{max} in minute^{-1} (a more useful unit for planet hunters) where R is the stellar radius, and T_{eff} the effective temperatures. The authors showed that the relationships for v_{osc} holds for almost two orders of magnitude in L/M:

$$\Delta\nu_0 = \left(\frac{M}{M_\odot}\right)^{1/2}\left(\frac{R}{R_\odot}\right)^{-3/2} 134.9\ \mu\text{Hz}, \tag{10.9}$$

$$\nu_{max} = \frac{M/M_\odot}{(R/R_\odot)^2\sqrt{T_{eff}/5777\ \text{K}}} 0.183\ \text{m}^{-1}, \tag{10.10}$$

$$v_{\mathrm{osc}} = \frac{L/L_{\odot}}{M/M_{\odot}}(0.234 \pm 1.4)\ \mathrm{m\ s^{-1}}. \tag{10.11}$$

These scaling relationships are very useful for estimating the RV contribution due to stellar noise. For example, an F_0 main-sequence star with a mass of 1.6 $M_{\odot}$ and luminosity of 6.5 $L_{\odot}$ will have $v_{\mathrm{osc}} \approx 1$ m s^{-1} or four times larger than the Sun. The oscillation periods would be about 9 minutes ($T_{\mathrm{eff}} = 7300$ K, $R = 1.5\ R_{\odot}$), or almost twice the solar value.

The noise contribution of stellar oscillations to the RV measurement is generally small, and in many cases, this would be much less than the typical measurement error. However, the velocity amplitude scales as the stellar luminosity and for evolved giant stars, this can approach amplitudes and periods that are characteristic of planetary signals. A relatively small K-giant star like β Gem with a radius $R = 8.3\ R_{\odot}$, mass, $M = 1.9\ M_{\odot}$, luminosity, $L = 43\ L_{\odot}$, and effective temperature, $T_{\mathrm{eff}} = 4850$ K, is expected to have a domimant pulsation period of $P_1 \approx 3$ hr with a velocity amplitude $v_{\mathrm{osc}} \approx 6$ m s^{-1}. On the other hand, an evolved K giant like γ Dra ($R = 44\ R_{\odot}$, $M = 1.7\ M_{\odot}$, $L = 510\ L_{\odot}$, $T_{\mathrm{eff}} = 3990$ K) will have a dominant pulsation period of $P_1 \approx 3.6$ days and a velocity amplitude of $v_{\mathrm{osc}} \approx 70$ m s^{-1}. For both stars this is largely consistent with what is observed for these stars (Figure 10.27). Figure 10.28

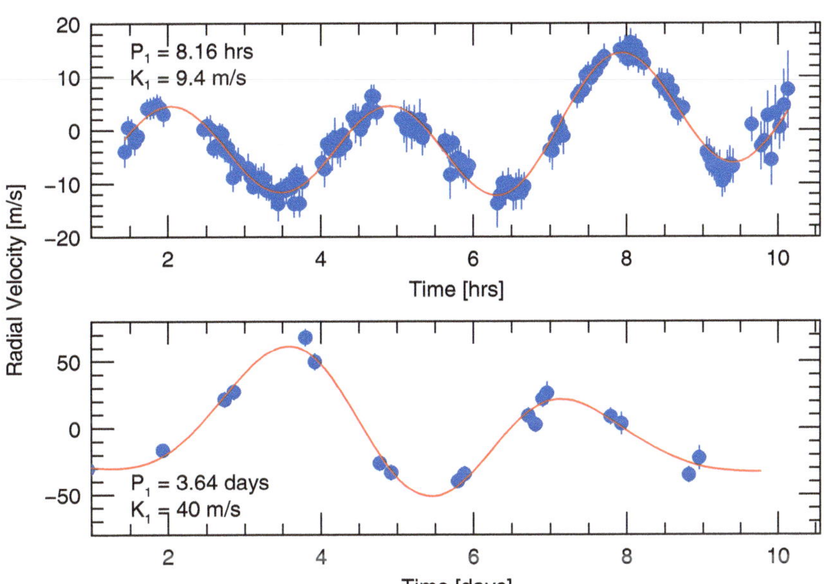

Figure 10.27. (Top) The radial velocity variations due to stellar oscillations in the K-giant star β Gem ($L = 48.3\ L_{\odot}$, $M = 19.3\ M_{\odot}$, $R = 8.3\ R_{\odot}$). The dominant mode has a period of 8.1 hr and a velocity K amplitude of 9.4 m s^{-1}. (bottom) The radial velocity variations due to stellar oscillations in the K-giant star γ Dra ($L = 510\ L_{\odot}$, $M = 1.7\ M_{\odot}$, $R = 44\ R_{\odot}$). The dominant mode has a period of 3.64 days and a velocity K amplitude of 40.4 m s^{-1}. The amplitudes and periods for the oscillations in both stars are consistent with the scaling relations.

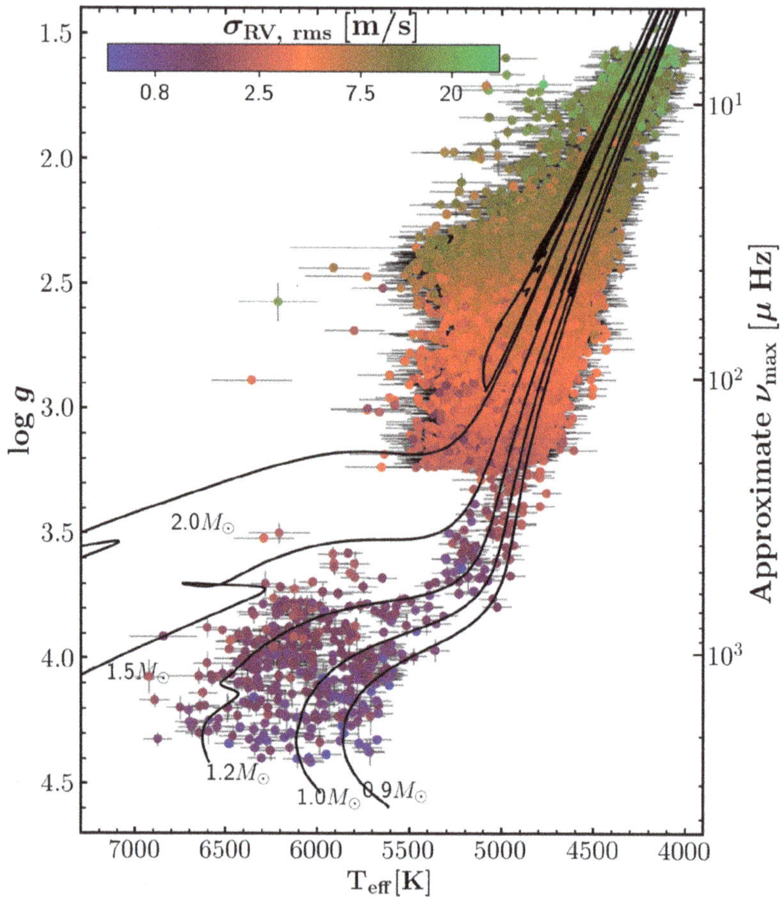

Figure 10.28. The calculated RV jitter, $\sigma_{\rm RV}$, in the surface gravity (log g)–effective temperature ($T_{\rm eff}$) diagram. The approximate $\nu_{\rm max}$ is labeled on the right vertical axis. The solid lines show evolutionary tracks (Bressan et al. 2012) for stellar masses in the range 0.8 to 2.0 $M_{\odot}$. (Reproduced from Yu et al. 2018. Copyright of OUP Copyright '2018'.)

summarizes the expected RV jitter due to stellar oscillations throughout the H-R diagram (Yu et al. 2018).

The influence of stellar oscillations on your RV measurement can be mitigated by averaging several measurements spanning a pulsation cycle. The top panel of Figure 10.29 shows RV measurements for α Cir. This star is a rapidly oscillating Ap (roAp) star showing up to 36 pulsation modes (Mkrtichian & Hatzes 2013) centered on a main period of about 6.83 minutes. The dominant mode has an RV amplitude of $\approx$20 m s^{-1}. The rapid time series shows an rms scatter of 14.7 m s^{-1}. If one takes binned averages covering five pulsation cycles, then the rms scatter reduces to a mere 0.83 m s^{-1}. An observing strategy of either having exposure times spanning the pulsational period or averaging several measurements taken in sequence can greatly improve the RV precision.

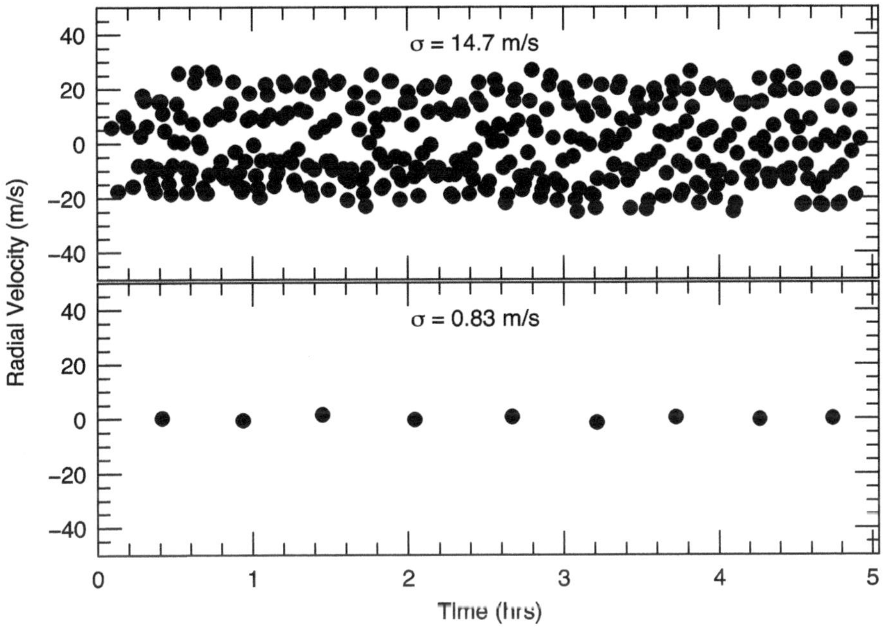

Figure 10.29. (Top) RV variations of the rapidly oscillating Ap star α Cir. The dominant variability is due a mode pulsating with a 6.8 minute period. The rms scatter is 14.7 m s^{-1}. (Bottom) The same RV data that have been averaged in bins with a time length corresponding to five pulsation periods. The rms scatter is 0.83 m s^{-1}.

10.6.4 Activity Jitter

Stellar magnetic activity is generally the dominant source of RV jitter for measurements on late-type main-sequence stars. This activity manifests itself in a wide range of phenomena that include dark spots, plage, faculae, flares, changes to the convection pattern, etc. We have seen that some of the variations from these can be tied to the rotation of the star and thus can be periodic (e.g., spots, plage) while others can be more stochastic in nature (e.g., flares).

Our current understanding of stellar activity is that it results from stellar rotation coupled with an outer convection zone—both are needed in order to generate a magnetic dynamo. The faster the rotation and the deeper the convection zone, the higher the level of activity. For the most part, activity noise becomes important for main-sequence stars later than the spectral type of about F6. Early-type (B to early F) stars do have high rotation rates, but they have a shallow outer convection zone resulting in low or no activity. For early-type stars, stellar oscillations (e.g., δ-Scuti type) rather than activity may be the major source of jitter. WASP-33 b is an example of a transiting planet around an A-type star whose RV confirmation was hindered by the presence of stellar oscillations (Collier Cameron et al. 2010; Herrero et al. 2011; Lehmann et al. 2015). For late-type (late F, G, K, and M), main-sequence stars with outer convection zones, RV jitter from stellar activity will be problematic.

An extreme case of activity jitter is young, active stars for which magnetic braking has not had time to take effect. You thus have rapid rotation which already results in a poorer RV measurement precision, plus high levels of activity jitter, which increases your scatter. It is challenging to find exoplanets around these stars, and it is for these reasons that the number of exoplanets around young, active stars is few.

As expected, for F, G, and K stars, the higher the stellar rotation rate, the larger the RV jitter σ_A (left panel of Figure 10.30). The jitter is roughly proportional to the projected rotatonal velocity of the star, $v \sin i$. The same behavior is seen as a function of the inverse rotation period of the star (right panel of Figure 10.30).

This increase is due to two effects. First, higher stellar rotation translates into a higher level of magnetic activity and thus larger spot coverage. The left panel of Figure 10.31 shows simulations of how the RV amplitude from a spot is roughly proportional to the filling factor of the spot.

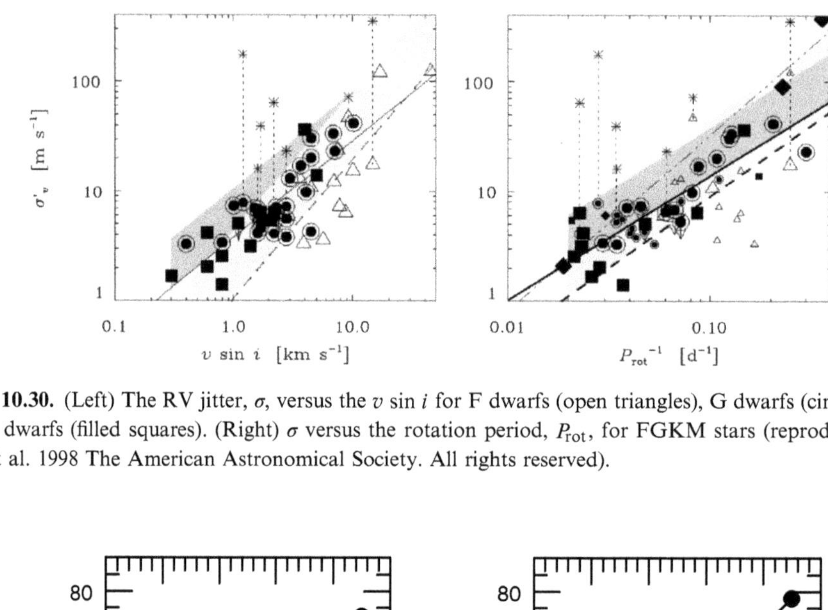

Figure 10.30. (Left) The RV jitter, σ, versus the $v \sin i$ for F dwarfs (open triangles), G dwarfs (circled dots), and K dwarfs (filled squares). (Right) σ versus the rotation period, P_{rot}, for FGKM stars (reproduced from Saar et al. 1998 The American Astronomical Society. All rights reserved).

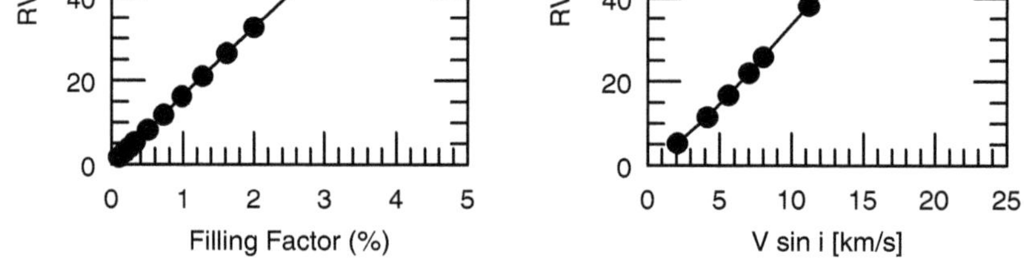

Figure 10.31. (Left) The RV amplitude due to cool spots with different filling factors and with fixed rotational velocities for the star. The simulations were performed using SOAP 2.0. (Right) SOAP 2.0 simulations of the RV amplitude versus $v \sin i$ of the star from a spot with a fixed filling factor.

Second, a higher $v \sin i$ means the stellar lines are more broadened, which accentuates the distortions due to surface structure. This, in effect, gives you more "leverage" in detecting the RV signal due to the spot. Thus, for a given spot size, this will cause an RV amplitude that is proportional to the $v \sin i$ of the star (right panel of Figure 10.31).

RV Jitter from Spots

A useful expression for approximating the RV jitter due to spots was given by Saar & Donahue (1997):

$$A_{\mathrm{RV}}\ [\mathrm{m\ s^{-1}}] \approx 6.5 v \sin i\, f^{0.9}, \tag{10.12}$$

where f is the spot filling factor in percent and $v \sin i$ the projected rotational velocity. This result was largely confirmed by other simulations by Hatzes (2002). This expression is useful for estimating the expected RV jitter of an active star when analyzing space-based light curves (e.g., from the *CoRoT*, *Kepler*, *TESS*, etc. missions). The amplitude of the light variations can give you a good estimate of the spot filling factor which can then be used to estimate the RV jitter. For example, CoRoT-7 shows photometric variations of up to ≈2%. This star has a rotational velocity of about 2 km $\mathrm{s^{-1}}$, so the expected RV amplitude for spots is approximately 24 m $\mathrm{s^{-1}}$, which is largely what is seen in the RV variations for this star (see Chapter 11).

Estimating RV Jitter from the S *Index and* R'_{HK}

If you have a measurement of the S index or R'_{HK}, these can be used to estimate the expected RV jitter from a star. Pioneering work on this was done by Saar et al. (1998) followed by a number of investigations (Santos et al. 2000; Paulson et al. 2002; Wright 2005; Martínez-Arnáiz et al. 2010; Isaacson & Fischer 2010).

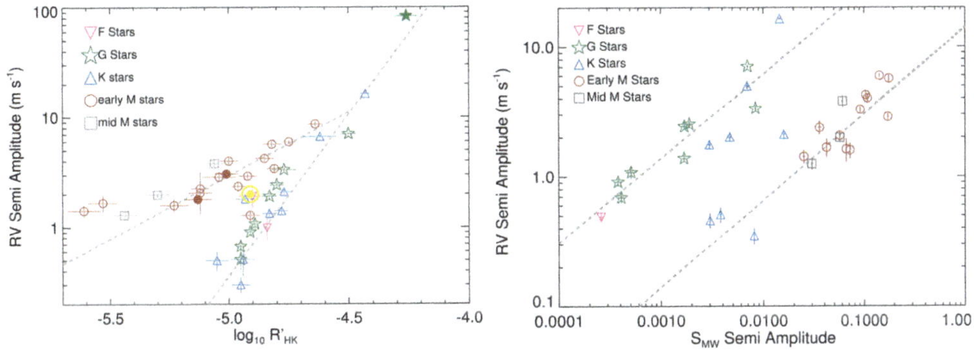

Figure 10.32. (Left) The RV semiamplitude of the rotation-induced signal versus log R_{HK}. The gray lines are the best fit according to Equations (10.13) and (10.14). (Right) The RV semiamplitude of activity versus the semiamplitude of the Mt. Wilson S index, S_{MW} . The lines are the fits according to Equations (10.15) and (10.16). (Reproduced from Suárez Mascareño et al. 2017, reproduced with permission © ESO).

Figure 10.32 shows the RV amplitude of rotation-induced signals versus the $R'_{\rm HK}$ (left panel) and Mt. Wilson S index, $S_{\rm MW}$, (right panel) from Suárez Mascareño et al. (2017). These form tight relationships that can be well fitted by linear regressions.

The $R'_{\rm HK}$ relationship for G–K dwarf stars is

$$\log_{10}(K) = (2.93 \pm 0.03)K + (14.23 \pm 0.12). \tag{10.13}$$

The $R'_{\rm HK}$ relationship for M-dwarf stars is

$$\log_{10}(K) = (1.15 \pm 0.02)K + (6.23 \pm 0.08). \tag{10.14}$$

The $S_{\rm MW}$ relationship for FG and early K-dwarf stars is

$$\log_{10}(K_{\rm v}) = (0.54 \pm 0.07)K_{S_{\rm MW}} + (1.61 \pm 0.17). \tag{10.15}$$

The $S_{\rm MW}$ relationship for late K and M-dwarf stars is

$$\log_{10}K_{\rm v} = (0.49 \pm 0.06)K_{S_{\rm MW}} + (0.63 \pm 0.06). \tag{10.16}$$

Note that these predicted RV amplitudes are due to the $R'_{\rm HK}$ and S-index variations due to rotational modulation.

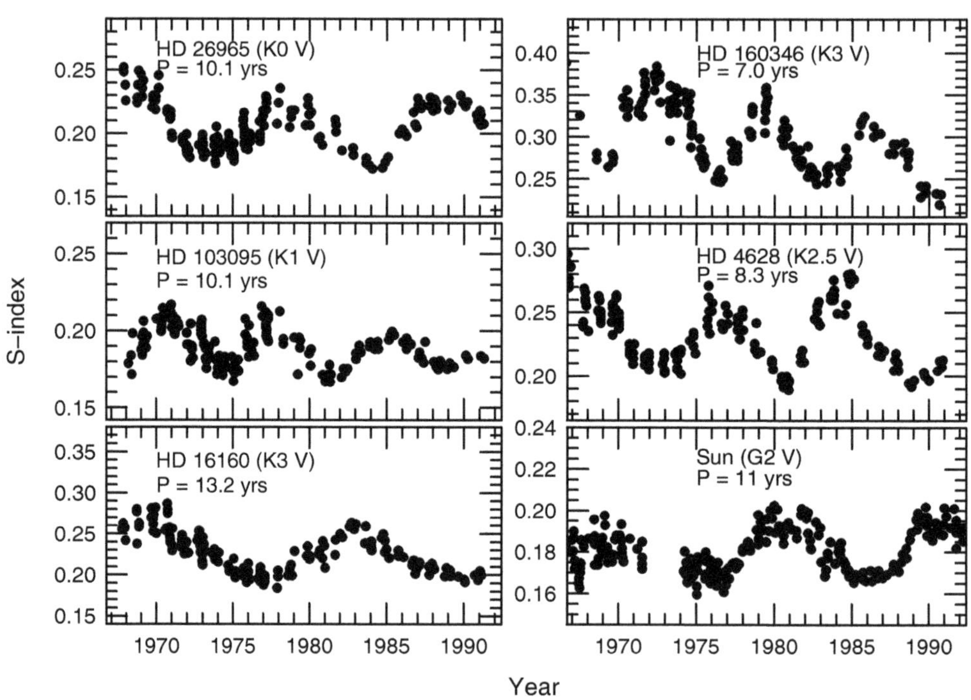

Figure 10.33. S-index measurements of stars from the Mt. Wilson S-index survey showing activity cycles. The lower right shows measurements for the Sun. (Reproduced from Baliunas et al. 1995. The American Astronomical Society. All rights reserved.)

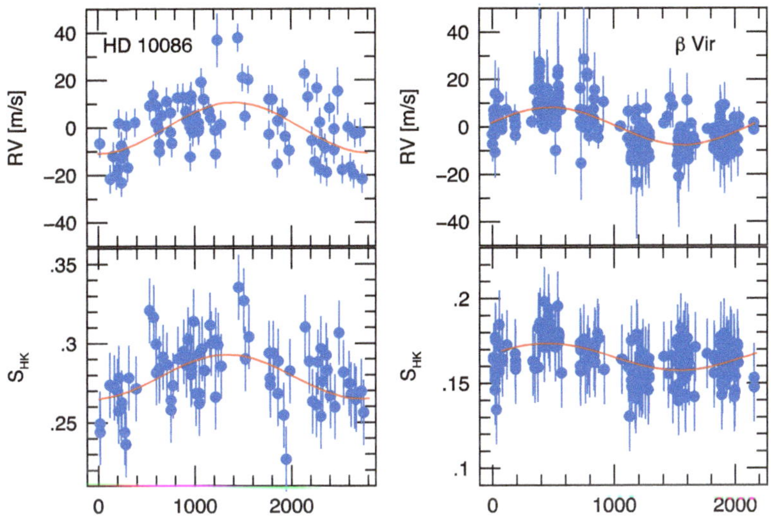

Figure 10.34. (Top panels) RV measurements of HD 10086 (right) and β Vir (left) from the McDonald Observatory Planet Search Program. (Bottom panels) The Ca II S-index measurements for HD 10086 (left) and β Vir (right). The same period is seen in the S-index measurements as in the RVs (from Endl et al. 2016).

10.7 Activity Cycles

The S index can be a powerful diagnostic to discern activity cycles in solar-type stars. The most extensive use of this index has been the Mt. Wilson survey (see Baliunas et al. 1995 and references therein). Figure 10.33 shows long-term S-index measurements of some solar-like stars, including the Sun, from the Mt. Wilson survey (Baliunas et al. 1995). Sun-like stars reveal cycle periods of a few years to decades, comparable to the solar cycle.

Activity cycles should produce low-amplitude, but measurable, RV variations. Because these periods are so long, they will mimic the signal of a long-period giant planet (i.e., Jovian analogs). Figure 10.34 shows the Ca II S-index and RV measurements from two stars from the McDonald Observatory Planet Search Program (Endl et al. 2016). HD 10086 shows RV variations with a period of 7.7 years and a K amplitude consistent with an $m \sin i = 0.74\ M_{\rm Jup}$ companion. The star β Vir shows RV variations with a period of 5.6 years, which could result in a giant planet with a minimum mass of 0.65 $M_{\rm Jup}$. However, in both cases, the same period is seen in the S-index measurements, indicating that activity is the source of the RV variations.

The RV K amplitude due to stellar activity cycles seems higher than the value expected from the S-index variations. Equation (10.15) gives a $K_{\rm v} = 3$ m s^{-1} and $K_{\rm v} = 4$ m s^{-1} for β Vir and HD 10086, respectively, given their respective $K_{S_{\rm SM}}$ amplitudes. These are about a factor of 2 smaller than what is observed. Also, the approximate slope of the log $K_{\rm v}$–log $K_{S_{\rm SM}}$ for the activity cycles is about a factor of 4 larger. Admittedly this is based on very small statistics, but it could point to a higher RV amplitude for a given S-index variation due to activity cycles, possibly

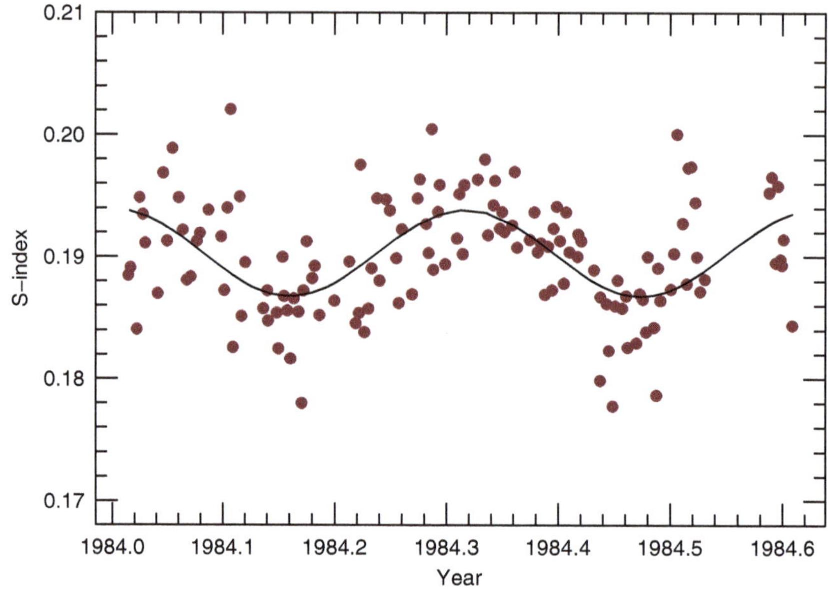

Figure 10.35. Mount Wilson *S*-index measurements of the F6 V star τ Boo. The curve represents a sine wave fit with a period of 120 days (from Schmitt & Mittag 2017).

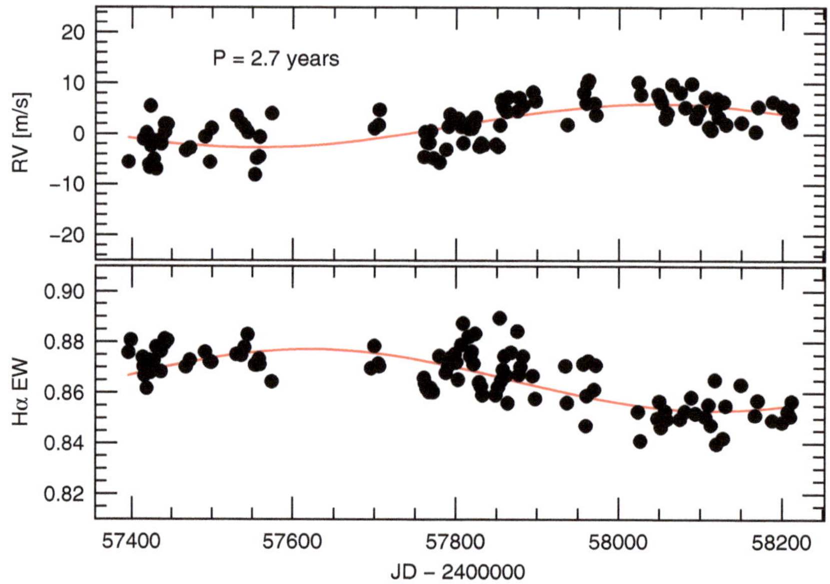

Figure 10.36. (Top) Long-term RV variations with a 2.7 year period of an M dwarf from the CARMENES sample. (Bottom) The same variations are seen in the equivalent width of Hα, indicating an activity cycle.

related to changes in the convection pattern on the star. Equation (10.15) may be more appropriate for surface inhomogeneities (e.g., spots).

Our greatest knowledge about cycles is for Sun-like stars due to the focus of the Mt. Wilson survey. Slowly we are learning about magnetic cycles on other types of stars, and these can be shorter than the 11 year solar cycle. Schmitt & Mittag (2017) found evidence for a ~120 day cycle in the planet-hosting star τ Boo, coincidentally using the S-index measurements from the Mt. Wilson survey (Figure 10.35).

Our knowledge of magnetic cycles in M-dwarf stars will increase due to the number of programs surveying these stars with RV measurements. Díez Alonso et al. (2019) took extensive photometric time series of 337 bright M stars and found evidence for long term activity cycles of 3 to 11.5 years, comparable to the period of the solar cycle. Figure 10.36 shows the RV and Hα equivalent width variations of a star from the CARMENES program (Reiners et al. 2018). It shows both RV and Hα variations with a period of about 2.7 years. This most likely is an activity cycle as this period seems too long (but not impossible) to be the rotation period of the star. The K amplitude of the star is 4.3 m s^{-1}, which could be misinterpreted as arising from a planet with mass $M \sin i = 0.2\ M_{\rm Jup}$ mass orbiting 2.6 au from the star.

10.8 Concluding Remarks

In this chapter, we presented a number of diagnostics or activity indicators that one can use for discerning the true nature of a signal in an RV time series. You should use as many indicators that you have access to. Some indicators are better than others, and this can change from star to star. Is there a perfect activity indicator? Photometric measurements come close, but even these represent just one piece of the puzzle.

Unless the star is very active, it is unlikely that you see variations in all indicators (Ca II, FWHM, bisector, Hα, etc.). You will stumble across cases where you see no variations in a large number of indicators only to find one that varies with the RV period. You only need to find the RV period in a single indicator to cast doubt on the planet hypothesis. The rule for bisector measurement can be applied to all other indicators: a lack of variations is a necessary, but not a sufficient, condition to prove the planet hypothesis.

References

Allende Prieto, C., & Garcia Lopez, R. J. 1998, A&AS, 129, 41
Anglada-Escudé, G., & Tuomi, M. 2015, Sci., 347, 1080
Baliunas, S. L., Donahue, R. A., Soon, W. H., et al. 1995, ApJ, 438, 269
Bedding, T. R., Butler, R. P., Carrier, F., et al. 2006, ApJ, 647, 558
Bedding, T. R., Kjeldsen, H., Arentoft, T., et al. 2007, ApJ, 663, 1315
Bressan, A., Marigo, P., Girardi, L., et al. 2012, MNRAS, 427, 127
Collier Cameron, A., Guenther, E., Smalley, B., et al. 2010, MNRAS, 407, 507
Díaz, R. F., Cincunegui, C., & Mauas, P. J. D. 2007, MNRAS, 378, 1007
Díez Alonso, E., Caballero, J. A., Montes, D., et al. 2019, A&A, 621, 126
Dravins, D. 1987a, A&A, 172, 200

Dravins, D. 1987b, A&A, 172, 211
Dumusque, X., Boisse, I, & Santos, N. C. 2014, ApJ, 796, 132
Dumusque, X. 2018, A&A, 620, A47
Endl, M., Brugamyer, E. J., Cochran, W. D., et al. 2016, ApJ, 818, 34
Gomes da Silva, J., Santos, N. C., Bonfils, X., et al. 2011, A&A, 534, A30
Gray, D. F. 1983, PASP, 95, 252
Gray, D. F. 1994, PASP, 106, 1248
Gray, D. F. 1997, Natur, 385, 795
Gray, D. F. 1998, Natur, 391, 153
Gray, D. F., & Hatzes, A. P. 1997, ApJ, 490, 412
Gray, D. F., & Johanson, H. L. 1991, PASP, 103, 439
Gray, D. F., & Nagel, T. 1989, ApJ, 341, 421
Hatzes, A. P. 1996, PASP, 108, 839
Hatzes, A. P. 2002, AN, 323, 392
Hatzes, A. P. 2016, A&A, 585, A144
Hatzes, A. P., Cochran, W. D., & Bakker, E. J. 1998a, Natur, 391, 154
Hatzes, A. P., Cochran, W. D., & Bakker, E. J. 1998b, ApJ, 508, 380
Hatzes, A. P., Cochran, W. D., & Johns-Krull, C. M. 1997, ApJ, 478, 374
Hatzes, A. P., & Mkrtichian, D. E. 2005, A&A, 430, 279
Herrero, E., Morales, J. C., Ribas, I., & Naves, R. 2011, A&A, 526, L10
Huélamo, N., Figueira, P., Bonfils, X., et al. 2008, A&A, 489, L9
Isaacson, H., & Fischer, D. 2010, ApJ, 725, 875
Kanaan, A., & Hatzes, A. P. 1998, ApJ, 503, 848
Käufl, H. U., Amico, P., Ballester, P., et al. 2008, Proc. SPIE, 7014, 70140
Kjeldsen, H., & Bedding, T. R. 1995, A&A, 293, 87
Kürster, M., Endl, M., Rouesnel, F., et al. 2003, A&A, 403, 1077
Larson, A. M., Irwin, A. W., Yang, S. L. S., et al. 1993, PASP, 105, 332
Lehmann, H., Guenther, E., Sebastian, D., et al. 2015, A&A, 578, L4
Livingston, W. C. 1983, in IAU Symp. 102, Solar and Stellar Magnetic Fields: Origins and Coronal Effects, ed. J. O. Stenflo (Cambridge: Cambridge Univ. Press), 149–52
Martínez-Arnáiz, R., Madonado, J., Montes, D., Eiroa, C., & Montesinos, B. 2010, A&A, 520, A79
Mayor, M., Bonfils, X., Forveille, T., et al. 2009, A&A, 507, 487
Mkrtichian, D. E., & Hatzes, A. P. 2013, in ASP Conf. Proc. 479, Progress in Physics of the Sun and Stars: A New Era in Helio- and Asteroseismology, ed. H. Shibahashi, & A. E. Lynas-Gray (San Francisco, CA: ASP), 115
Mkrtichian, D. E., Hatzes, A. P., & Kanaan, A. 2003, MNRAS, 345, 781
Mkrtichian, D. E., Hatzes, A. P., Saio, H., & Shobbrook, R. R. 2008, A&A, 490, 1109
Neff, J. E., O'Neal, D., & Saar, S. H. 1995, ApJ, 452, 879
Paulson, D. B., Saar, S. H., Cochran, W. D., & Hatzes, A. P. 2002, AJ, 124, 572
O'Neal, D., Neff, J. E., Saar, S. H., & Mines, J. K. 2001, AJ, 122, 1954
O'Neal, D., Saar, S. H., & Neff, J. E. 1996, ApJ, 463, 766
Queloz, D., Henry, G. W., Sivan, J. P., et al. 2001, A&A, 379, 279
Ramsey, L. W., & Nations, H. L. 1980, ApJ, 239, L121
Reiners, A., Mrotzek, N., Lemke, U., Hinrichs, J., & Reinsch, K. 2016, A&A, 587, A65
Reiners, A., Zechmeister, M., Caballero, J. A., et al. 2018, A&A, 612, A49

Robertson, P., Endl, M., Cochran, W. D., MacQueen, P. J., & Boss, A. P. 2013, ApJ, 774, 147
Robertson, P., Endl, M., Henry, G. W., et al. 2015, ApJ, 801, 79
Robertson, P., Mahadevan, S., Endl, M., & Roy, A. 2014, Sci., 345, 440
Saar, S. H., Butler, R. P., & Marcy, G. W. 1998, ApJ, 498, L153
Saar, S. H., & Donahue, R. A. 1997, ApJ, 485, 319
Santos, N. C., Mayor, M., Naef, D., et al. 2000, A&A, 361, 265
Schmitt, J. H. M. M., & Mittag, M. 2017, A&A, 600, A120
Setiawan, J., Henning, T., Launhardt, R., et al. 2008, Natur, 451, 38
Soderblom, D. R., Duncan, D. K., & Johnson, D. R. H. 1991, ApJ, 375, 722
Suárez Mascareño, A., Rebolo, R., González Hernández, J. I., & Esposito, M. 2017, MNRAS, 468, 4772
Tassoul, M. 1980, ApJS, 43, 469
Tassoul, M. 1990, ApJ, 358, 313
Vogt, S. S. 1979, PASP, 91, 616
Vogt, S. S. 1981, ApJ, 247, 975
Vogt, S. S., Butler, R. P., Rivera, E. J., et al. 2010, ApJ, 723, 954
Vogt, S. S., & Penrod, G. D. 1983, PASP, 95, 565
Wright, J. T. 2005, PASP, 117, 657
Yu, J., Huber, D., Bedding, T. R., & Stello, D. 2018, MNRAS, 480, L48
Zechmeister, M., Reiners, A., Amado, P. J., et al. 2018, A&A, 609, A12

A P Hatzes

Chapter 11

Dealing with Stellar Activity

We have seen that for late-type stars, the radial velocity (RV) variations caused by activity can dominate the measurement uncertainty from photon statistics and instrumental errors. At best, these add an RV "jitter" term to your measurement error; at worse, they can masquerade as a planet signal. When trying to find small ($m < 50\ M_\oplus$) planets around very active stars, the RV signal due to activity can dominate the signal of the planet by a factor of 10 or more. If you could filter out the activity signal, you can (1) be sure of the planet signal you have detected, (2) get more accurate orbital parameters, and (3) search for even lower mass planets in the data.

In this chapter, we discuss various methods for removing the activity signal. No single method can be applied universally to all active stars. Depending on the periods of the planet, stellar rotation, and activity cycles, as well as the activity level of the star, some methods will work better than others. Ideally, more than one method should be employed. If these arrive at the same answer, you can be more confident of your results.

Most of these methods work for stars that you know you have an exoplanet, i.e., you wish to confirm a detection. A classic application is for transiting planets. Photometry has yielded a discovery and the planet period, now all you need is the orbital K amplitude to confirm the nature of the transit and to get the companion mass. Knowing the planet orbital period as well as the stellar rotation (often seen in the light curves) affords you the luxury of tailoring your filtering method to that specific discovery. You can of course apply these tools to RV data where there is no known planet, but you are still left grappling with the nature of the signal. Is it really a planet?

11.1 Fourier Filtering

A classic and effective method for removing the activity signal is Fourier filtering. The mathematical basis for this is the fact that sines and cosines are orthogonal

doi:10.1088/2514-3433/ab46a3ch11

(or equivalently "basis") functions. As such, any continuous function can be represented by a sum of trigonometric functions. This is often forgotten when using periodograms to search for primarily a single periodic signal. One might argue that there is no physical basis for arbitrarily using sine functions to model the activity. However, there is a mathematical basis for this, the expansion in a Fourier series. You can represent a linear function by $y = mx + b$—all you need find is the slope and intercept. Equivalently, one can represent the line by a trigonometric series, $y = \sum A_i \cos(\omega t)$. Now all you need to find are the coefficients to each cosine term, admittedly a more difficult task.

Figure 11.1 shows the basis of Fourier filtering and the two ways of fitting an underlying trend. The left panel of the figure shows simulated RV measurements consisting of a planet signal with a period of 2.85 days superimposed on the linear trend. Random noise with a standard deviation of 2 m^{-1} is also present. The sampling is typical for ground-based measurements.

The obvious method is to remove the trend by simply fitting a straight line to the data (blue line in figure), subtracting it, and then performing a period analysis on the residual RV data. The result is shown in the top panel on the right.

Fourier filtering should also work. You calculate the sum of the dominant Fourier components present in the data (red line in Figure 11.1), subtract this, and look at the periodogram of the residuals. The lower-right panel shows that you arrive at the same result. Clearly, the linear fit removes all of the higher Fourier components, which is not done with standard prewhitening. There may be additional "wiggles" due to the residual sine components that were not removed in the

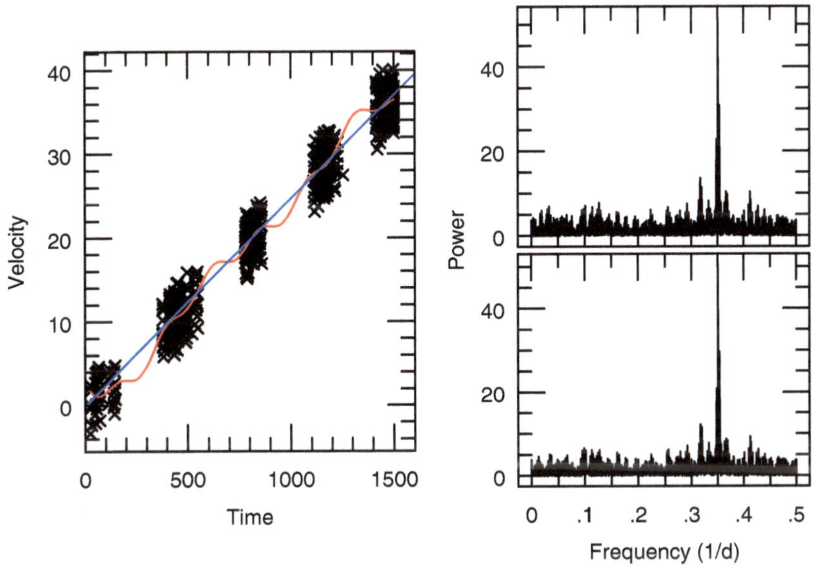

Figure 11.1. (Left) Synthetic RVs consisting of a planet signal superimposed on a long-term trend. The blue line shows the fit to the trend from a linear fit, the red line from the prewhitening sine coefficients. The LS periodogram of the RV residuals after removing the trend by fitting a straight line (top) or by prewhitening the data (bottom). Both methods for removing the trend produce consistent results.

multisine fit to the trend. In this case, the amplitude of these was obviously below the measurement errors. One caveat is that if the underlying trend has Fourier components with the same frequency as your planet, then the signal will be hidden—not the case for the linear trend fitting. So, if the trend fitting method produces a different result, you should be skeptical.

You can find the Fourier components to time series data by finding the highest peak in the amplitude spectrum, fitting a sine function with this frequency, subtracting it, and looking for the additional peaks in the residuals. This is essentially the prewhitening procedure, but in this case, we are not finding multiple planet systems, but rather the Fourier representation of the underlying activity signal. The trick is finding enough Fourier coefficients to adequately represent your data.

We shall apply Fourier filtering to an analysis of CoRoT-7. This star hosts the first discovered transiting rocky planet (Léger et al. 2009), whose mass was confirmed with RV measurements (Queloz et al. 2009; Hatzes et al. 2010). We first start with the space-based light curve measured by the *CoRoT* satellite, rather than the RVs. We do this because space data have exquisite time sampling and a clean window function. There are also few gaps in the data so the data is as close to a continuous function as observations can give you. Finally, the data are of high precision, and thus all the peaks you see in the amplitude spectrum are most likely due to signal and not noise. In short, it presents us with a "textbook case."

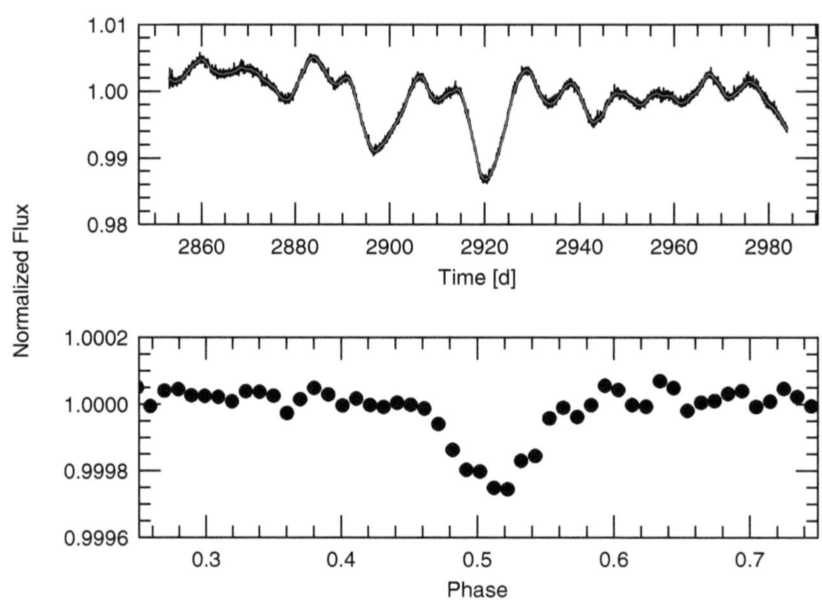

Figure 11.2. Fourier filtering of the CoRoT-7 light curve. (Top) The light curve as measured by *CoRoT* (black line). The 40 highest Fourier components in the frequency range 0–0.5 day^{-1} were used to provide the fit shown by the red line. (Bottom) The resulting transit curve for CoRoT-7b after subtracting the 40 sine component fit to the underlying activity signal.

The top panel of Figure 11.2 shows the *CoRoT* light curve of CoRoT-7. The large variations are due to rotational modulation from spots that clearly evolve with time.

We fit the CoRoT-7 light curve using the first 40 Fourier components found in the amplitude spectrum. This provides an adequate fit to the light curve (red line in Figure 11.2). Subtracting this yields the transit curve for CoRoT-7b (lower panel of Figure 11.2).

For ground-based observations, you never get such exquisite sampling, but in many instances the method works well. However, let us see if the same technique can extract the planet signal from ground-based RV measurements that cover a comparable time span, but with much poorer sampling and time gaps.

The top panel of Figure 11.3 shows the RV measurements for CoRoT-7 taken with the HARPS spectrograph (Queloz et al. 2009). These consist of 106 measurements spanning 109 days. The typical measurement error is about 2 m s^{-1}. A Fourier analysis (prewhitening) was performed to find the dominant frequencies in the data. The stopping point for the procedure was the 1% false-alarm probability (FAP) criterion, i.e., when a peak in the residuals had an amplitude less than 3.6 times the surrounding peaks. The line shows the multisine component fit coming from both activity and planets found by finding the Fourier coefficients shown in Table 11.1.

The challenge now is to interpret the signals found in Table 11.1. The dominant signal with a period of $P \approx 22$ days is most likely due to rotational modulation of activity as confirmed by the light curve. Other signals, such as the one at approximately 10 days and 5 days, are most likely harmonics. Note that the activity signal produces several frequencies; this is just the Fourier way of fitting a

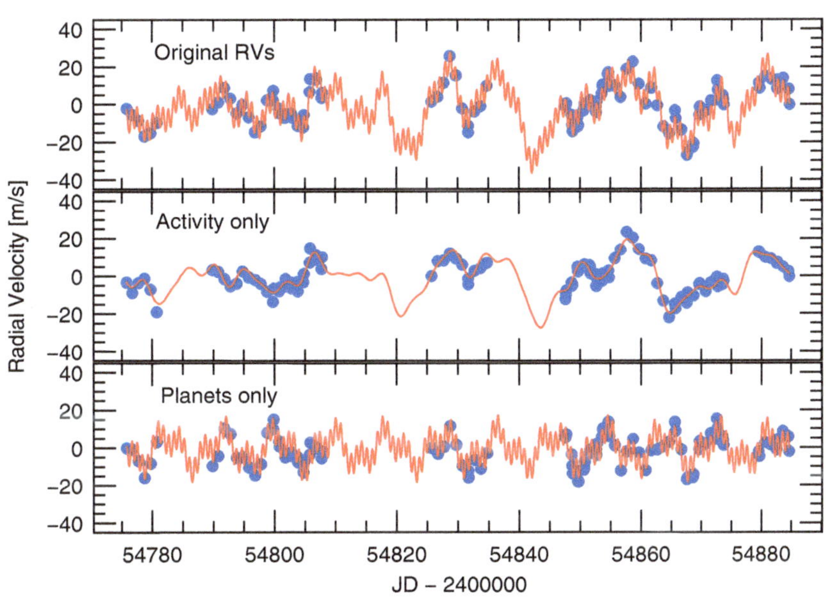

Figure 11.3. Fourier filtering of the CoRoT-7 RVs. (Top) The RVs (points) for CoRoT-7 and the multisine component fit (curve) using the frequencies in Table 11.1. (Middle) The RV and fit due to activity after removing the signals of all planets (including f_2). (Bottom) The RV and fit due to planet signals.

Table 11.1. Frequencies Found in the CoRoT-7 RV Data with Prewhitening

	ν (day^{-1})	P (days)	K (m s^{-1})	ν_{activity} (day^{-1})	Comment
f_1	0.045	22.22	9.18	0.045	Activity
f_2	0.111	9.01	7.02		Planet?
f_3	0.096	10.42	6.07	0.097	Activity
f_4	0.271	3.68	5.21		CoRoT-7c
f_5	1.170	0.85	5.75		CoRoT-7b
f_6	0.181	5.52	3.28	0.167	Activity
f_7	0.035	28.57	5.49	0.033	Activity
f_8	0.087	11.49	3.05	0.091	Activity

complicated light curve. But other signals may be planetary in origin. How can you tell which signal is due to activity versus planets? A look at activity indicators will help. Also shown in the table are the frequencies found in at least one of the activity indicators (Ca II, FWHM, bisectors). In this way, we can identify which signals are due to planets and those due to activity. Note that the analysis found the additional planet, CoRoT-7c, with a period of 3.7 days, which is largely accepted by the community (Queloz et al. 2009; Hatzes et al. 2010).

The signal at 9 days may also be due to a planet, but the question is still open. There are no obvious periods in the activity indicators coincident with this signal. It also appears not to be associated with a harmonic of the rotation period. Hatzes et al. (2010) argued that this was evidence for a third planet in the system, although this is still disputed (Haywood et al. 2014).

The upper panel of Figure 11.4 shows the phased RV curve for CoRoT-7b produced by removing the contribution of all other signals in Table 11.1. One important comment: the Nyquist frequency for the RV data is 0.5 day^{-1}, so normally one does not examine frequencies beyond this. In this case, we know that there is a planet signal at a frequency of 1.17 day^{-1} ($P = 0.85$ days), which appears at the alias of 0.17 day^{-1}. In this case, we chose the appropriate frequency since we know the orbital frequency of the transiting planet. The lower panel of Figure 11.4 shows the phased RV variations of CoRoT-7c ($P = 3.7$ days) also found by the prewhitening process.

When dealing with the Fourier amplitude spectrum the value of a peak (in m s^{-1}) should be approximately the amplitude of the periodic signal in the data. However, the spectral window, or leakage from other periodic signals in the data, may influence this amplitude. You should worry that by removing signals you may over- or underestimate the true amplitude of your signal. How do you know if you have removed too many frequencies, or not enough?

As a check, let's take the CoRoT-7 RVs and prewhiten these beyond the frequencies show in Table 11.1, that is, even including peaks that are not significant according to our 1% FAP criterion. Figure 11.5 shows the evolution of the K amplitude of CoRoT-7b as a function of the number of Fourier components removed (excluding the orbital frequency of CoRoT-7b). The first seven frequencies correspond to those in Table 11.1, excluding f_5 (the 0.85 day period). One can see

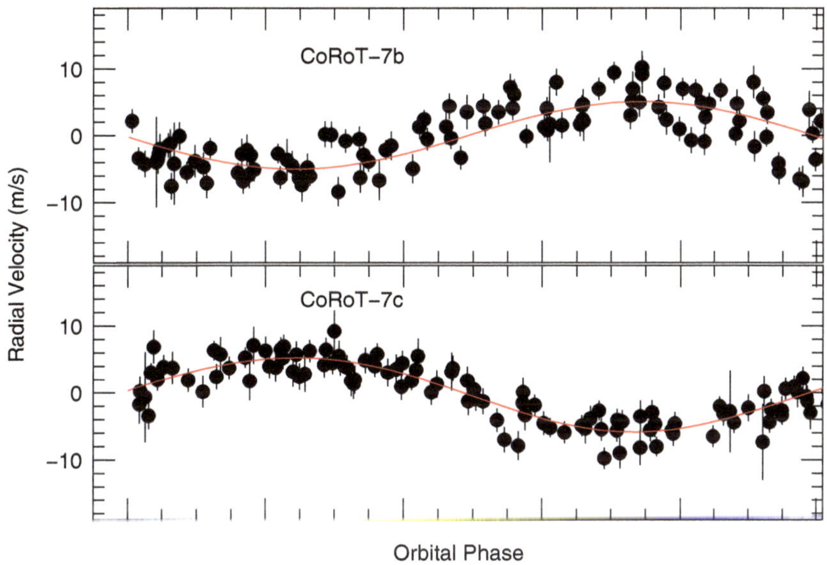

Figure 11.4. The RV variations of CoRoT-7b (top) and CoRoT-7c (bottom) phased to the respective orbital periods (0.85 days for CoRoT-7b and 3.7 days for CoRoT-7c). The contribution of all other signals due to activity has been removed via the prewhitening process.

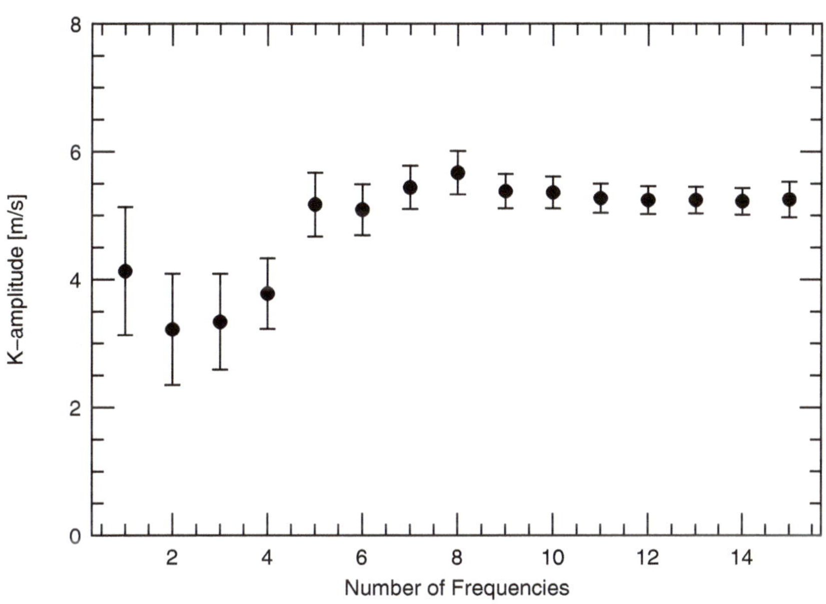

Figure 11.5. The K amplitude of CoRoT-7b as a function of the number of frequencies removed in the prewhitening process (excluding the signal of CoRoT-7b).

that the largest influence on the amplitude is from the rotational frequency, followed by the 9 day signal (CoRoT-7d), the first rotational harmonic, and finally the signal of CoRoT-7c. After that, the K amplitude settles down to $K = 5.1\ \mathrm{m\,s^{-1}}$, regardless of the number of frequencies one removes. Removing just the dominant components gets you close to the correct amplitude. Real cases may of course be different and will depend on the spectral window, the periods of stellar activity, and the number of planets and their respective K-amplitudes. At least for this case, the prewhitening procedure worked extremely well.

A variant of prewhitening, or Fourier filtering, is the so-called harmonic analysis (Queloz et al. 2009). In standard prewhitening, one chooses the highest peak regardless of the frequency. In harmonic analysis, you only use those frequencies tied to the rotation of the star, f_{Rot}, namely f_{Rot}, $2f_{\mathrm{Rot}}$, f_{Rot}, etc. (or equivalently P_{Rot}, $P_{\mathrm{Rot}}/2$, $P_{\mathrm{Rot}}/3$, etc). The rotational frequencies of the roAp star α Cir (Table 9.1) give support to this procedure. However, as we have seen in the case of the Sun, the signal due to activity can have frequencies unrelated to the rotational period.

11.1.1 The Pitfalls of Prewhitening

Before we leave Fourier filtering and prewhitening, we would like to emphasize two important points:

(1) Prewhitening can be a powerful tool for removing the activity signal and it will always give you an answer—you will find significant frequencies in the data! It is easy to use and can give you an answer in minutes, if you have the appropriate tools. The difficult part, as in analyzing any Fourier transform, is interpreting the signals that you do find. This can take weeks, if not months, of hard analyses to ferret out the true nature of the signal—user beware! It is reckless to simply use the prewhitening, quote an FAP, and then be done.

(2) Prewhitening is an excellent "quick look" tool. It can find potential planetary signals relatively easily. In cases where the K amplitude is large and dominates signal from stellar variability, or when this variability is a clean signal (e.g., single period), it works extremely well. Pitfalls and false conclusions, as we shall soon see, often arise when the K amplitude is comparable to the activity signal, as well as measurement error. If you find a signal with prewhitening, it is always wise to check the results with other methods, or use more sophisticated techniques tailored to extracting this known period.

11.2 High Pass Filtering

Photometric transit surveys have uncovered a large population of planets with periods less than a few days. In particular, the ultrashort-period planets have periods less than one day (Sanchis-Ojeda et al. 2014). The orbital frequency of these planets are typically much higher than the frequency of the rotation of the star or activity cycles. Because the orbital frequency of the planet is far removed from the low-frequency components due to stellar activity, high pass filtering—i.e., the suppression of low-frequency signals in the time series—is an effective way of isolating the RV variations of the planet. Here we discuss two methods of high pass filtering.

11.2.1 Local Trend Fitting

The danger when using Fourier filtering is that the process may accentuate noise peaks, making them look like real signals, especially if there is spectral leakage from other signals, the sampling window, etc. Remember that by removing dominant peaks you are reducing the mean amplitude of peaks in a certain frequency range and thus boosting the significance of the remaining peaks in the periodogram. You may get fooled into thinking that a noise peak is actually a signal. In these cases, it is always good to check your results with an independent analysis method.

One such method is local trend fitting (LTF), best demonstrated by Figure 11.1. You simply fit an underlying trend, but in this case over short time intervals of the measurements (a local trend). To do this, you subdivide the RV time series into time chunks, ΔT, whose lengths are constrained by two timescales. The first is the orbital period of the planet, $P_{\rm p}$, The second is the timescale defined by the rotational period of the star, $P_{\rm Rot}$. Note that for $P_{\rm Rot}$ one should also consider the harmonics if these make a significant contribution to the RV rotational modulation.

Therefore, the criterion for using LTF is

$$P_{\rm p} < \Delta T < P_{\rm Rot}. \tag{11.1}$$

The trick is in the exact choice of ΔT. It should be long enough so that you have at least a full orbital cycle or more of the planet. It should be short enough that the underlying activity signal is slowly changing and can be represented by a low-order polynomial. In other words, the planet signal should be varying much more rapidly than that of the activity.

So in essence, we are applying a high pass filter. By removing low-frequency, slowly changing components of the activity signal, we let through the higher frequency changes of the orbit. Coincidentally, this is the method often employed in filtering light curves to search for transits. For the light curve in Figure 11.2, we could have also broken the photometric data into chunks with time spans much larger than the planet orbital period of the planet, fitted low-order polynomials to these, and then searched for the transit in the combined residuals. The transit shape may be slightly different (and more correct) but to first order, the results will be the same.

We will apply LTF to the case of the purported planet–planet α Cen Bb. This demonstrates how two different approaches can come up with inconsistent results.

Dumusque et al. (2012) reported the presence of a planet with a mass of 1.13 ± 0.09 $M_\oplus$ and orbital period 3.23 days in RV data taken with the HARPS spectrograph. The K amplitude of the planet signal was only about 0.5 m s^{-1}, which was completely dominated by the activity of the star (Figure 11.6). To remove the activity signal, the authors used harmonic analysis—a version of Fourier filtering but

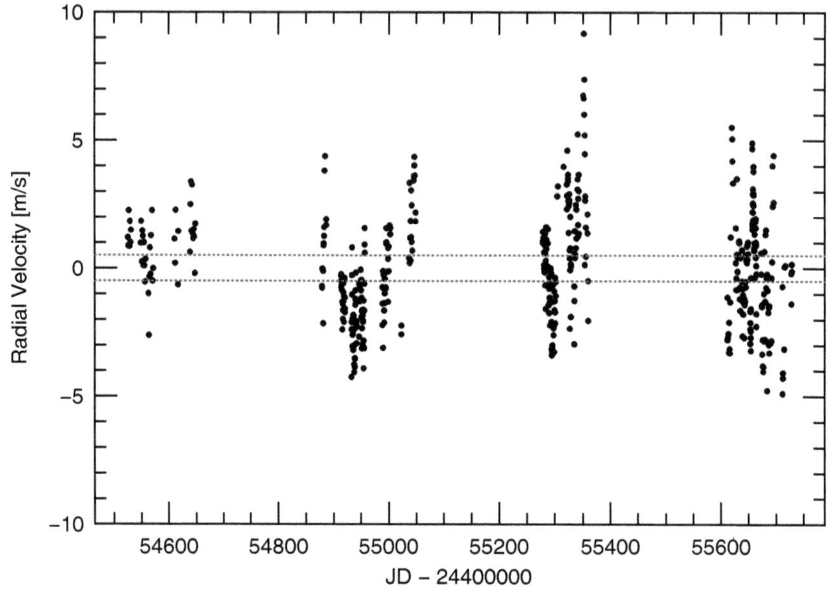

Figure 11.6. RV measurements of α Cen B taken with the HARPS spectrograph (Dumusque et al. 2012). The variations are almost entirely due to stellar activity. The horizontal dashed lines mark the amplitude of the presumed planet α Cen Bb.

restricted to the components of rotational period (≈38 days) and its harmonics. The resulting planet signal after removing the activity had a convincing FAP of 0.02%.

This result was confirmed using standard prewhitening by Hatzes (2013). The top panel of Figure 11.7 shows the Lomb–Scargle (LS) periodogram of the α Cen B RV residuals from all time segments after removing eight significant frequencies found in the prewhitening procedure. The resulting peak coincides with the planet orbital frequency of 0.309 day^{-1} ($P = 3.23$ days). A bootstrap analysis of this signal showed that the FAP is convincingly low at 0.004, consistent with the Dumusque et al. (2012) result.

This Fourier filtering (harmonic analysis and prewhitening) is the same as fitting sine functions to an underlying trend to a signal (lower panel of Figure 11.1). If the planet signal is robust, then using LTF to fit a simple function to the underlying local activity trend should produce the same result. For example, as shown in Figure 11.1, the two approaches arrived at the same answer when applied to a simulation with a simple sine function. Is this the case for α Cen B?

Figure 11.8 shows LTF applied to the α Cen B RV data. The data were divided into 21 time chunks. Figure 11.8 shows only six representative chunks. Clearly, the orbit of the planet has much higher frequency variations than the underlying activity signal. Removing the activity signal from each chunk and combining the RV residuals produces an LS periodogram with greatly reduced power (middle panel of Figure 11.7). In fact, the FAP of this signal is ≈0.4, or a factor of 1000 greater than the prewhitening result. Applying LTF to simulated data consisting of the activity signal and the planet signal shows that it would have detected a planet signal if it

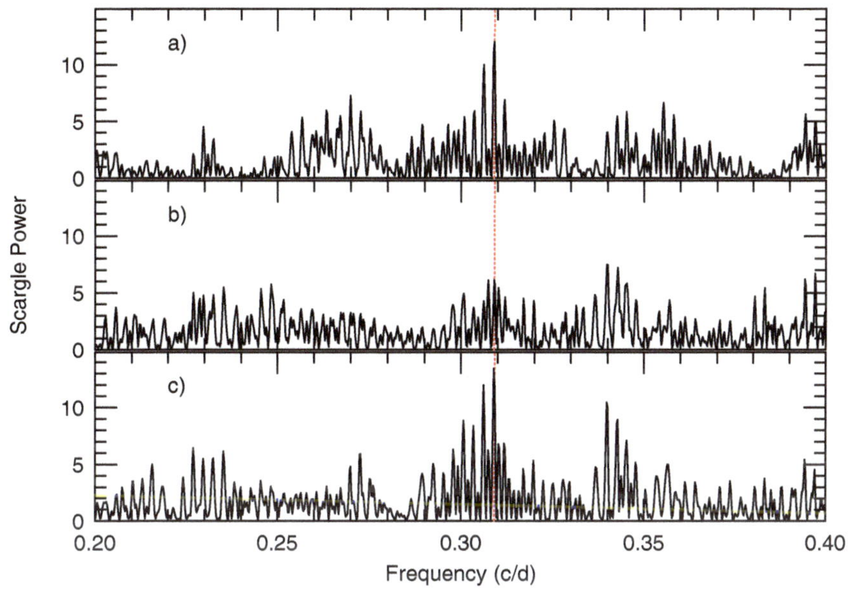

Figure 11.7. (a) The LS periodogram of the α Cen B residual RV measurements after removing all dominant frequencies via the prewhitening process. The vertical red line marks the orbital frequency of the planet, and it has a false-alarm probability (FAP) of 0.4%. (b) The LS periodogram of the α Cen B residual RV measurements produced via the LTF process. The FAP in this case is 40%. (c) LSP of residual RV measurements using the LTF on simulated data consisting of a model activity signal for α Cen B and an artificial signal for the planet inserted. The FAP of this "false" signal is 0.04%.

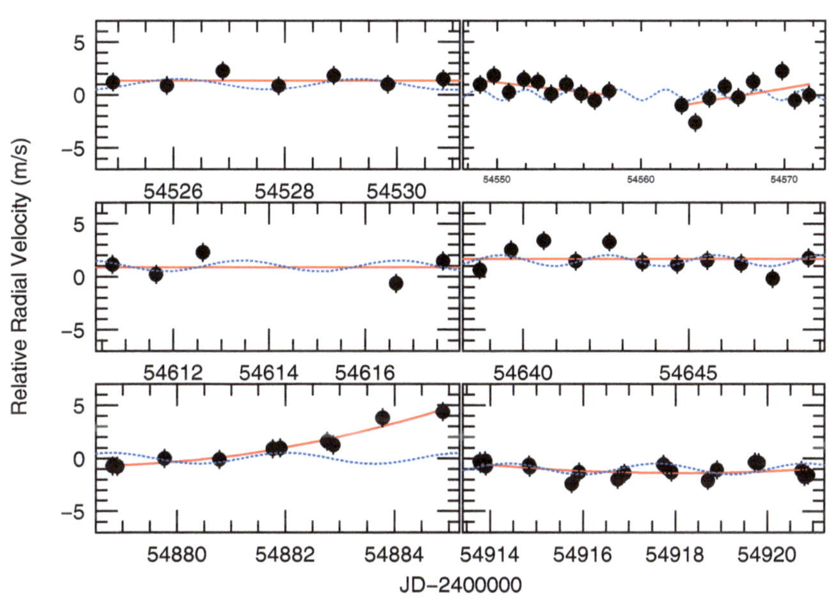

Figure 11.8. Subsets of the α Cen B RVs showing local trend fits (solid line) of the activity variations. The dashed line shows the expected RV variations from the planet α Cen Bb.

were there (lower panel of Figure 11.7). So two approaches to filtering the activity signal produce radically different results, which should cast uncertainty as to the reality of the signal. But how can Fourier filtering be so off?

A simple simulation shows how noise may masquerade as a planet after the filtering process. We created simulated data consisting of a model of the activity for Alp Cen Bb generated using the dominant Fourier components of the RV data, but without the planet signal. The planet signal found by Dumusque et al. (2012) was then added to the data. These simulated data were sampled in the same way as the real data, and random noise at a level of 2 m s^{-1}, the typical HARPS error for the α Cen B data, was finally added. As with the real data, prewhitening was used to remove the activity signal. The LS periodogram of the residuals shows a peak at 0.309 day^{-1} (Figure 11.9) at the same orbital frequency of the planet. A bootstrap yields an FAP of 0.04%. The problem is that no planet signal at this frequency was present in the data; it is an artifact.

This demonstrates that the combination of the activity signal, noise, and sampling window can create false signals. After filtering, these can appear to have a formally high statistical significance using the standard tools to assess the FAP. Hatzes (2013) suggested that the signal of α Cen Bb was most likely due to a combination of (1) the activity signal, (2) sampling of the data, (3) spectral leakage, and (4) the peculiarities of the filtering process. Rajpaul et al. (2016) conclusively demonstrated that the planet signal indeed most likely arises from "ghosts" due to spectral leakage.

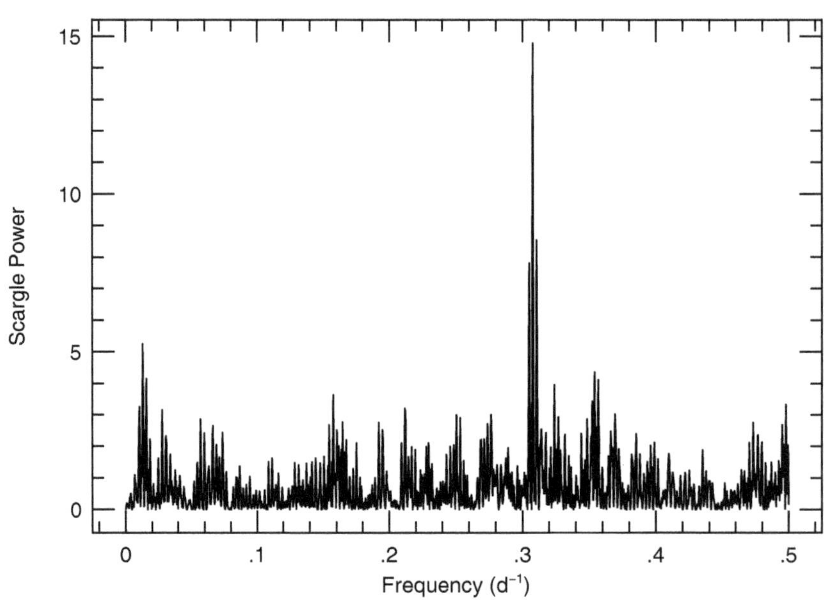

Figure 11.9. LSP of a synthetic data set consisting of a simulated activity signal that has been prewhitened. The peak at $\nu = 0.309$ day^{-1} is an artifact due to prewhitening as no signal at this frequency was inserted in the data.

11.2.2 Floating Chunk Offset

Ultrashort-period planets with periods less than one day offer us another way to filter out the activity signal. The first such planets were CoRoT-7b (Léger et al. 2009) and Kepler-10b (Batalha et al. 2011), both with periods of 0.82 days. One of the shortest-period planets discovered with a measured mass is Kepler-78b, with an orbital period of a mere 0.35 days or 8 hr (Sanchis-Ojeda et al. 2013).

Once again, for these short-period planets, we can exploit the fact that the orbital period of the planet is much shorter than the rotation period of the star and thus the timescale for stellar activity. Figure 11.10 shows how we can exploit this. For short-period planets, over the course of a night ($\Delta T \approx 8$ hr), you will observe a significant fraction of an orbital phase. Assuming a circular orbit, this will be a short segment of a sine wave (blue segments in figure). If the rotation period of the star is much longer than the planet's orbital period, then the RV contribution from spots is constant because there has not been enough time for the star to rotate significantly or for spots to evolve. The spot distribution is essentially frozen-in on the stellar surface, and it creates a velocity offset, v_1, for that first night. The observed RV variations are thus coming almost entirely from the orbital motion of the planet.

Note that we do not care what is causing this velocity offset. It could be spots, it could also be systematic errors, even additional long-period planets. All we care about is that during the course of the night, the RV contribution from all these other phenomena remains more or less constant.

We then observe the star on another night when stellar rotation has moved the spots or these have evolved so that we have a different view of the stellar activity. This will create a different velocity offset, v_2. If we do this for several nights and phase the data to the orbital period of the planet, we will see segments of sine waves, each having a different velocity offset v_i. All we have to do is calculate the best offset v_i for each segment, in a least-squares sense, that causes all these segments to align

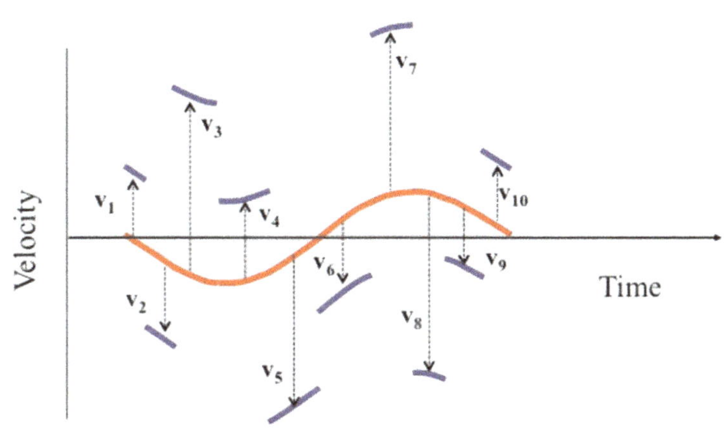

Figure 11.10. Schematic of the floating chunk offset method (FCO). Observations of a short-period planet on a given night appear as short segments of a sine function for circular orbits. The segment on a given night has a velocity (v_i) offset due to activity, long-period planets, systematic errors, etc. By finding the appropriate offset, v_i, and subtracting it, one can line up the segments to recover the planet's orbit (red curve).

on the orbital RV curve. This method was used to provide a refined measurement of the mass of CoRoT-7b (Hatzes et al. 2011). Because the offsets in each time chunk is allowed to "float," the technique is referred to as the floating chunk offset (FCO) method.

Kepler-78b provides us with a nice application of the FCO method. This is an ultrashort-period (P = 0.35 day) transiting planet found by the *Kepler* mission (Sanchis-Ojeda et al. 2013). Two independent groups used RV measurements made with Keck HIRES (Howard et al. 2013) and HARPS-N (Pepe et al. 2013) to measure a planet mass of $\approx 1.8\ M_{\oplus}$. Kepler-78 shows RV variations of ± 20 m s^{-1} due to activity, which is a factor of 10 larger than the RV amplitude of the planet (top panel of Figure 11.11). Furthermore, the two instruments had different zero-point offsets. Both are not a problem for the FCO method because the orbital period of Kepler-78b is about a factor of 30 smaller than the rotational period of the star of 10.4 days.

The lower panel of Figure 11.11 shows the RV orbital curve due to Kepler-78b after applying the FCO method to the combined RV data sets. When one compares the LS periodogram of the RV data before and after application of the FCO method, one can see that it is very effective at suppressing the low-frequency noise due to activity (Figure 11.12).

The method can also be used as a periodogram to search for unknown, short-period planets in your RV data. Basically, you take a trial period and find the

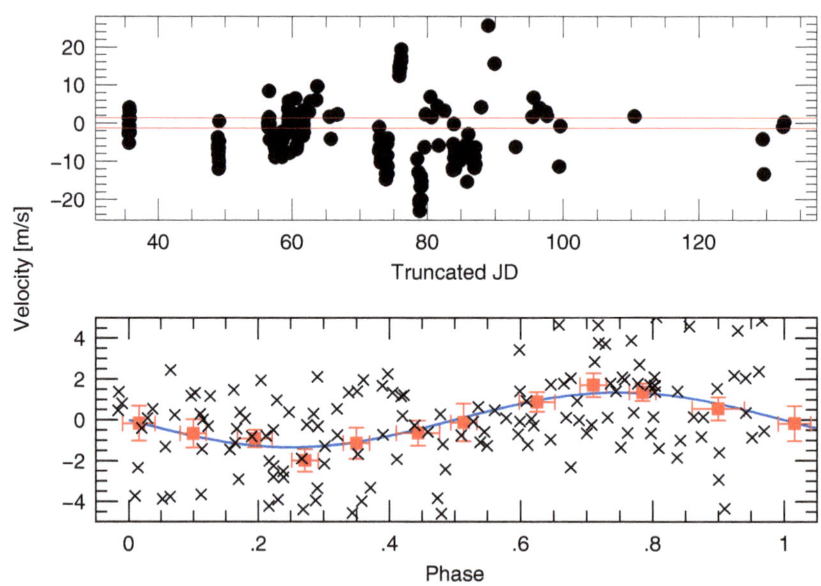

Figure 11.11. (Top) The RV measurements for Kepler-78 from the combined HARPS-N and Keck data. Both data sets have the same zero point. The two horizontal lines show the RV extrema of Kepler-78. Most variations are due to stellar activity. (Bottom) The RV measurements of Kepler-78b phased to the orbital period after using the FCO method to calculate and remove the nightly offsets due to activity and instrumental effects. Dots represent binned values.

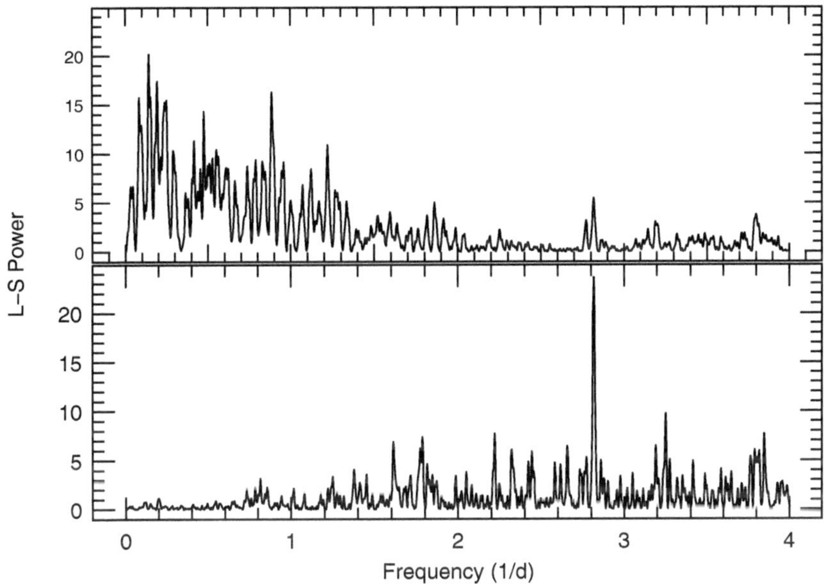

Figure 11.12. The LS periodogram of the RV measurements for Kepler-78b before (top) and after (bottom) applying FCO filtering. Note that all of the low-frequency components due to stellar activity have been filtered out, which enhances the planet signal. For the unfiltered periodogram, the two data sets were put on the same zero-point scale.

velocity offsets that provide the best sine fit to the data and calculate the reduced χ^2. Another trial period is used and an new χ^2 calculated. A plot of the reduced χ^2 as a function of input periods will show the one that best fits the data.

The FCO periodogram for the Kepler-78 RV data is shown in Figure 11.13.[1] The reduced χ^2 is plotted with decreasing values along the ordinate so that the minimum value appears as a peak, like in the standard periodogram. The best-fit period to the Kepler-78 data is indeed at 0.35 days. Even if we did not know a transiting planet was present, the FCO periodogram would have detected it.

It is also possible to use the FCO periodogram on eccentric orbits. One simply uses a Keplerian orbit with nonzero eccentricity rather than a sine wave (zero eccentricity).

The FCO method is also useful for finding periodic signals even if these have much longer orbital periods than ultrashort-period planets. All that is required is that the orbital period of the planet is at least two to three times shorter than the period of other phenomena (rotation, activity, etc.) and that you have good sampling (three to four points) for a significant fraction of the orbit within a time chunk, or about $>0.1\ P_{\mathrm{orbit}}$.

To demonstrate this, we apply FCO to a not so obvious case, the planet around Proxima Centauri. This planet has an orbital period of 11.2 days and the host star a

[1] The reader will note that in this case, we plot the period along the abscissa in spite of my exhortations to the reader to use frequency in the periodogram.

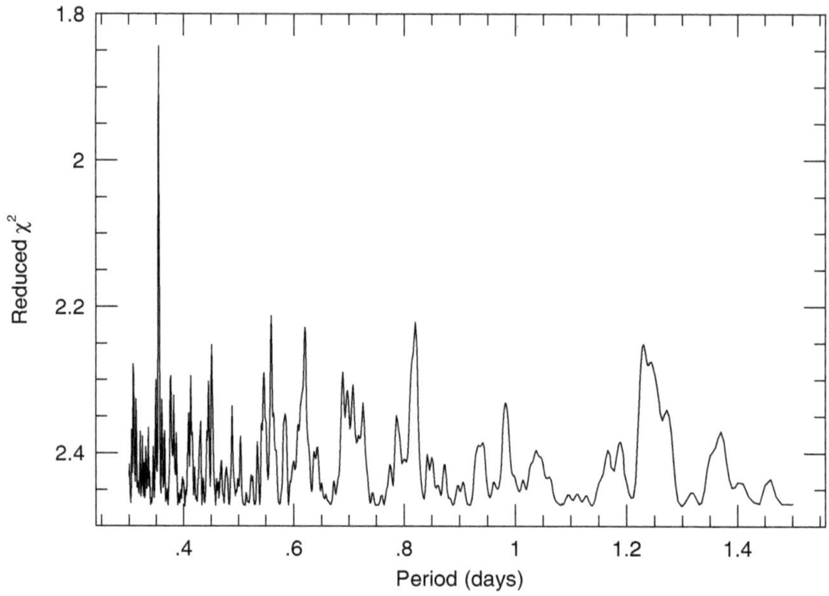

Figure 11.13. The FCO periodogram of the Kepler-78 RV measurements. Note how the filtering enhances the detection of Kepler-78b.

rotational period of ≈80 days (Anglada-Escudé et al. 2016). There are no obvious harmonics of the rotational period seen in the RV data. We divide the RV data of Anglada-Escudé et al. (2016) into subsets spanning four to nine days, or up to one orbital cycle of the planet, yet less than 10% of a rotation period.

Figure 11.14 shows the results of the FCO analysis. The best-fit orbital parameters are period $P = 11.184 \pm 0.001$ days and amplitude $K = 1.29 \pm 0.23$ m s^{-1}. These are in excellent agreement with the published values $P = 11.186^{+0.001}_{-0.002}$ days and $K = 1.38 \pm 0.21$ m s^{-1}. Allowing the eccentricity to vary shows no significant nonzero eccentricity, consistent with the poorly determined published value of $e < 0.35$.

How does FCO work as a periodogram on this data? The FCO periodogram does show the highest peak at $P = 11.18$ days (Figure 11.15), but not as strongly as does the generalized Lomb–Scargle (GLS) periodogram. All other peaks can be identified with those in the GLS.

In summary, the FCO method can be an effective tool for finding planets in the presence of activity noise. However, with one strong caveat, you must know the timescales of other phenomena. It is most effective in determining the K amplitude of transiting planets where you have a known orbital period, but the amplitude is distorted by activity jitter.

11.3 Gaussian Processes

Gaussian processes[2] have become a popular method for modeling the underlying activity RV signal from stars. In most applications, to fit observational data one

[2] For an excellent tutorial, see ftp:ftp.tuebingen.mpg.de/pub/ebio/chrisd/GPtutorial.pdf by M. Ebden.

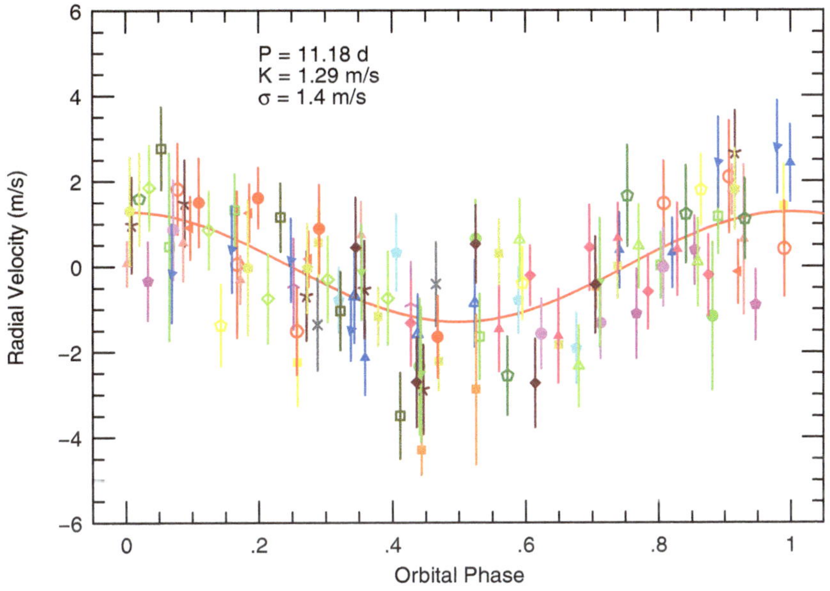

Figure 11.14. The FCO method applied to the RV data of Proxima Cen. Colored symbols represent the RV data in each subset. The period of 11.18 days and K amplitude of 1.29 m s^{-1} are consistent with published values.

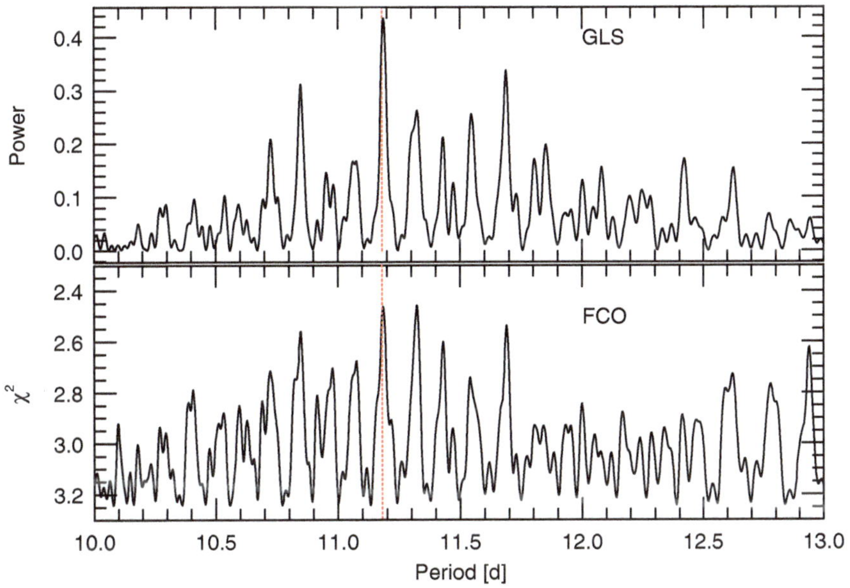

Figure 11.15. (Top) The GLS periodogram of the Proxima Cen RV data. (Bottom) The FCO periodogram of the same data. The vertical dashed red line marks the published orbital period of the planet.

chooses a model, either physical or mathematical. For example, if you expect the underlying function of $f(x)$ to be linear, we employ standard linear regression to fit the data. In Fourier filtering, we assume that the activity signal can be fit by a sum of trigonometric functions.

What if you have no good knowledge of what the appropriate model is for your data? Gaussian processes (GP) are a nonparametric way to fit your data that essentially "let's the data speak for itself" (M. Ebden), but in a mathematically rigorous manner. Gaussian processes are not completely free-form as you have to make some basic assumptions regarding $f(x)$. In a sense, it is a form of supervised learning.

The Gaussian process extends multivariate Gaussian distributions to infinite dimensions. It generates data in some domain such that any subset of the data follows a multivariate Gaussian distribution. What relates one function to another is the covariance matrix, $k(x, x')$, and a popular choice is

$$k(x, x') = \sigma_f^2 \exp\left[\frac{-(x - x')^2}{2l^2}\right], \tag{11.2}$$

$$k(t, t') = \eta_1^2 \exp\left[\frac{-(t - t')^2}{2\eta_2^2} - \frac{2\sin^2\left(\frac{\pi(t - t')}{\eta_3}\right)}{\eta_4^2}\right]. \tag{11.3}$$

The hyperparameters of $k(t, t')$ (ηs) describe various timescales for the activity:
η_1: amplitude of the Gaussian process.
η_2: timescale for the growth and decay of active regions.
η_3: recurrence timescale for active regions.
η_4: smoothing parameter.

Typically, the recurrence timescale, η_2, can be set to the rotation period of the star. The other hyperparameters can be estimated through a Monte Carlo Markov Chain (MCMC) and training the GP by maximizing the likelihood, $\mathcal{L}$, fit to, say, the photometry or RV data (Haywood et al. 2014). For a data set y, the log of the likelihood is given by (Rasmussen & Williams 2006)

$$\log \mathcal{L} = \frac{-n}{2}\log(2\pi) - \frac{1}{2}\log\left(\left|\left(\mathbf{K} + \sigma_i^2\mathbf{I}\right|\right) - \frac{1}{2}\underline{y}^T\left(\mathbf{K} + \sigma_i^2\mathbf{I}\right)^{-1}\underline{y}, \tag{11.4}$$

where $|\mathbf{K}|$ is the determinant of the covariance matrix which serves to penalize complex models. The first term is merely a normalization constant, and the χ^2 of the fit is represented by the third term. The term $\sigma_i^2\mathbf{I}$ is an additional white-noise component, where σ_i is the error on each data point y_i and $\mathbf{I}$ is the identity matrix. Sometimes this term is added to Equation (11.3).

The timescale for the growth and decay of active regions (η_2) is often comparable to the rotation period of the star (Haywood 2014; Dai 2017). Figure 11.16 shows an example of a Gaussian process model fit to the RV variations of EPIC 228732031 (Dai et al. 2017).

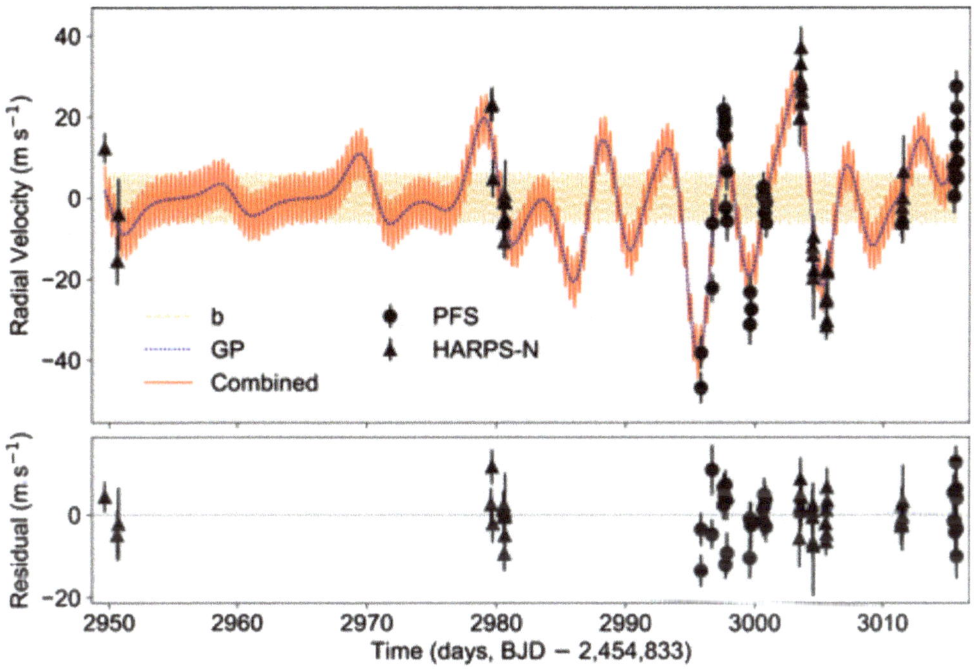

Figure 11.16. (Top) The Gaussian process fit to the radial velocities of EPIC 228732031 (blue dotted line). The red solid line is the best-fit model including the signal of the transiting planet alone with correlated stellar noise. The yellow dashed line is the signal of the planet. (Bottom) The RV residuals to the fit. (Reproduced from Dai et al. 2017. The American Astronomical Society. All rights reserved.)

11.4 A Short Comparison of Filtering Methods

In this chapter, we have presented several methods for filtering out the RV signal due to rotational modulation of stellar activity. It is of interest to see how the results of some of these compare for the same star. Ultrashort-period planets are good test cases as the RV variations are short and usually distinguishable from rotation frequencies. We will see the results of Gaussian processes, prewhitening, and the FCO method on the ultrashort-period planets CoRoT-7b, Kepler-78b, and K2-131b.

Haywood et al. (2014) used Gaussian processes on RV measurements of CoRoT-7 spanning 26 consecutive nights using the HARPS spectrograph. They also had the benefit of incorporating simultaneous photometric measurements in the GP. They derived $K = 3.10 \pm 0.68$ m s^{-1} for CoRoT-7b. Applying prewhitening to these RV data results in $K = 4.01 \pm 1.05$ m s^{-1}. The FCO method yields $K = 4.39 \pm 1.00$ m s^{-1}.

Grunblatt et al. (2015) applied Gaussian processes to the HIRES (Howard et al. 2013) and HARPS-N (Pepe et al. 2013) RVs for Kepler-78 and found $K = 1.86 \pm 0.23$ m s^{-1}. Prewhitening yields $K = 1.88 \pm 0.44$ m s^{-1}, whereas the FCO method yields $K = 1.63 \pm 0.23$ m s^{-1} (Hatzes 2014).

Finally, K2-131b is an ultrashort-period planet in a 0.37 day orbit. Dai et al. (2017) applied GP to RV data and photometric measurements for this star and derived $K = 6.55 \pm 1.48$ m s^{-1}. The FCO method yields $K = 6.77 \pm 1.50$ m s^{-1}. The

Table 11.2. *K*-amplitudes of exoplanets found via different filtering methods.

Planet	Prewhitening ($m\,s^{-1}$)	FCO ($m\,s^{-1}$)	GP ($m\,s^{-1}$)
CoRoT-7b	4.01 ± 1.05	4.39 ± 1.00	3.10 ± 0.68
Kepler-78b	1.88 ± 0.44	1.63 ± 0.23	1.86 ± 0.23
K2-131b	6.95 ± 0.66	6.77 ± 1.50	6.55 ± 1.48

RV data are not ideal for prewhitening as they used data from two instruments, HARPS-N and PFS, and prewhitening cannot deal, in a consistent way, with different instrumental offsets. Using only the more numerous HARPS-N RV data, prewhitening yields $K = 6.95 \pm 0.66\ \mathrm{m\,s^{-1}}$. We should note that the formal error for the prewhitening amplitude is most likely underestimated. Table 11.2 summarizes the *K*-amplitudes found in our choice of active stars using the various filtering methods.

The upshot is that all methods yield *K* amplitudes that are consistent with one another to the 1σ level. The preferred method comes down to the philosophical choice of the user and the appropriateness of the data. For example, FCO cannot be applied in all cases. GP tends to have smaller errors on the *K* amplitude, but these are almost always within 1σ of other methods. In many cases you can use your preferred method, but it does not hurt to compare these to the results of others just make sure your results are robust. If the planet is only found by one method, you should be cautious of the result.

Finally, it is worth mentioning the FCO and prewhitening do not incorporate the photometric data as GP did in these cases. The fact that these give consistent results mean they can be used as a "quick look" estimate of the *K* amplitude before applying the more computationally intensive GP.

11.5 The RV Challenge

The detection of small planets that produce low-amplitude RV variations is one of the more challenging problems in the detection of exoplanets with the Doppler method. If the *K* amplitude is high, then virtually all period search methods described in Chapter 7 will find planetary signals. However, if the RV amplitude is small, it will be buried in the Fourier noise of the periodogram. Complicating matters is the forest of peaks caused by stellar activity, which can masquerade as planetary signals.

To address this, Xavier Dumusque (2016) issued a "Radial-velocity Fitting Challenge" to several groups who search for planets with the Doppler method. Dumusque (2016) generated a set of RV curves from planetary systems in the presence of stellar noise using the SOAP 2.0 code. Most of the planets had a wide range of masses, but most were small (Earth, super-Earth, or Neptune). Aside from the planetary systems, the data included RV signals from instrumental noise, stellar oscillations, granulation, supergranulations, and magnetic activity. The data were sampled in a manner typical for most RV programs. These data were then handed to teams whose job it was to find the planets in the system (Dumusque et al. 2017).

Eight teams participated in the challenge. Teams 1–4 used a Bayesian framework incorporating different methods to account for the red noise. Team 5 also used a Bayesian framework but with white noise. Team 6 used traditional prewhitening (Fourier filtering) as was described in the previous chapters. Teams 7 and 8 used filtering in frequency space, but the latter with compressed sensing.

Figure 11.17 summarizes the results from the eight teams. Green regions indicate the claimed and probable planets that were true and red/orange regions the false positives, negatives, or mistaken planets. Methods including a Bayesian approach appear to be more successful at finding real planets over traditional methods based on Fourier filtering, finding about twice as many planets. The reader should see Dumusque et al. (2017) and references therein for a more detailed description of these methods.

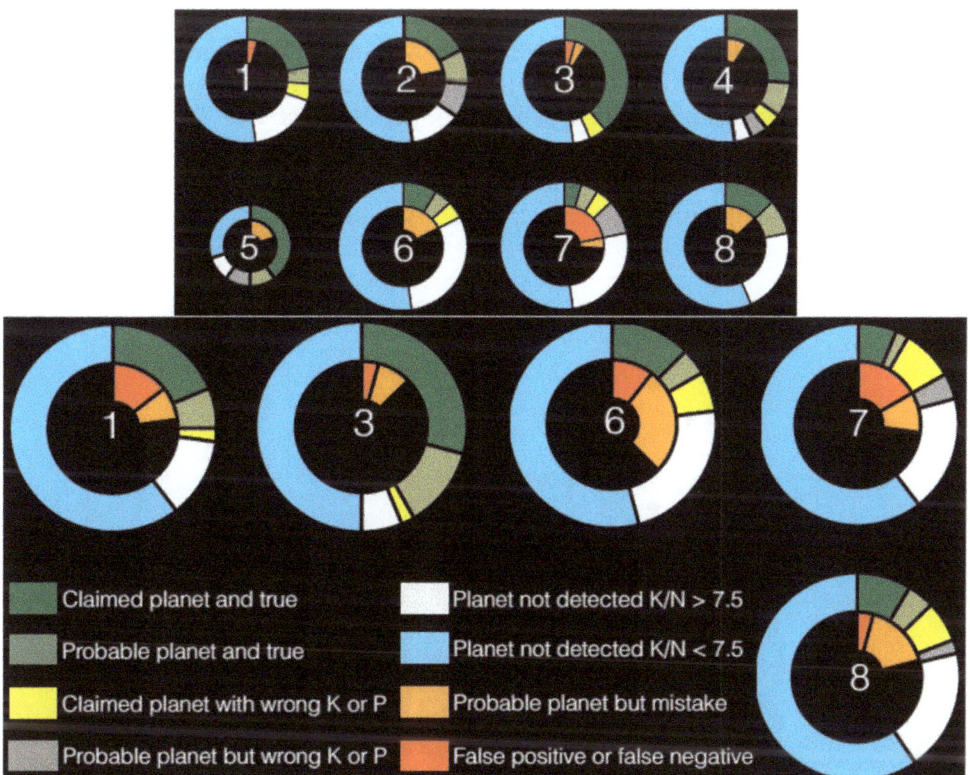

Figure 11.17. Summary of the RV challenge. (Top) The first five systems examined by all teams. (Bottom) The same for all systems, but only performed by five teams. The outer circle represents true signals that were in the data showing how each team performed. The inner circle represents planets announced by the team but were not in the data. The size of the circle represents the number of systems analyzed: large for all 14 systems, medium for five systems, and the smallest for two systems. For more information, see Dumusque et al. (2017). (From Dumusque et al. 2017, reproduced with permission © ESO.)

This RV challenge highlighted the difficulty in detecting small planets in the presence of activity noise. Even the most successful Bayesian approach (Team 3) was only able to find one-third of the planets that were present in the data. Activity noise undoubtedly represents the greatest obstacle to finding small planets with the Doppler method. The methods described in this chapter are the first steps; more work needs to be done in reliably extracting planetary signals from RV data dominated by the activity signal.

11.6 Toward Earth Analogs

The detection of Earth analogs, i.e., Earth-mass planets in the habitable zone of G-type stars, represents the largest obstacle for the Doppler method. The RV challenge demonstrated the difficulty in detecting the K amplitude of small planets in the presence of intrinsic stellar variability, even if these are in relatively short-period orbits with amplitudes of $\approx$0.5 m s^{-1}. On the other hand, an Earth analog will produce a K amplitude of $\approx$10 cm s^{-1} with a period of approximately one year. The current RV precision of the best instruments is about 0.5–1 m s^{-1}. Even if new techniques manage to bring this down to a few cm s^{-1} via improved wavelength calibration of superstable instruments, the detection of an Earth analog will still be challenging simply because the star will not cooperate. Even the "quietest" stars will show an intrinsic variability no smaller than 0.5–1 m s^{-1}.

The RV detection of an Earth analog requires that its peak in the periodogram rise above the surrounding noise peaks (i.e., Fourier noise floor) due to activity,

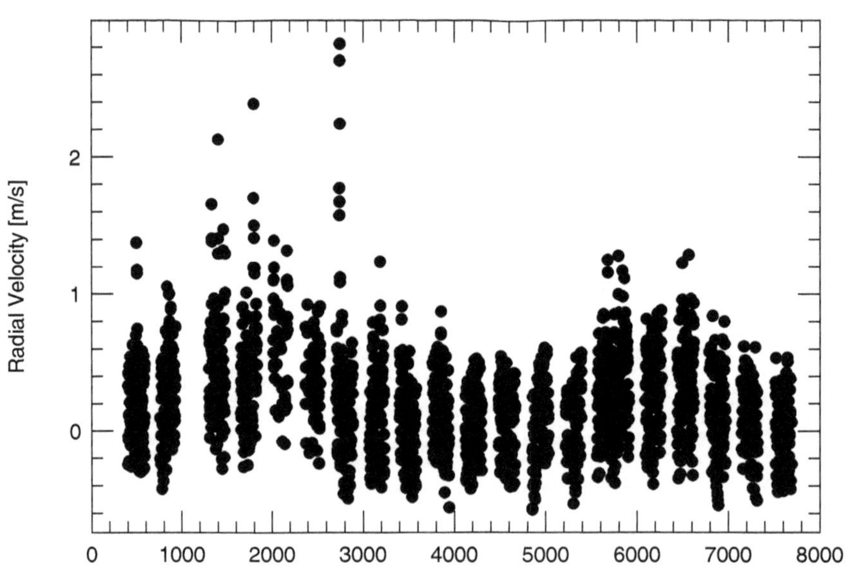

Figure 11.18. Simulated time series of Sun-like RV variations due to activity generated from the real areal coverage of sunspots. The signals of an Earth-like planet (P = 410 days, K = 10 cm s^{-1}) and a Venus-like planet (P = 280 day, K = 10 cm s^{-1}) have been added as well as random noise with σ = 0.25 m s^{-1}.

instrumental, and photon noise. There are two approaches to suppressing the Fourier noise floor. (1) You can devise clever filtering methods to suppress the contribution of the activity. Such approaches, however, may have a limited effect. (2) You can boost the planet signal above the noise level simply by taking more measurements. The RV signal due to noise is at some level stochastic—activity features come and go through their evolution, or due to the activity cycle. It is true that some periodic signals due to activity may be present, but hopefully the activity indicators described in Chapter 10 can be used to identify these.

How many measurements will it take to detect an Earth analog? As a simple experiment, we took the areal coverage of actual sunspots with time (Balmaceda et al. 2009) and converted this to an RV signal. The time series was then sampled using a typical pathology for a real RV program—measurements every five to seven days once a month and with random gaps to account for weather. We also added random noise (measurement error) of $\sigma = 0.25$ m s^{-1}. The RV time series is shown in Figure 11.18. The final time series had 2300 measurements spanning 20 years. We then added the signal of a Venus-like planet at $P = 280$ days and an Earth-like plant at $P = 410$ days. We chose a slightly longer period for the Earth-like planet to avoid the obvious problems with a one-year period (alias effects, etc.).

The top panel of Figure 11.19 shows the LS periodogram of the time series. The raw periodogram (top panel) shows a dominant peak at low frequencies due to the activity signal, although the peak due to Venus is readily apparent. Removing

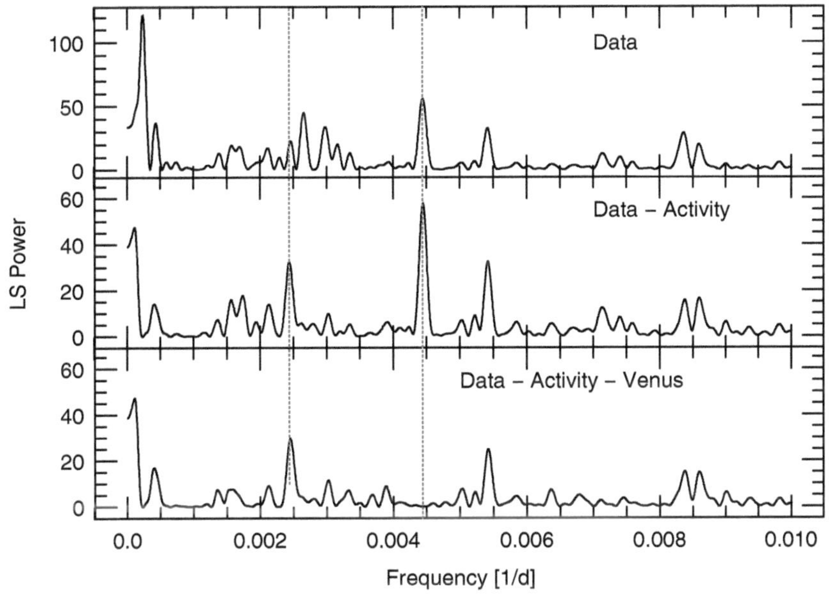

Figure 11.19. (Top) LS periodogram of simulated RV data (2300 measurements over 20 years) containing an Earth-like and a Venus-like planet in the presence of activity noise. The vertical red dashed lines mark the orbital frequency of the planets. (Middle) The LS periodogram after removing the dominant signal due to activity. The signal of Venus is readily visible. (Bottom) The LS periodogram after also removing the signal due to Venus. The Earth-like planet is just visible.

the activity signal via simple prewhitening shows Venus as the dominant peak, but the hint of an Earth signal is apparent. This becomes stronger upon removing the signal due to Venus (lower panel).

So, with 2300 measurements, one should be able to detect Earth analogs. What if the measurement error or the activity signal is larger? Simply take more measurements. For example, if your measurement error is 0.5 $\mathrm{m\,s^{-1}}$ instead of 0.25 $\mathrm{m\,s^{-1}}$, then you will need approximately four times the measurements, or almost 10,000 data points. This simple experiment demonstrates that even with superb measurement precision, it will take many thousands of measurements to detect an Earth-mass planet in orbit 1 au from a G-type star.

Any program to detect Earth analogs with the Doppler method requires:

1. A measurement technique with a stability on a timescale of decades. This precludes ones based on hollow cathode lamps.
2. Dedicated resources, i.e., a telescope facility where all of the time is dedicated to the program.
3. A relatively small sample of stars in order to ensure sufficient measurements per star.
4. An independent confirmation of the signal with a different instrument.

It will be difficult but possible with a long-term commitment of resources.

References

Anglada-Escudé, G., Amado, P. J., Barnes, J., et al. 2016, Natur, 536, 437
Balmaceda, L. A., Solanki, S. K., Krivova, N. A., & Foster, S. 2009, JGRA, 114, A07104
Batalha, N. M., Borucki, W. J., Bryson, S. T., et al. 2011, ApJ, 729, 27
Dai, F., Winn, J. N., Gandolfi, D., et al. 2017, AJ, 154, 226
Dumusque, X. 2016, A&A, 593, A5
Dumusque, X., Borsa, F., Damasso, M., et al. 2017, A&A, 598, 133
Dumusque, X., Pepe, F., Lovis, C., et al. 2012, Natur, 491, 207
Grunblatt, S. K., Howard, A. W., & Haywood, R. D. 2015, ApJ, 808, 127
Hatzes, A. P., Dvorak, R., Wuchterl, G., et al. 2010, A&A, 520, 93
Hatzes, A. P., Fridlund, M., Nachmani, G., et al. 2011, ApJ, 743, 75
Hatzes, A. P. 2013, ApJ, 770, 133
Hatzes, A. P. 2014, A&A, 568, A84
Haywood, R. D., Collier Cameron, A., Queloz, D., et al. 2014, MNRAS, 443, 2517
Howard, A. W., Sanchis-Ojeda, R., Marcy, G. W., et al. 2013, Natur, 503, 381
Léger, A., Rouan, D., Schneider, J., et al. 2009, A&A, 506, 287
Pepe, F., Cameron, A. C., Latham, D. W., et al. 2013, Natur, 503, 377
Queloz, D., Bouchy, F., Moutou, C., et al. 2009, A&A, 506, 303
Rajpaul, V., Aigrain, S., & Roberts, S. 2016, MNRAS, 456, L6
Rasmussen, C. E., & Williams, C. K. I. 2006, Gaussian Processes for Machine Learning (Cambridge, MA: MIT Press)
Sanchis-Ojeda, R., Rappaport, S., Winn, J. N., et al. 2014, ApJ, 787, 47
Sanchis-Ojeda, R., Rappaport, S., Winn, J. N., et al. 2013, ApJ, 774, 54

Chapter 12

Contributions to the Error Budget

The theoretical limit of the radial velocity (RV) precision is given by the photon noise. As we have seen, other errors, such as instrumental shifts, largely prevent you from reaching this limit. However, even if you have a superstabilized instrument, there are other, more subtle errors that can creep in, especially if you want to get to a precision well below 1 m s^{-1}. You should keep in mind that the final RV error results from a budget with every component from the front end of the spectrograph to the focal plane where the detector resides contributing to this budget (Figure 12.1). Wavelength calibration is important, but it only represents one component of the error budget. It makes no sense to invest resources to improving this one component when other factors can cause much larger uncertainties.

In Chapter 4, we discussed instrumental shifts and how simultaneous wavelength calibration can minimize these, and in Chapter 6, we discussed how changes in the instrumental profile can induce RV measurements and how these can be modeled, at least using the iodine method. Here we discuss other contributions to the "error budget" when measuring precise RVs.

12.1 Guiding Errors

As we mentioned in Chapter 2, the spectrograph is merely an optical system that produces a dispersed image of the entrance slit (or fiber) at the detector. Changes in this image due to atmospheric seeing or guiding errors can have a large influence in the measured RV.

Figure 12.2 shows a simple slit, the kind often used in classic spectrograph designs. In this example, the slit has a width that projects to 2″ on the sky. No telescope guide system is perfect, and there will be slight motions of the stellar image during the exposure; this results in a displacement of the (dispersed) slit image (i.e., the spectrum) at the focal plane. If the seeing is a poor 2″ (left panel of figure), the central portion of the Gaussian profile defining the stellar image and where most of

doi:10.1088/2514-3433/ab46a3ch12

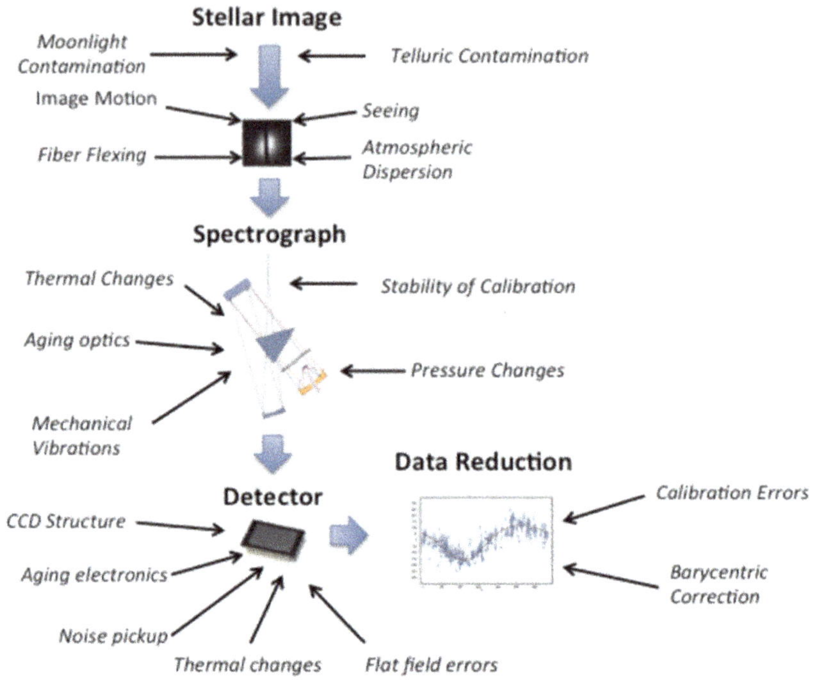

Figure 12.1. Schematic of the some of the contributions to the RV error budget.

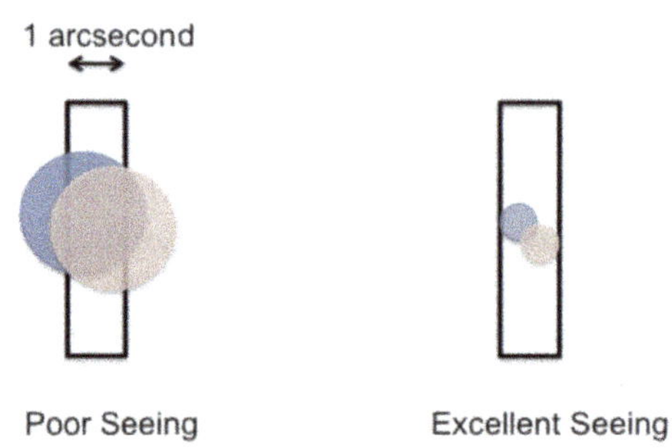

Figure 12.2. Poor guiding under different seeing conditions for a spectrograph slit with a projected width of 1″. (Left) For poor seeing conditions (≈2″), the stellar image is much larger than the slit width. In spite of poor guiding, the centroid of the stellar image is more or less in the same position. (Right) In excellent seeing conditions (≈0.5″)m there can be large motions of the star within the slit, which translates into a larger fake Doppler shift of the star.

the light is will largely stay in the center of the slit. In this case, the effect on the measured Doppler displacement of the lines will be relatively small.

This is not the case if one has excellent seeing of, say, 0.5″. Here the image can undergo large displacements relative to the image size, all the while staying within the slit. This can result in large RV displacements (right panel of Figure 12.2). Exacerbating things is that, due to the small size of the stellar image, the likelihood for bad guiding increases. Virtually all guide cameras use the reflected light from the slit jaws to determine the stellar image position. In superb seeing, there is very little reflected stellar light—it is all going down the slit! The image can wander around inside the slit, and it is only when there is a displacement large enough to bring the stellar image back on the reflective part of the slit will the guide camera react and move the telescope accordingly. Unlike for most astronomical observations, good seeing is not good for precise stellar RV measurements!

Figure 12.3 demonstrates with real data how bad guiding can affect your RV measurement. This shows a time series of RV measurements of 51 Peg using an iodine cell taken by the students of the Tautenburg Observing School. Midway through the time series, the students turned off the autoguider and moved the stellar image halfway off the slit to mimic bad guiding. One can clearly see that this produced a substantial RV displacement of $\approx$40 m s^{-1}. Note that this Doppler shift is the equivalent of the RV amplitude due to the planet 51 Peg b. A similar effect would occur if the telescope dome would occult the mirror. This is obviously a "worst case scenario," but even subtle guiding errors could introduce errors a significant fraction of 1 m s^{-1}.

To minimize the effects of guiding errors, most spectrographs designed for precise RV work feed the spectrographs with optical fibers which serve to "scramble" the stellar image. In spite of image motion at the fiber entrance (often called the near field), the output (far field) is spatially stable. There are other tricks that astronomers employ to improve the scrambling of optical fibers and thus the stability of the input into the spectrograph. These include a double scrambler—you feed the light through

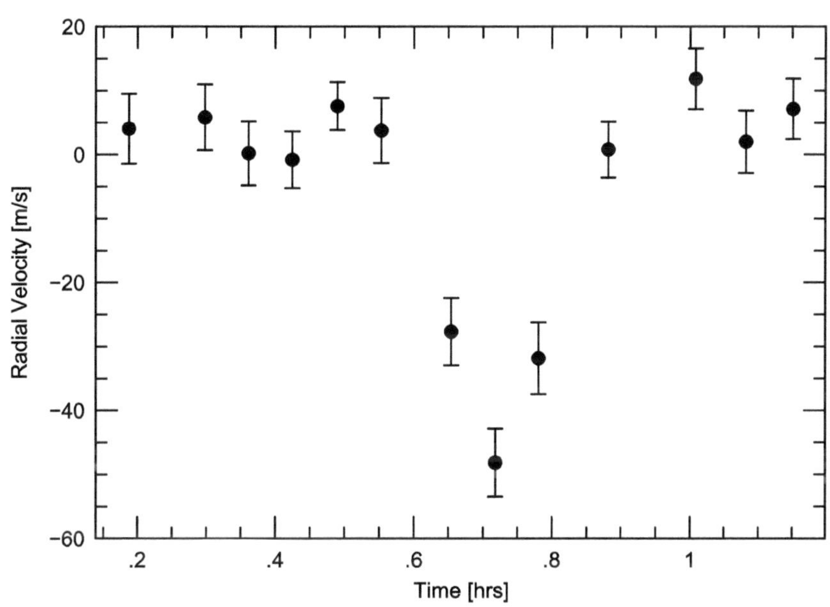

Figure 12.3. A time series of RV measurements of the star 51 Peg taken with the TCES spectrograph. For 0.6 hr, the autoguider was turned off, and the star position manually moved to the edge of the slit before returning to the nominal position.

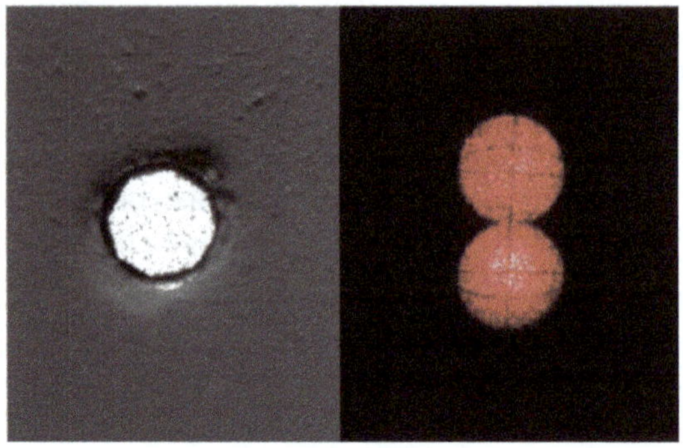

Figure 12.4. (Top) The near-field image of the octagonal fiber observed under a microscope. (Bottom) The far-field image of the back-illuminated fibers (the object and sky fibers). (From Lo Curto et al. 2015. Image courtesy of the European Southern Observatory.)

not one, but two, fibers. The scrambling ability can also be improved by "shaking" the fiber, that is, moving with a mechanism during the exposure.

Recently, spectrographs for precise RVs have started to employ hexagonal fibers as these provide superior scrambling to traditional circular fibers. Figure 12.4 shows the hexagonal fibers employed by the HARPS spectrograph (Lo Curto et al. 2015).

Even optical fibers can have guiding errors when you have to deal with atmospheric dispersion. The Earth's atmosphere acts as a refracting optical component that produces an extremely low-dispersion image of your star, particularly if you are observing at high air mass. This means that the red coming from the star could hit a different part of the optical fiber (or slit) than the blue light.

Figure 12.5 shows RV measurements from the fiber-fed FIES spectrograph at the Nordic Optical Telescope. Each point is an RV measured using a single spectral order. The RV decreases linearly by about 150 m s^{-1} from red to blue spectral orders. This is due to the fact that each order has a different central wavelength, and the stellar image for that wavelength is hitting a different region of the fiber. So, if you want to have increased RV precision, it is wise to include an atmospheric dispersion corrector as part of your spectrograph.

12.2 Changes in the Instrumental Setup

The cardinal rule of precise RV measurements is, "don't change anything!" It is best to have no moving parts in your spectrograph. Do not move gratings, cross-dispersers, slit or fiber assemblies, etc. Any change in the instrumental setup will translate into a systematic offset in Doppler shift. This is particularly true with the choice of slit width. If you have taken a time series of RV measurements using a particular slit width, then all measurements should be made with the exact same width. A different slit width results in a different resolution, a different instrumental profile, and most likely a different location of the slit "image" on the detector. All of these will produce a systematic offset in the RV measurements. Measurements taken with a different setup should be treated as if it were taken with a different instrument.

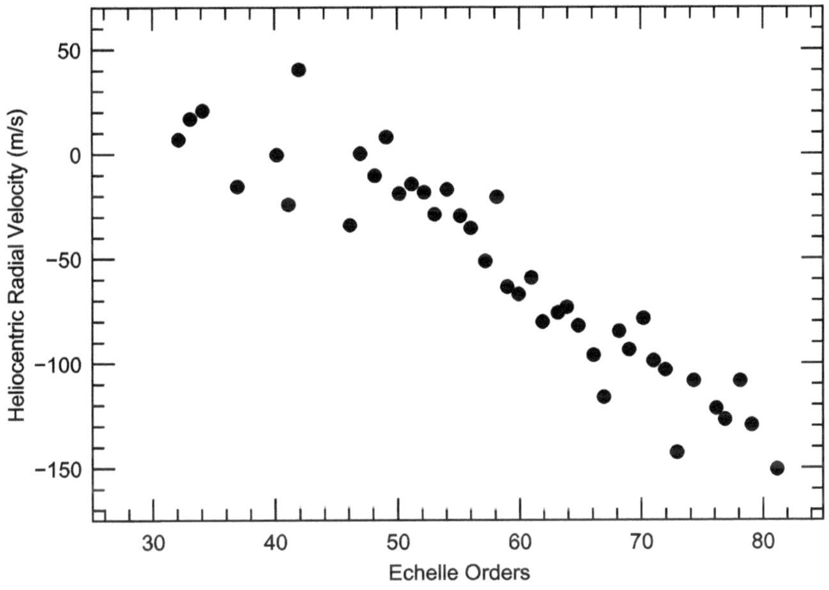

Figure 12.5. The radial velocity of a star as a function of echelle orders from the FIES spectrograph. Smaller number echelle orders correspond to longer wavelengths. (Figure courtesy of Davide Gandolfi.)

The magnitude of such an offset can also depend on how the slit assembly was designed. Do the slit jaws move in unison such that the centroid of their image is more or less on the same location on the detector? Or does one side stay fixed while the other moves? In this case, one would expect a much larger systematic RV shift.

An example of how a slight change in the instrumental setup can produce a velocity offset is the case of the planet reported around the M-dwarf star VB 10. Astrometric measurements detected a possible planet with a true mass of 6.5 M_{Jup} in a 0.74 yr orbit (Pravdo & Shaklan 2009). The detection seemed real. The false alarm probability (FAP) was a convincing[1] 3×10^{-1}, and measurements of control stars could exclude an instrumental origin for the signal.

RV measurements for VB 10 were made in the near-infrared using the telluric method by Zapatero Osorio et al. (2009), one that should, in principle, eliminate instrumental shifts; a total of five measurements[2] were made. The left panel of Figure 12.6 shows these measurements phased to the orbital period along with the orbital solution. The orbit is eccentric, but the solution is clearly driven by only one point. Remove this and there are no convincing RV variations with the orbital period.

Anytime an orbital solution is driven by one measurement, one should be cautious. You should investigate possible causes for this being an outlier. If you

[1] This only highlights that a low FAP does not provide sufficient evidence to confirm the presence of a companion.
[2] This author is of the opinion that the number should be a factor of 2–3 higher.

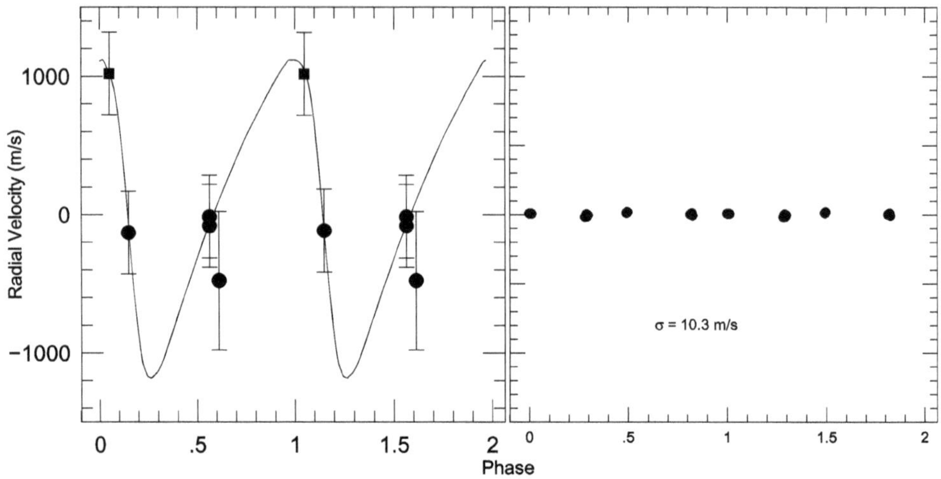

Figure 12.6. (Left) RV measurements of VB 10 phased to the orbital period of the presumed planet (Zapatero Osorio 2009). The outlier (square) was a measurement taken with a different slit width. (Right) RV measurements taken with the CRIRES spectrograph (Bean et al. 2010), again phased to the orbital period. No planet signal is seen in these data.

find none, always play it safe and get more data, particularly at key phases of an eccentric orbit.

Is there a possible cause for this discrepant point? A careful reading of the Zapatero Osorio et al. (2009) paper reveals that this one measurement was taken with a wider slit and therefore different spectral resolution. On this particular night, observing conditions were probably poor, so the observers decided to open the slit so as to get a higher signal-to-noise ratio. This almost certainly resulted in a systematic RV offset for this point. The zero-point offset between this measurement and the ones taken at higher resolution has to be determined, which is problematic when using one data point. The use of the telluric method may minimize instrumental shifts, but it still will not correct for any changes in the instrumental profile that is introduced by using data of different spectral resolutions.

Subsequent RV measurements confirmed that there were indeed problems with this one measurement. Bean et al. (2010) obtained RV measurements using the the CRyogenic high-resolution InfraRed Echelle Spectrograph (CRIRES) and an ammonia gas absorption cell to provide the simultaneous wavelength calibration. These measurements phased to the purported orbital period of the planet to VB are constant to 10.3 m s^{-1} (right panel of Figure 12.6) and thus refute the planet hypothesis.

Many fiber-fed spectrographs offer different spectral resolutions via fibers of different diameters. For these, all RV measurements should be made with the same optical fiber. Change fibers, and you have a different data set that cannot be easily combined with others. This also holds if you replace fibers, even if you keep the same diameter. You may have the same spectral resolution, but it is a change in the setup, and you should expect zero-point offsets.

12.3 Detector Errors

Great efforts often go into making the spectrograph as stable as possible, employing the most sophisticated wavelength calibration and exploiting the state-of-the-art optical fiber technology, but the detector is an afterthought. Detector performance can also introduce RV errors.

Twenty years ago, major observatories had their own CCD laboratories with dedicated technical staff whose job it was to provide detectors for all instruments used at the observatory and to ensure that these were performing properly. Currently, CCD technology has advanced to the stage where one can get high-quality, science-grade CCD cameras, including control electronics, directly from the commercial enterprises. For astronomy, CCD detectors have become turnkey devices. Today, you simply buy your CCD detector, mount it on your spectrograph, and start collecting data. No thought it given to ensuring that the CCD is a stable device that delivers consistent long-term performance.

CCD detectors can introduce RV errors in a variety of ways. We have already discussed how such things as flat-field errors and errors due to fringing may affect your RV measurements. However, there are other ways that the detector can influence your RV uncertainty in ways that are not always obvious.

12.3.1 Electronic Noise Pickup

CCD detectors literally "live in a vacuum," but they do not live in an isolated environment. At a telescope facility, motors, computers, electronics, and assorted cabling are nearby. The CCD readout is done at the spectrograph, but these have to be transferred via cables to the data storage computer. There is an infinite number of places where electronic noise can creep into your signal.

An example of this is Figure 12.7, which shows shows the time series of RV measurement errors for a star that was part of the Tautenburg Observatory Planet Search Program. For more than three years, the star had a mean measurement error of about 5 m s^{-1}. However, starting at JD = 3,454,500, the measurement error increased to more than 15 m s^{-1}. An inspection of the reduced data showed nothing out of the ordinary.

The bias level of the CCD gave the first indication that the problem was with the CCD detector. Normally, the bias level for this CCD should be about 100 analog-to-digital units (ADUs). For these anomalous measurements, the CCD bias level was about 1000, despite no changes in the control electronics. It took the technical staff of the observatory more than a week of investigations to track down the problem.

Sine-wave signal generators were used to drive the motors for the telescope motion. Recall that a sine function has a very clean Fourier spectrum represented by a δ function at the frequency of the sine wave. New motors were purchased and were driven by the old signal generators. The manufacturer of the motors urged changing the drivers to square-wave generators or risk damaging the motors. The Fourier spectrum of a square wave function is much more complicated, with many components over a wide range of frequencies. Evidently, one of these frequencies hit a resonance with the CCD electronics and introduced noise into the control

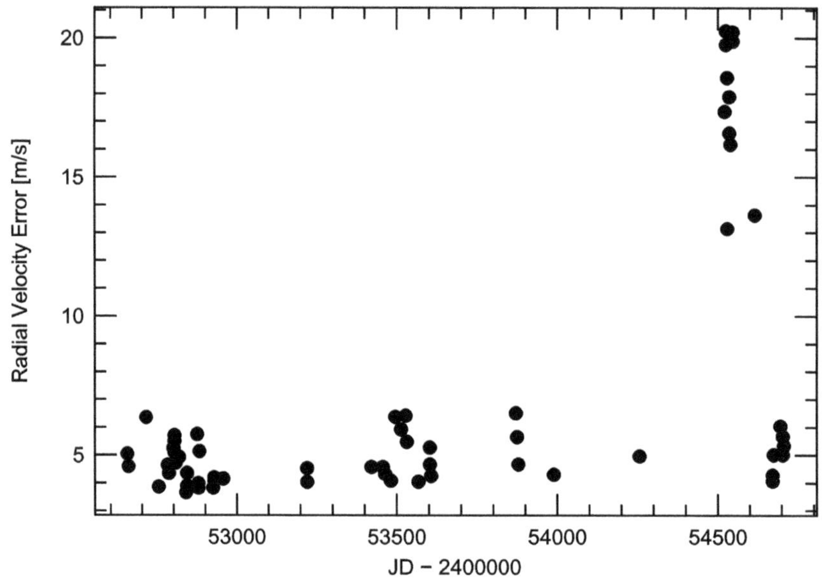

Figure 12.7. The RV measurement error as a function of time for a star from the Tautenburg Observatory Planet Search Program. The outliers were measurements taken when there were CCD electronics picking up noise from the signal generators that drove the motors for the telescope drive.

system. The "pure" frequency of the sine-wave generators had a single frequency that caused no interference. Once the signal from the motors were isolated from the RV measurements, the errors returned to their normal values, as shown by the last measurements.

The lesson learned is that if you are using an instrument for precise RV measurements, you should be concerned with what the technical day crew are doing to the telescope far from your spectrograph. In this example, the influence of the detector noise was large (tens of m s^{-1}) so it was easily noticed. But what if you are making measurements at the sub-m s^{-1} level? It is important to shield your detector and control electronics from sources of noise. No instrument is truly isolated.

12.3.2 CCD Inhomogeneities and Discontinuities

The low number density of Th–Ar lines (Figure 4.9), and all hollow cathode lamps for that matter, means that we often cannot get a good global solution to the wavelength calibration. In spectral regions where there are no thorium emission lines, one must rely on large-scale interpolations provided by the global fit. Because of this, a Th–Ar hollow cathode lamp cannot map out discontinuities in the CCD detector.

The fine line density provided by the laser frequency comb (LFC), on the other hand, enables one to map out such discontinuities. Figure 12.8, reproduced from Wilken et al. (2010), shows the residuals of the wavelength calibration for the LFC using piecewise fourth- and eight-order polynomials with a pattern having a period of 512 CCD pixels. This period arises from variations in the pixel size, shape, and

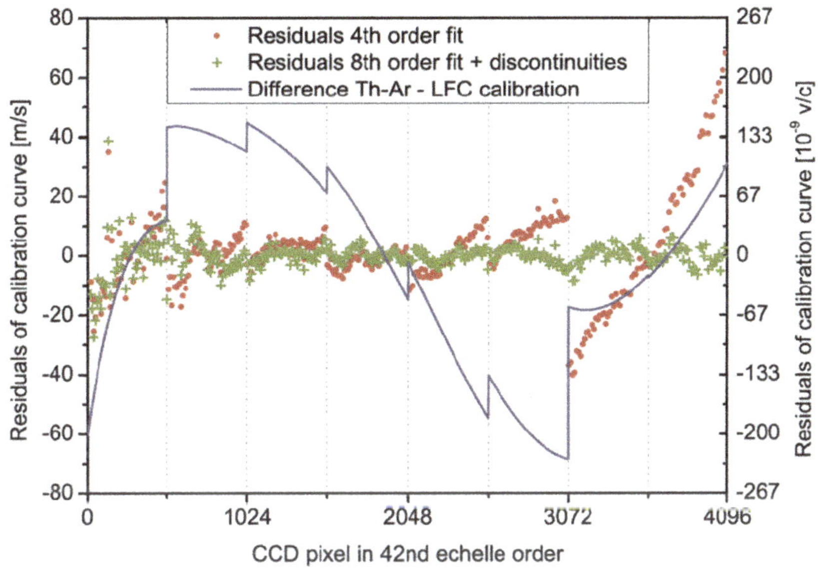

Figure 12.8. CCD detector inhomogeneities revealed by the residuals in the wavelength calibration curve using Th–Ar lines and the LFC. For the Th–Ar, a global fourth-order polynomial was used. The red dots show the residuals to the LFC calibration using a fourth-order polynomial, whereas the green cross are the residuals from fitting the data with an eighth-order polynomial. A pattern with a 512 pixel period is due to variations in the pixel size, shape, and location, due to the manufacturing process of the CCD. The solid blue line shows the difference between the Th–Ar calibration curve (fourth-order polynomial) and the LFC calibration. The deviations of the Th–Ar calibration deviates strongly due to the low line density of Th–Ar, which cannot detect CCD discontinuities. (Reproduced from Wilken et al. 2010. Copyright of OUP Copyright '2010'.)

position due to the manufacturing process. The LFC calibration improves the absolute wavelength calibration by an order of magnitude (Wilken et al. 2010).

Of particular interest is comparing the LFC to that of traditional Th–Ar calibration (fourth-order polynomial). The blue line in Figure 12.8 shows the differences between the Th–Ar and LFC calibrations. Th–Ar has insufficient line density to map out the discontinuities of the CCD, which the LFC does quite well. The differences in calibration can lead to errors of 60 m s^{-1}.

With errors of 60 m s^{-1}, you may well ask, "why does Th–Ar calibration achieve RV precisions considerably better, or at $\approx$2 m s^{-1} (Figure 4.30)?" The answer is that we are making relative RV measurements so that the absolute wavelength calibration is not as important. It only matters if the wavelength solution changes with time. Of course, if you wish to achieve RV precision below 1 m s^{-1}, treating the CCD discontinuities will be important.

Structure in the CCD wafer in the form of interpixel gaps that are introduced as part of the manufacturing process can also be the source of RV errors when coupled with other effects such as Earth's barycentric motion. One can see such structure in Figure 2.19 as a pattern of horizontal lines, which are removed (at least to the eye) in the flat-fielding process.

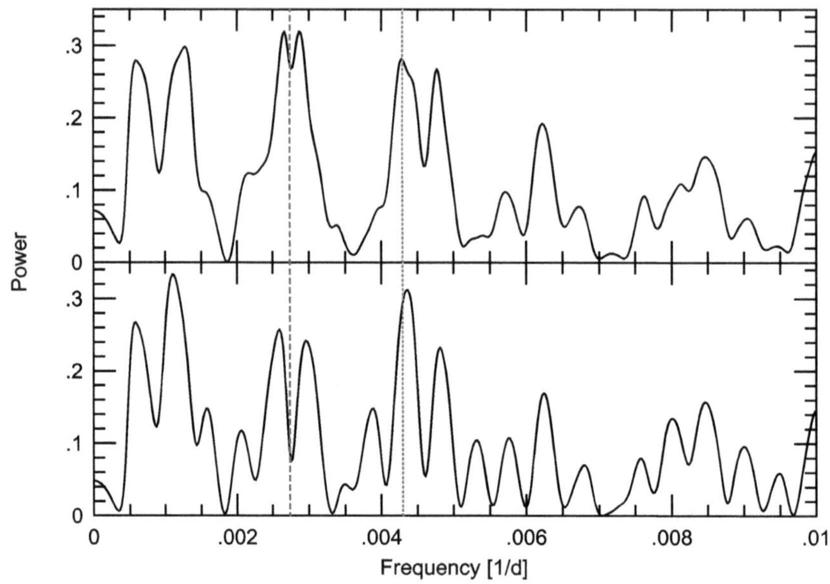

Figure 12.9. (Top) The GLS periodogram of HARPS RV measurements for Barnard's star. The vertical red dashed line to the right marks the location of the planet, whereas the maximum amplitude appears at a frequency corresponding to a period of 365 days (vertical blue dashed line at left). The RVs were calculated using a wavelength solution that included the "stitches" of the CCD. (Bottom) The HARPS RVs for Barnard's star calculated using a wavelength solution that avoided the stitch boundaries of the CCD. (The RV data was courtesy of G. Angalada-Escudé and I. Ribas.)

The CCD used for the HARPS spectrograph has such a structure of "stitch" pattern. Dumusque et al. (2015) noticed that several stars in the HARPS program were showing an RV signal of a few m s^{-1} at the suspicious period of one year. They found that a few spectral lines were crossing the stitch patterns of the CCD due to the barycentric motion of Earth. These had an RV amplitude of up to a hundred m s^{-1}, with a period of one year. When computing the RV signal using all lines, this amplitude was reduced to a few m s^{-1}, but still a significant contribution.

Figure 12.9 shows this effect for Barnard's star, an M-dwarf that was shown to host an planet with a 233 day orbit (Ribas et al. 2018). The top panel shows the generalized Lomb–Scargle (GLS) periodogram of HARPS RVs calculated using all of the spectral lines. There is a strong double peak centered at a frequency of 0.00274 day^{-1}, i.e., a period of one year. One can also see the peak due to the planet, but it is not as strong. The lower panel shows the RV calculated from the same HARPS data, but after masking out those lines near the CCD stitches. The power at a frequency of yr^{-1} is greatly reduced. If your photon errors produce an RV precision of ≈5–10 m s^{-1}, this effect would not be noticed.

12.3.3 Charge Transfer Effects

In Chapter 2, we discussed the charge transfer efficiency (CTE) for CCDs. Recall that this is the loss of charge as it is transferred from pixel to pixel in the readout

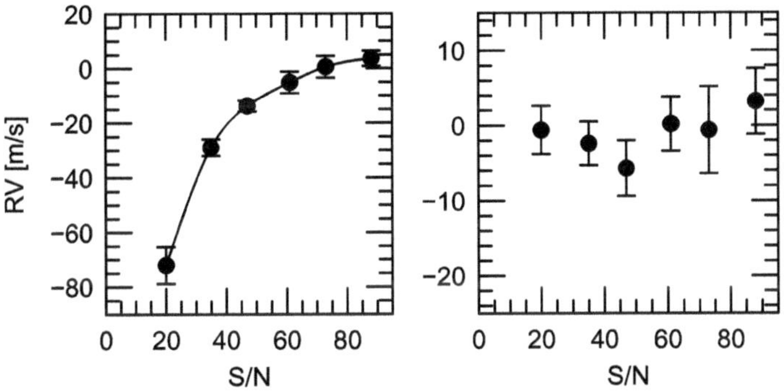

Figure 12.10. (Left) RV versus signal-to-noise from an average of three stars from Bouchy et al. (2009). The velocities are strongly correlated with S/N due to charge transfer inefficiency. (Right) Doppler shift versus S/N after a software correction.

process. Even though this is rather high (99.9997%), the decrease in S/N can have an influence in the measured RV.

As an example of the CTE effects on the RV, we use the investigation of Bouchy et al. (2009) on the CCD used for the SOPHIE spectrograph. They found a strong correlation between the RV and the signal for S/N < 70. The left panel of Figure 12.10 shows the binned average of data for three stars measured by Bouchy et al. (2009).

Bouchy et al. (2009) used the calibration of the charge transfer inefficiency (CTI = 1 − CTE) derived for the STIS CCD on the *Hubble Space Telescope* by Goudfrooij et al. (2006) to include a correction term for CTI to the measured RV. If $I(y)$ and $B(y)$ are the measured signal and background on pixel y, respectively, then

$$\mathrm{CTI}(I,\, B) = \alpha I^{-\beta} \exp\left(-\gamma\left(\frac{B}{I}\right)^{\delta}\right), \tag{12.1}$$

where I is the signal level and B is the background level.

For the SOPHIE CCD, similar parameters to Goudfrooij et al. (2006) were found, namely $\alpha = 0.056$, $\beta = 0.82$, $\gamma = 0.205$, and $\delta = 3.00$.[3] As expected the CTI decreases with increasing signal and decreasing background (at a given signal),

$$\mathrm{CTI}(y) = \alpha I(y)^{-\beta} \exp\left(-\gamma\left(\frac{B(y)}{I(y)}\right)^{\delta}\right), \tag{12.2}$$

$$I_0(y) = I(y)/(1 - \mathrm{CTI})^{y}, \tag{12.3}$$

$$I_0(y + 1) = I(y + 1) - (I_0(y) - I(y)). \tag{12.4}$$

Applying these corrections provided a noticeable improvement in the RV measurement. Figure 12.10 shows the RV shift versus S/N measured on thorium spectra before

[3] Parameters may depend on the CCD.

and after the CTI correction. Reducing the flux by a factor of 30 results in an RV shift of about 35 m s^{-1}. The RV drift is removed after applying the correction.

Should one apply the CTI correction? First of all, one should measure how large the effect is for a given CCD. Bouchy et al. (2009) also investigated the CTI effect on the HARPS CCD and found it to be approximately a factor of 10 lower than for the SOPHIE CCD. This puts it within the uncertainty of the photon noise uncertainty, so no CTI correction is needed. It may be be important for low-S/N data, but one should make sure that the effect is not simply buried in the photon noise. However, if one ultimately needs to achieve RV precisions at the level of cm s^{-1}, then the CTI should be taken into account.

12.4 Errors in the Barycentric Correction

As we have seen, the orbital motion and rotation of the Earth can cause Doppler shifts as large as ±30 km s^{-1} and as high as ±460 m s^{-1}, respectively. Although this motion can be removed with exquisite precision when applying standard tools for barycentric correction, there are a number of other ways for the errors in the barycentric correction to creep into the measurements.

12.4.1 Inaccurate Time of Observations

The largest error on the barycentric correction comes from an inaccurate time for the observations. Typical exposure times for RV measurements are a few minutes up to half an hour, depending on the brightness of the target. In order to do proper barycentric correction, you need to have an accurate time for the observation. Often, one simply takes the midpoint in time of the exposure. This is valid so long as there are no transparency variations during the exposure.

Figure 12.11 is a schematic of an observation where there are large sky transparency variations during the exposure—e.g., clouds are moving in. Most of the photons arrive before the midpoint in time (referred to as the geometric midpoint) of the observation. In this case, the intensity median occurs at a time Δt before the midpoint of the exposure. This time difference will result in a slight error in the barycentric correction. All precise RV programs use the flux-weighted

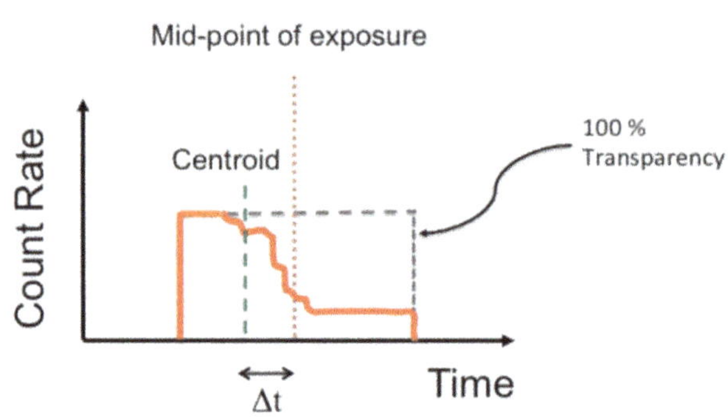

Figure 12.11. For 100% transparency, the time of the observation is defined as the midpoint of the exposure. A variable transparency shifts the centroid of the time by an amount Δt.

time as the time of the observations. This is typically done with an exposure meter that monitors the count rate and either records the entire count rate versus time, or simply calculates and records the intensity midpoint of the observations.

Tronsgaard et al. (2019) investigated the RV errors due to differences between using the geometric and the photon-weighted midpoints in time. If v_0 is the barycentric velocity correction based on the geometric midpoint and v_p the photon-weighted midpoint, then the difference in time can be approximated by

$$v_0 - v_\mathrm{pm} \approx -2.0 \text{ m s}^{-1} \cos(lat)\cos(\delta)\cos(\psi(t_\mathrm{p}))\frac{t_0 - t_\mathrm{pm}}{1 \text{ min}}, \quad (12.5)$$

where *lat* is the latitude, δ the declination of the object, and $\psi(t)$ the local hour angle. The latter can be expressed in radians in terms of the local sidereal time (LST):

$$\psi(t) = \frac{2\pi}{24 \text{ h}}\mathrm{LST}(t, lon) - \alpha. \quad (12.6)$$

A systematic error of up to 2 m s^{-1} in the barycentric correction can occur for every minute of time difference between using the geometric and photon-weighted midpoints in time.

There is also a second-order effect due to the fact that the barycentric correction does not change linearly, but has some curvature due to the change in the barycentric velocity from the diurnal rotation of Earth. This is highlighted in

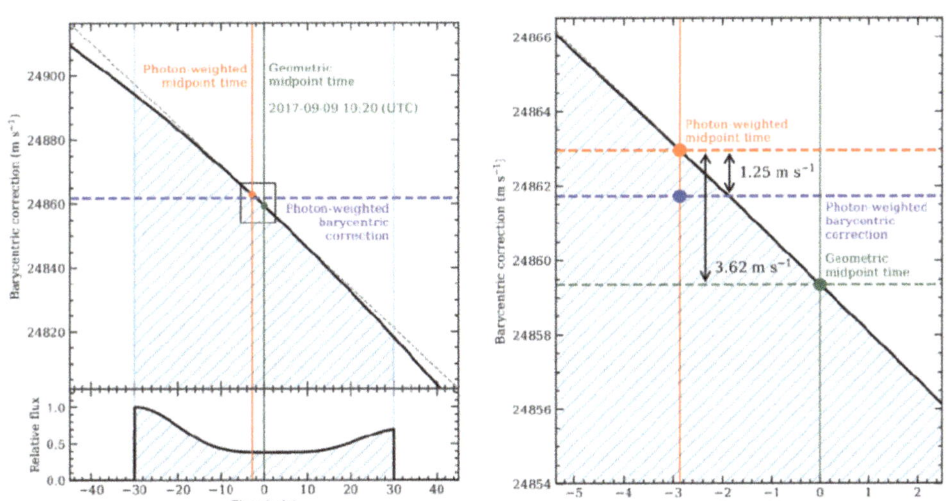

Figure 12.12. (Left panel) The barycentric correction (black curve) for a fictional star observed during a 60 minute exposure from Maunakea. The dashed line is the tangent to the barycentric correction (BC) function and highlights the curvature of this function. The lower panel shows the flux recorded by an exposure meter. The decrease can be due to a thin cloud passing over the telescope or from seeing changes. The red dot and line mark the photon-weighted midpoint in time while the blue line the photon-weighted BC. The green dot and line mark the geometric midpoint in time. The right panel is an enlarged view around the center of the exposure. In this case, the difference is 1.25 m s^{-1} between the photon-weighted BC (blue dot) and the BC computed using the photon-weighted midpoint (red dot). (Reproduced from Tronsgaard et al. 2019. Copyright of OUP Copyright ‘2019’.)

Figure 12.12, from Tronsgaard et al. (2019), which shows the instantaneous barycentric correction for a fictional star observed from Maunakea. The error in the barycentric correction between using the geometric midpoint as opposed to the photon-weighted midpoint in time results in an error of 3.62 m s^{-1}. However, if one uses the photon-weighted barycentric correction, which accounts for the curvature of the barycentric motion, then this results in a time difference of 1.25 m s^{-1} between it and the correction using just the photon-weighted midpoint in time.

If v_{pa} is the photon-weighted average of the barycentric correction, then the difference between this and the velocity calculated using the photon-weighted midpoint in time (v_{pm}) can be approximated by

$$v_{pm} - v_{pa} \approx -2\pi^2 V_0 \sin(\psi(t)) \frac{t_{pm}^2 - t}{(24\ \mathrm{h})^2}. \tag{12.7}$$

12.4.2 Inaccurate Telescope Coordinates

If you need to apply barycentric corrections to your RV measurements, the first step is to obtain accurate coordinates of your telescope in terms of latitude, longitude, and height (the latter needs the radius to Earth's center). Wright & Eastman (2014) showed that a positional error of 100 m in either height, latitude, or longitude produces a velocity error of about 1 cm s^{-1} in barycentric correction (Figure 12.13). This is about 10% of the amplitude for a terrestrial planet in the habitable zone of a Sun-like star.

One should be cautious in using coordinates of telescopes provided by observatories. What is the location of these coordinates? The telescope pier? If the observatory has multiple telescopes, for which telescopes were the coordinates taken? Finally, what is needed are the coordinates of the spectrograph. For traditional coudé spectrograph rooms, these can be 5–10 m below the telescope pier. With fiber-fed spectrographs, these can be located quite far from the nominal

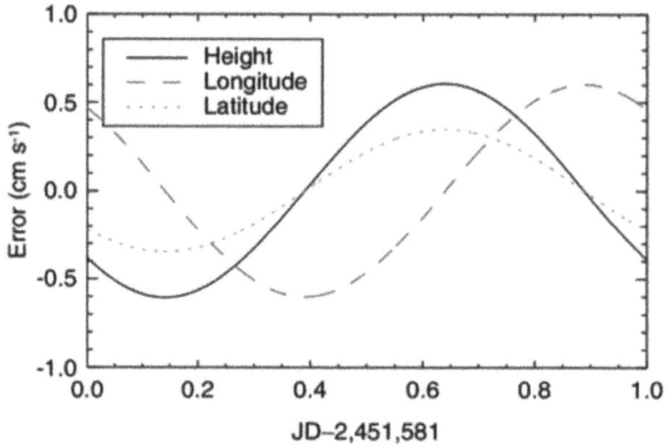

Figure 12.13. Barycentric error due to a positional error of 100 m in height (solid), longitude (dashed), and latitude (dotted) in the coordinates for the CTIO 1.5 m telescope. (Reproduced from Wright & Eastman 2014. The American Astronomical Society. All rights reserved.)

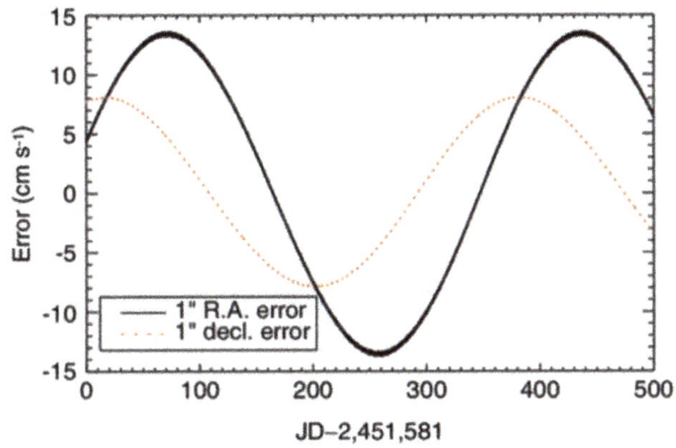

Figure 12.14. The error in the barycentric correction incurred for a 1″ error in declination (red dashed curve) and right ascension (solid black curve) in the position for τ Ceti. The stellar position must be known to better than 1″ to achieve an RV precision below 10 m s^{-1} (Reproduced from Wright & Eastman 2014. The American Astronomical Society. All rights reserved.)

position of the telescope. If one is interested in ultraprecise RV measurements down to several cm s^{-1}, you have get precise coordinates of your instrument. If one has a 1 arcsecond error in either R.A. or declination this can produce an RV error of up to 10 cm s^{-1} (Figure 12.14).

12.4.3 Inaccurate Stellar Positions

In order to have an accurate barycentric correction, you need to know the position of the star quite accurately. Unfortunately, stars have a proper motion, and over time, the stellar coordinates will be wrong. The worst case is Barnard's Star with the highest proper motion of 14″ yr^{-1}. If you do not update this star's coordinates after one year, the wrong barycentric correction will introduce an RV error of about 10 m s^{-1}.

For long-term RV surveys, it is essential to correct the stellar coordinates for the proper motion of the star. Clearly, the error becomes larger the longer you monitor stars. This is also essential for ultraprecise RVs. For example, if a star has a modest proper motion of one-tenth of an arcsecond per year, the barycentric error will be several tens of cm s^{-1} after a decade. Thanks to *Gaia*, we now have very precise positions and proper motions for all stars accessible by precise RV measurements.

Figure 12.15 shows the barycentric error introduced by a 10 mas yr^{-1} error in the proper motion of τ Ceti (from Wright & Eastman 2014). For this star, a 10 mas yr^{-1} error in the proper motion amounts to an error in the barycentric correction of $\approx$5 m s^{-1}.

12.4.4 Differential Barycentric Motion

The left panel of Figure 12.16 shows the barycentric motion of Earth for a typical star. If you take an exposure of 30 minutes, the spectral lines will have shifted by about 50 m s^{-1} from the start of the observation. This results in a blurring of the spectral lines due to the barycentric motion—the "BM-blurring function." This will

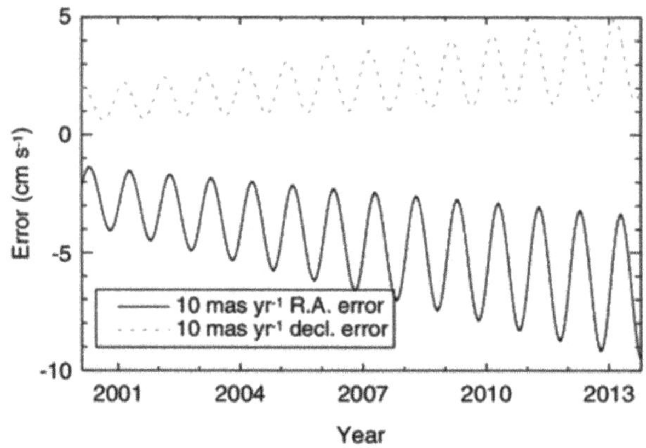

Figure 12.15. The barycentric error introduced by a 10 mas yr^{-1} error in the proper motion of τ Ceti. (Reproduced from Wright & Eastman 2014. The American Astronomical Society. All rights reserved.)

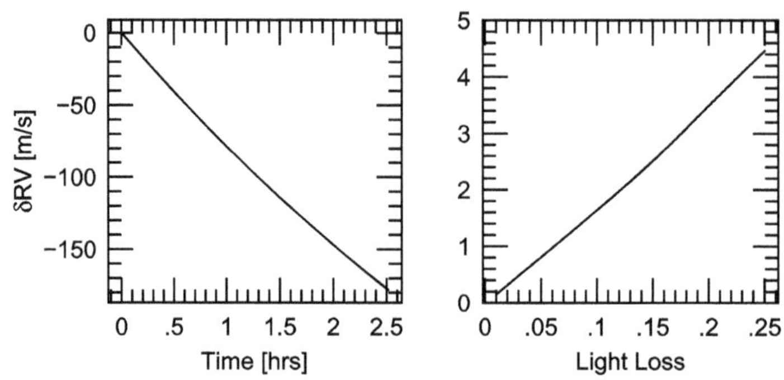

Figure 12.16. (Left) The typical barycentric motion for a star. (Right) The Doppler shift due to barycentric motion of the spectral lines on the detector, which creates an asymmetric profile as a function of transparency losses during a 30 minute exposure. A loss of 0.2 means the count rate is 20% less at the end of the exposure with a linear trend.

affect the line shapes, and it will be different for each exposure due to different barycentric motions, transparency effects, etc. If you use a stellar spectrum as your template, this, too will have its own BM-blurring function. Differences in the BM-blurring function can introduce errors in your RV similar to those due to changes in the instrumental.

To estimate the magnitude of this, we performed a simple numerical experiment. We took a synthetic stellar line profile and shifted this at a rate of 50 m s^{-1} per hour but at the same time decreasing the flux for each shifted profile. This would mimic a situation where the sky transparency was decreasing during the exposure, resulting in a slightly asymmetric line profile. This integrated profile was then cross-correlated with a profile produced using the same BC-blurring function, but this time not decreasing the flux of each shifted profile.

The right panel of Figure 12.16 shows the induced Doppler shift for a 30 minute exposure and for different factors for the light loss. For example, a light loss of 0.20 means that the count rate from the star is 20% less at the end of the exposure, and with a linear decrease, For a light loss of 20% due to transparency, at the end of a 30 minute exposure, a Doppler shift of $\approx$3 m s^{-1} will be introduced. This is because the decreasing light level from the star has a slightly asymmetric spectral line profile as the lines move across the detector, and these are not present in the template.

12.5 The Secular Acceleration

Another error resulting from the proper motion of a star is the so-called secular acceleration. This acceleration is not actually an error, but a physical effect due to the motion of the star. However, if not accounted for, one may mistake the phenomenon for a long-term RV trend due to a companion in a long-period orbit. It arises from the different viewing angle of a high proper motion star (right panel of Figure 12.17). Imagine a high proper motion star that is approaching you. When it is at a large distance, you will measure a blueshifted velocity ($-v$). As the star approaches you, the tangential velocity of the star increases at the expense of the radial component. It then crosses your line of sight, where the radial velocity goes through zero and to positive values. Finally, when the star is far away, the radial velocity you measure is a redshifted value, $+v$. The secular acceleration depends on the proper motion of the star and your viewing angle. We include this as an "error" because it is an unwanted signal when searching for exoplanets.

The minimum distance, d_0, at closest approach is given by

$$t = t_0 - \frac{\sqrt{d^2 - d_0^2}}{v},$$

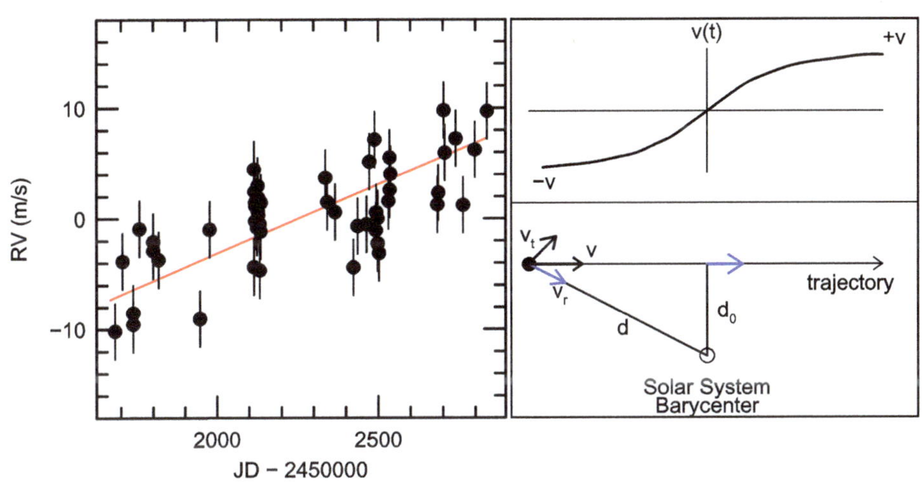

Figure 12.17. (Right) Schematic showing how the secular acceleration arises due to the proper motion of the star. Far from the observer, the star has a high radial velocity, $\pm v$. As the star crosses the line of sight, the radial velocity passes through zero. (Left) RV measurements for the high proper motion star, Barnard's Star (Kürster et al. 1999). The line represents the predicted secular acceleration.

where $v = \sqrt{v_r^2 + v_t^2}$ is the space velocity and t_0 is the time of closest approach. Setting t_0 to be the astrometric radial velocity (ARV) as a function of time,

$$v_r(t) = \frac{v^2 t}{\sqrt{v^2 r^2 + d_0^2}}.$$

The upper-right panel of Figure 12.17 shows a plot of this function. Differentiating yields the secular acceleration of the RV:

$$\frac{dv_r(t)}{dt} = \frac{v^2}{\sqrt{v^2 t^2 + d_0^2}} - \frac{v^4 t^2}{\left(v^2 t^2 + d_0^2\right)^{3/2}} \tag{12.8}$$

The left panel of Figure 12.17 shows the RV measurement of Barnard (Kürster et al. 1999), the star which has the highest proper motion. The predicted secular acceleration of the star (line) fits the observed trend in the RVs quite well (line). If this is not taken properly into account, you would think this linear trend was due to a companion (stellar or planetary) to the star.

12.6 Telluric Line Contamination

We have seen how telluric lines can be used as a reference for measuring relative stellar Doppler shifts, thus minimizing the effects of instrumental shifts. However, when not used as an RV "technique," these telluric lines are a nuisance that contaminate the stellar spectrum and degrades the RV precision. Telluric features are fixed in wavelength, but the stellar lines move across mostly due to Earth's barycentric motion. Telluric line contamination is recognized as one of the main contributors to the RV error budget (Halverson et al. 2016).

Figure 12.18 shows a synthetic spectrum of telluric features from 0.3–30 μm (Smette et al. 2015). Telluric features (H_2O, CO_2, N_2, etc.) completely dominate the wavelength region beyond about 1μm with virtually no windows where the stellar spectrum will be uncontaminated. It is hopeless to perform precise RV measurements in this region.

In the optical regions, the telluric features largely "kick in" beyond about 0.62 nm, and these regions should generally be avoided when calculating RVs. Only in the 0.3–0.4 nm region will the stellar spectrum be largely free from telluric contamination.

In calculating precise RVs, one solution is simply to mask out those spectral regions contaminated by telluric lines. If one is interested in achieving the highest RV precision possible, this is the method of choice.

Another approach is to divide out the contribution of the telluric lines. The drawbacks of the telluric division method have been extensively discussed in the literature (Vacca et al. 2003; Bailey et al. 2007; Seifahrt et al. 2010; Gullikson et al. 2014; Smette et al. 2015).

There are two approaches to telluric division. The first is to divide your science observation with a telluric standard star, which is usually a rapidly rotating B- to A-type stars main-sequence stars. These types of stars have relatively few spectral

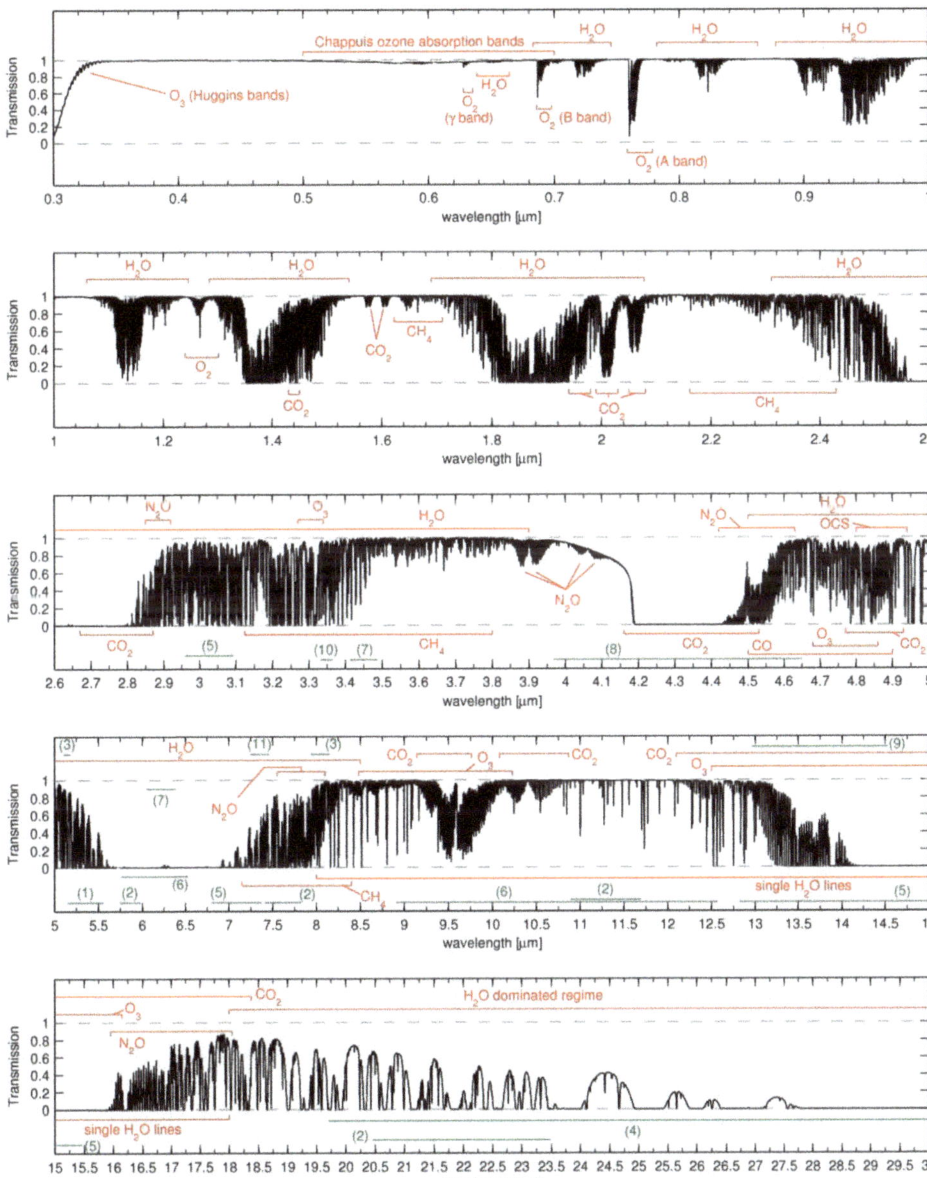

Figure 12.18. Synthetic spectrum of telluric features from 0.3 to 30 μm. (From Smette et al. 2015, reproduced with permission © ESO.)

lines, which are broad and shallow due to the high stellar rotation. There are several drawbacks to using standard stars:

1. Division by the telluric standard will increase noise in your science frame.
2. Even though the few spectral lines from the standard are broad, they will still alter the shape of the local continuum.

3. It may be difficult to get an observation of the standard star taken at the exact same air mass as the science target. Some scaling of the telluric line depths will most likely be required.
4. Increased overhead as observing time needed for science targets will be required for the standard star.

The second approach is to use a synthetic spectra of Earth's atmospheric transmission. The software tool Molecfit is commonly used to remove telluric absorption lines (Kausch et al. 2015; Smette et al. 2015, 2017). The main advantages of this method is that you will not add additional noise to your science spectrum, and you do not have to waste precious telescope time observing standard stars. The main drawback is that you may not have a perfect match to the observed telluric line absorption spectrum.

Seifahrt et al. (2010) investigated how well one can remove telluric lines using synthetic spectra. Figure 12.19 shows a fit to the water vapor feature near 1504.8 nm. They demonstrated that by using a synthetic spectrum, one could remove the contribution of the water lines to about 2%. This may be good for some applications, but maybe not for precise RV work.

How well must you remove the telluric lines? To answer this, we performed a simple simulation where we took a very narrow spectral line (our "telluric" feature)

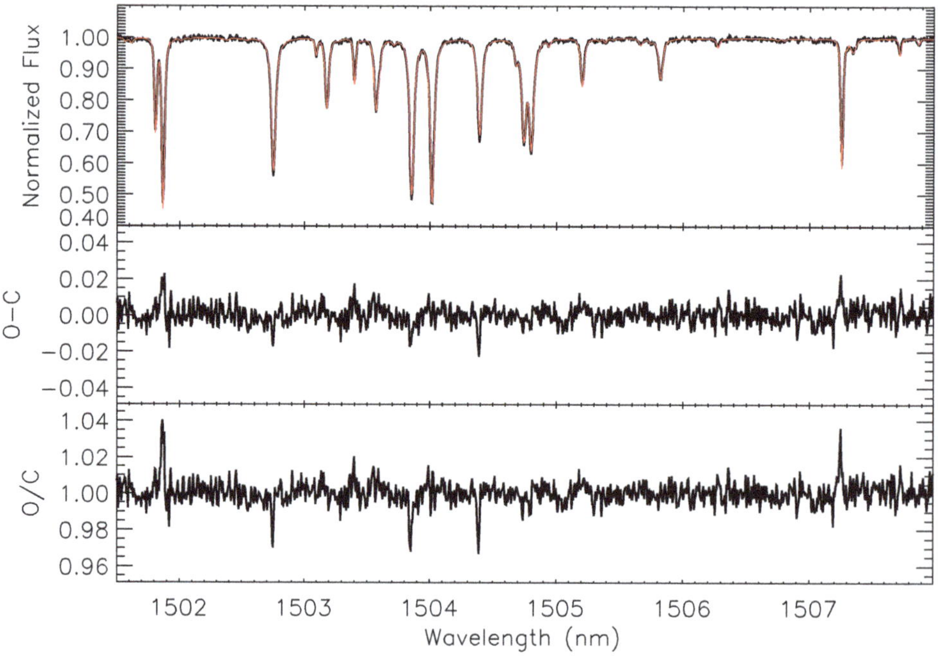

Figure 12.19. (Top) Water vapor near 1504.8 nm. Black is a spectrum observed with the CRIRES spectrograph and the synthetic transmission spectrum is shown in red. (Middle) The $O-C$ residuals of the fit. (From Seifahrt et al. 2010, reproduced with permission © ESO.)

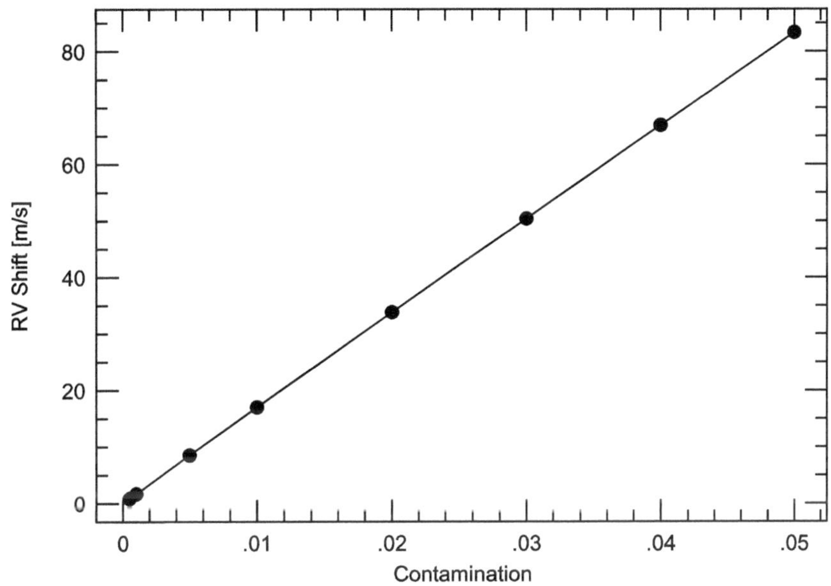

Figure 12.20. The RV shift of a spectral line due to a simulated "telluric" line (see text) as a function of the level of contamination. A contamination factor of unity implies a telluric line with the same depth as the stellar line.

and shifted it by 0.055 Å (resolving power, R = 100,000) with respect to a broader ($v \sin i = 3$ km s^{-1}) "stellar" line. Figure 12.21 shows the calculated RV shift as a function of the contamination factor, where a factor of 1.0 corresponds to the telluric line having the same depth as the stellar line. This shows that if you want to achieve an RV precision better than 1 m s^{-1}, you need to remove the telluric line contribution to better than 0.1%. As a rule, if you wish to use contaminated spectral regions, the telluric lines should be removed to a fraction of a percent.

If one is measuring RVs of very cool objects with lots of stellar lines (and flux) at longer wavelengths, you might gain by using telluric-contaminated regions. E. Nagel et al. (2019, in press) used modeling of the telluric spectrum to correct NIR spectra taken with the CARMENES spectrograph. Figure 12.20 shows the improvement in a time series of RV measurements in the VIS and NIR channels of CARMENES. In the NIR channel, where telluric line contamination is more severe, the rms scatter improves from 9.5 m s^{-1} to 5.7 m s^{-1} after the correction.

12.7 Moonlight Contamination

All precise RVs are taken with high-resolution spectrographs, which generally means you cannot observe faint stars. For a 4 m class telescope, a V-magnitude of ≈10–11 is a realistic limit. For 8 m class telescopes, you can probably observe stars with V = 12–13. Traditionally, stellar observations were always scheduled near the full moon; precious dark time should be devoted to imaging and spectroscopy of faint objects.

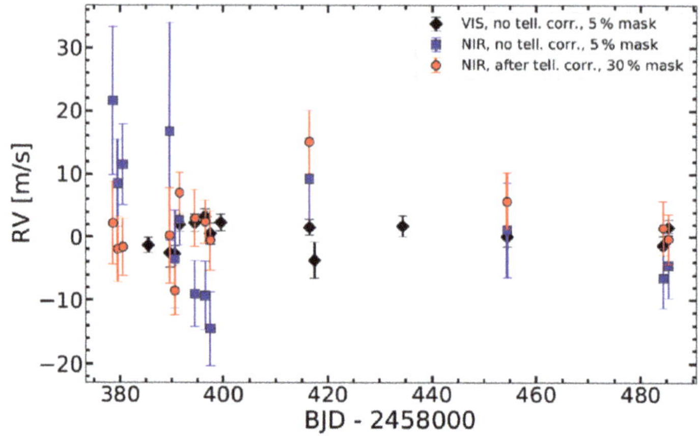

Figure 12.21. Time series of RV measurements in the VIS without telluric correction (black diamonds) and the NIR channel before (blue squares) and after (red circles) telluric correction. For the NIR channel RVs, the rms scatter is reduced from 9.5 m s^{-1} to 5.7 m s^{-1}. (From E. Nagel et al., 2019, in press.)

For precise stellar RV measurements, contamination by moonlight can be a serious source of error when observing faint stars. This is especially true for spectroscopic observations of late-type stars. You now have a spectrum of a G2 main-sequence star (the Sun!) contaminating your stellar spectrum. The problem becomes more acute for faint stars at low signal-to-noise levels taken with long exposures. You can try avoiding the Moon by requesting observing time during so-called "dark" (new moon) or "gray" (quarter moon) time, but that will decrease your chances of success. You are now competing with your extragalactic colleagues who simply must use dark time. So the chances are high that when you make RV measurements, the Moon will be up for a portion of the night, and you have to worry about contamination by moonlight.

Figure 12.22 shows how moonlight contamination can seriously affect your RV measurement in a worst case scenario. The top panel shows the cross-correlation function (CCF) of a faint star that was observed during new moon. A clean, single peak in the CCF is seen at an RV of ≈ -32 km s^{-1}.

The lower panel shows the CCF computed from an observation of the same star, but taken during full moon. In this case, the scattered light from the Moon (i.e., the solar spectrum) produces a CCF peak that is stronger than that from the target star. Fortunately, the CCFs from the star and the solar spectrum are clearly separated. If the RV displacement between the star and solar spectrum were smaller you may not even see the effects of moonlight contamination. However, it would still distort the CCF and produce an erroneous RV measurement.

If want to improve the RV precision of a moon-contaminated spectrum then the best way is to remove the solar spectrum from your observation. This requires an observation of the moonlight taken at the same time, in the same part of the sky, and with the same exposure time. If you are using a slit spectrograph, then you should use a slit that is longer than the stellar image so that you record the solar (moon)

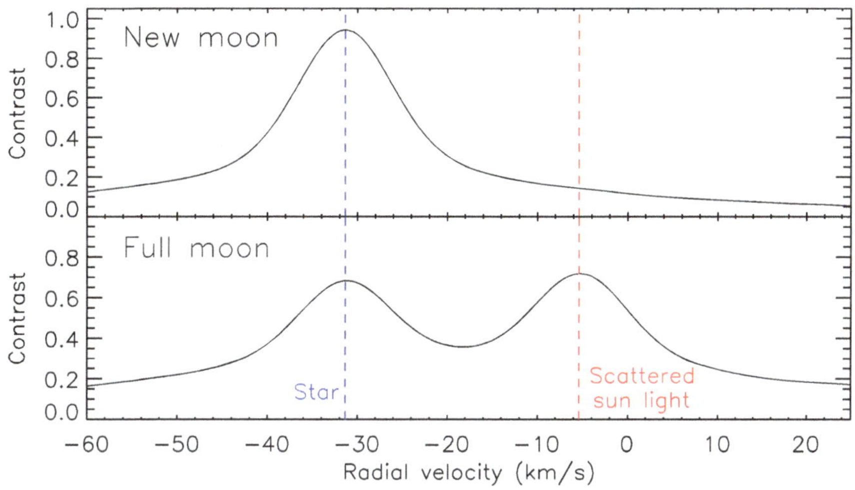

Figure 12.22. (Top) The cross-correlation function (CCF) of a program's star observed during new moon. (Bottom) The CCF of the same target observed during full moon. One can clearly see the peak at ≈ -5 km s^{-1} due to scattered moonlight. (Figure courtesy of Davide Gandolfi.)

spectrum on either side of the stellar one. Most fiber-fed spectrographs have a so-called sky fiber, which can be placed near the stellar image to record a spectrum of the background sky.

If you do not have either of these options, then one can try using a scaled version of the solar spectrum (e.g., an observation of the Moon or an asteroid). You then take a cross-correlation of your stellar spectrum and hope that the RV of the Sun does not coincide with your peak. If all works well, you will see a double peak in the CCF, which well tell you the relative Doppler shift and scaling factor you have to apply to the solar spectrum. However, if this is your only option for removing moonlight contamination, you should simply not observe stars near the Moon.

References

Bailey, J., Simpson, A., & Crisp, D. 2007, PASP, 119, 228

Bean, J. L., Seifahrt, A., Hartman, H., et al. 2010, ApJ, 711, L19

Bouchy, F., Isambert, J., Lovis, C., et al. 2009, in EAS Publications Series, Vol. 37, ed. P. Kern, 247–53

Dumusque, X., Pepe, F., Lovis, C., & Latham, D. W. 2015, ApJ, 808, 171

Goudfrooij, P., Bohlin, R. C., Maíz-Apellániz, J., & Kimble, R. A. 2006, PASP, 118, 1455

Gullikson, K., Dodson-Robinson, S., & Kraus, A. 2014, AJ, 148, 53

Halverson, S., Terrien, R., Mahadevan, S., et al. 2016, Proc. SPIE, 9908, 99086P

Kausch, W., Noll, S., Smette, A., et al. 2015, A&A, 576, A78

Kürster, M., Hatzes, A. P., Cochran, W. D., et al. 1999, A&A, 344, 5

Lo Curto, G., Pepe, F., Avila, G., et al. 2015, Msngr, 162, 9

Pravdo, S. H., & Shaklan, S. B. 2009, ApJ, 700, 623

Ribas, I., Tuomi, M., Reiners, A., et al. 2018, Natur, 563, 365

Seifahrt, A., Käufl, H. U., Zängl, G., et al. 2010, A&A, 524, A11
Smette, A., Sana, H., Noll, S., et al. 2015, A&A, 576, A77
Smette, A., Sana, H., Horst, H., et al. 2017, in ESO Calibration Workshop: The Second Generation VLT Instruments and Friends, 41
Tronsgaard, R., Buchhave, L. A., Wright, J. T., Eastman, J. D., & Blackman, R. T. 2019, MNRAS, 489, 2395
Vacca, W. D., Cushing, M. C., & Rayner, J. T. 2003, PASP, 115, 389
Wilken, T., Lovis, C., Manescau, A., et al. 2010, MNRAS, 405, L16
Wright, J. T., & Eastman, J. D. 2014, PASP, 126, 838
Zapatero Osorio, M. R., Martín, E. L., del Burgo, C., et al. 2009, A&A, 505, L5

IOP Astronomy

The Doppler Method for the Detection of Exoplanets

A P Hatzes

Chapter 13

The Rossiter–McLaughlin Effect

13.1 Introduction

Up until now, we have been discussing the Doppler method as it is applied to the detection of exoplanets with the goal of deriving the companion mass. The Doppler method can tell you more about the planet's orbit, in particular the alignment of the orbital axis to the spin axis of the star. This is done via the Rossiter–McLaughlin (R–M) effect.

This effect was discovered in the RV curves of eclipsing binaries almost 100 years ago by Rossiter (1924) and McLaughlin (1924). Although the effect is named after both astronomers, Rossiter's paper was dated 11 days prior to McLaughlin's. Furthermore, McLaughlin remarked in his paper that Rossiter first observed the effect in β Lyrae.[1]

Rossiter correctly noted that this was an effect of stellar rotation, remarking that "the secondary oscillation occurring in the residuals of the eclipse time are due not to orbital motion, but to the rotation of the more luminous star about its axis."[2]

Figure 13.1 shows the discovery of the effect in the RV variations of β Lyrae (Rossiter 1924). The RV measurements are impressively good with an rms of approximately 1.3 km s^{-1}, a remarkable RV precision for 1924! With such an RV precision, Rossiter would have been able to detect a short-period massive planet or brown dwarf. Figure 13.2 shows more modern measurements on an eclipsing binary (Lehmann & Mkrtichian 2008).

The grayscale in the right panel shows the RV variations of both components of the binary, and you can see that the R–M effect occurs during the eclipse.

[1] Coincidentally, both Rossiter and McLaughlin were colleagues at the Detroit Observatory.

[2] In fact, the title of his paper was "On the detection of an effect of rotation during the eclipse in the velocity of the brighter component of Beta Lyrae, and on the constancy of velocity of this system." A rather lengthy but descriptive title!

doi:10.1088/2514-3433/ab46a3ch13

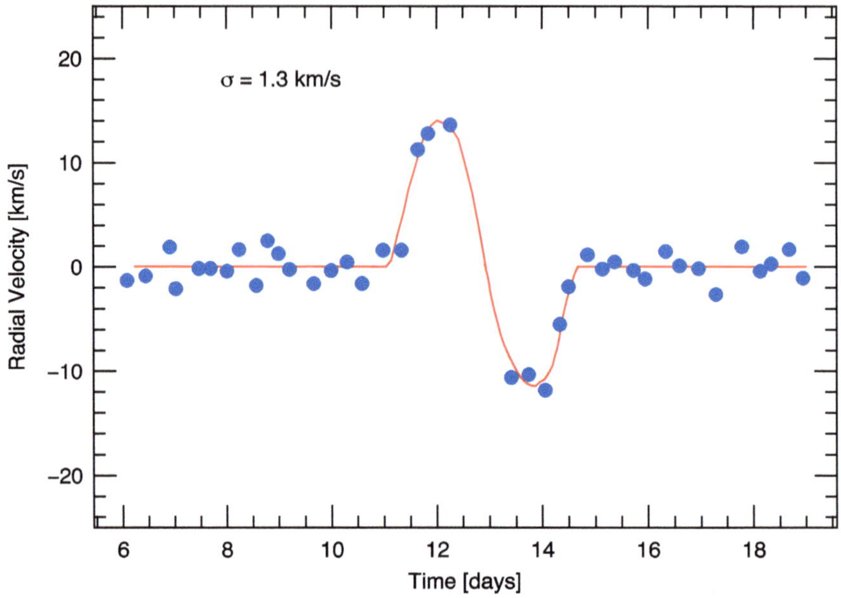

Figure 13.1. The discovery of the Rossiter effect. Rossiter's RV measurements for the binary star β Lyrae (Rossiter 1924). The Keplerian orbital motion has been removed.

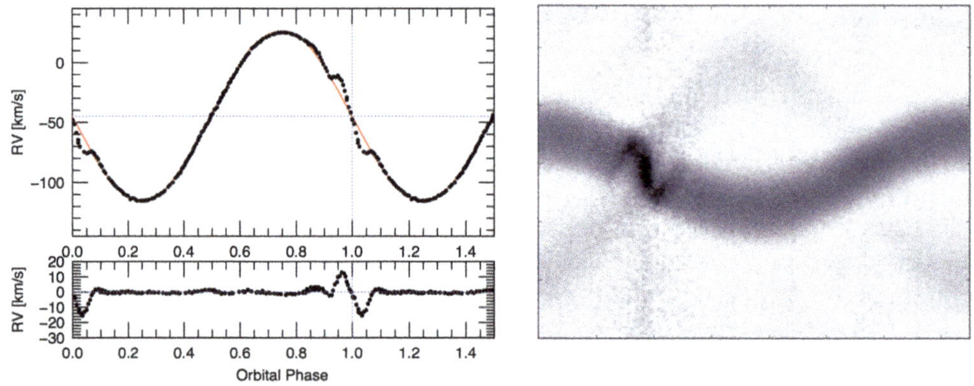

Figure 13.2. The R–M effect as seen in an Algol-type eclipsing binary, RZ. (Top left) The RV curve of the orbital motion. Note the distortion at phase 0 (1). (Lower left) The R–M distortion after subtracting the orbital motion of the binary. (Right) A grayscale image of the RV orbital motion of the primary (dark gray) and secondary (light gray). Note the R–M at mid-eclipse. (From Lehmann & Mkrtichian 2008.)

13.2 Origin of the Rossiter–McLaughlin Effect

The Rossiter–McClaughlin effect arises from the fact that the orbiting body blocks a portion of the light from the rotating stellar sphere. This produces a distortion in the spectral line that moves across the line profile, due to the orbital motion. For a rapidly rotating star, one can actually see the distortions due to the R–M effect by eye. As a planet crosses the blueshifted disk of the rotating star, it produces a

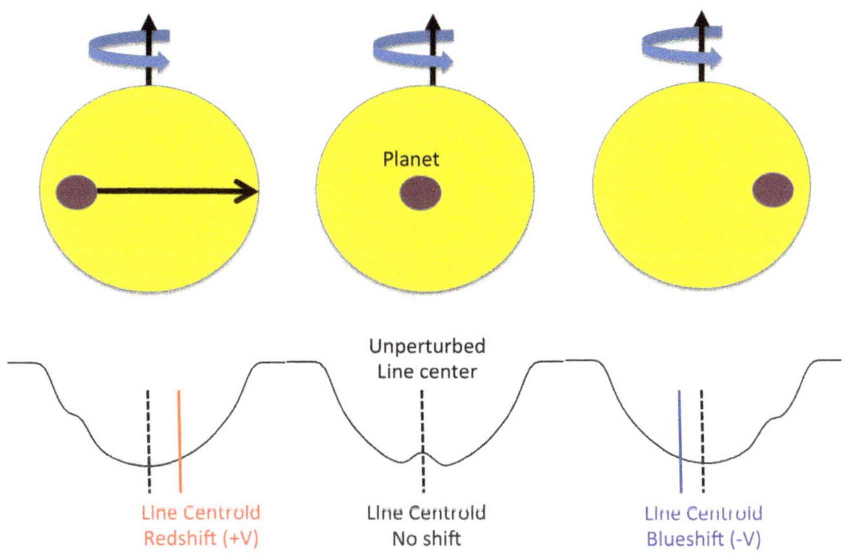

Figure 13.3. The R–M effect resulting from a distortion in the line profile. (Left) The planet blocks flux from the approaching limb of the star, producing a bump in the blue wing of the line. This shifts the centroid of the line to a positive velocity. (Center) As the planet crosses the line of sight, the bump is at line center, producing a symmetrical profile and thus zero Doppler displacement. (Right) As the planet moves to the redshifted limb of the star, the distortion appears in the red wing, thus producing a blueshifted line centroid.

"pseudo-emission" bump in the blue wing (Figure 13.3). This results in a slight displacement of the line centroid to positive velocities. As the planet crosses the line of sight, the distortion is a line center, and the profile is symmetrical, i.e., no Doppler perturbation. Finally, when the planet is in front of the receding limb, it now blocks redshifted light from the star. The distortion is now in the red wing of the line profile, and the centroid shifts to negative velocities.

We have seen this effect before in spotted stars. Figure 9.2 shows the distortion due to a stellar spot on the rotating star. In fact, the Doppler displacement is essentially an R–M effect due to starspots, and these are instructive in understanding the origin of the pseudo-emission bump—it is not so obvious. Vogt & Penrod (1983) nicely demonstrated how an emission bump appears in the rotationally broadened line profiles of spotted stars. For the R–M effect, a transiting planet will produce the same feature as a cool starspot, only that the planet moves with respect to its orbital motion, rather than to stellar rotation.

Figure 13.4 shows a rapidly rotating star with spectral lines that are dominated by the Doppler broadening of stellar rotation. In this case, the local line profiles of all regions of the star lying on the vertical chord (segment) have the same Doppler shift. All local line profiles in this segment get mapped into the same location in the spectral line. There is thus a one-to-one correspondence in Doppler shift between a location on the star (in a given segment) and position in the line profile. The constant Doppler-shift segments are marked by numbered zones. In this idealized case, the local line profile from each zone has a continuum value of 0.2 and zero flux in line

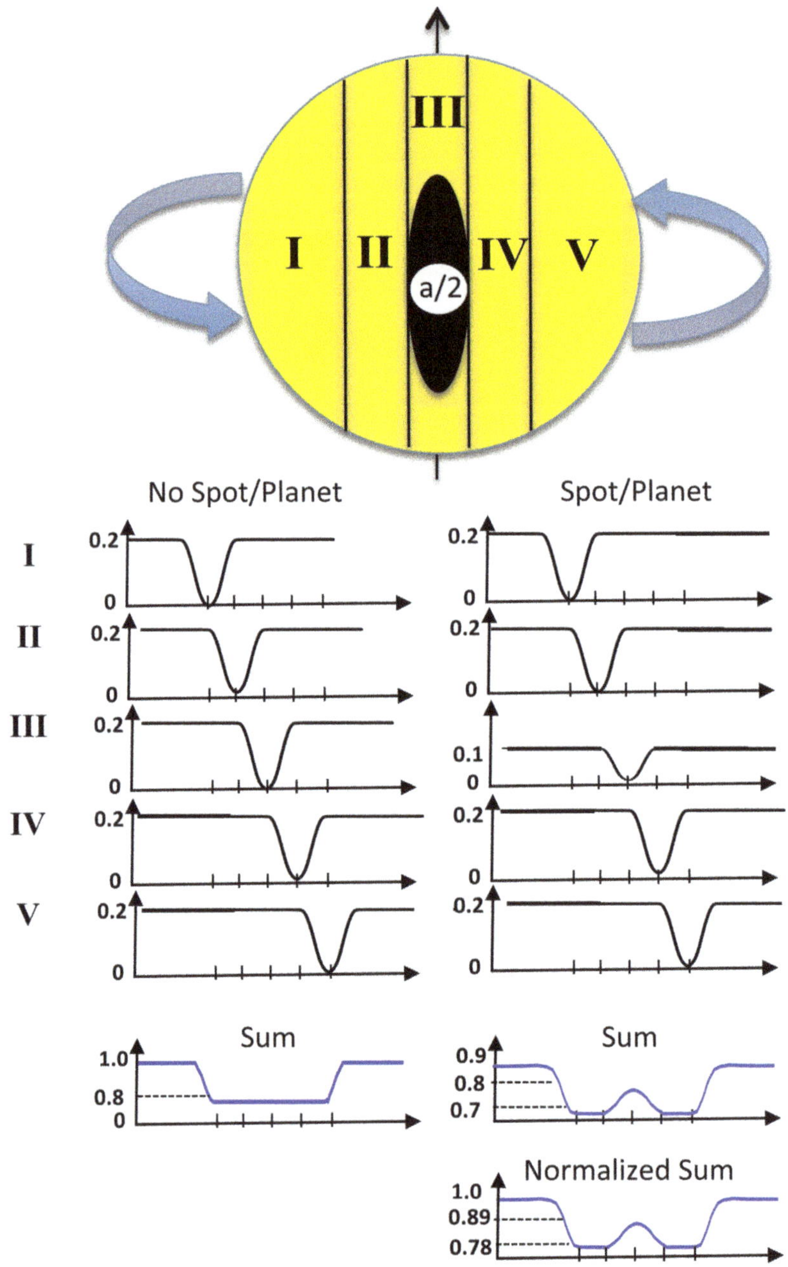

Figure 13.4. Illustration of the formation of the bumps in the line profiles due to the planet-blocking flux from the star or a surface spot. For illustrative purposes, we have made the planet oval-shaped so that it covers one-half of the area of the central stellar segment. (After Vogt & Penrod 1983. The American Astronomical Society. All rights reserved.)

center. The line profile of each zone is Doppler-shifted, and the sum produces a broad profile of continuum value 1.0 and a flux of 0.8 in the line core (left panels).

The right panels show what happens when half of the flux in the central zone is occulted by, in this case, a planet. For illustrative purposes, we have made the planet shape "oval" so that it covers roughly one-half of the area of a constant Doppler-shift segment. The local line profile of zone III has half of its former value, but still zero flux in the core. When the contributions of all zones are added, the line center still has a flux of 0.8, but either side (flux from the other zones) is now at 0.7 due to the reduced flux level. Thus, an emission bump appears. This is the flux effect, and one can consider that the line center, where the planet (spot) appears, stays at the same flux level, but the surrounding regions of the line profile drop in flux.

Figure 3.13 shows the appearance of the pseudo-emission bump due to the transiting planet WASP-33b. The host star has an RV of 90 km s^{-1}. The planet "bump" crosses the line profile from the red to the blue wing, indicating a planet in retrograde motion. For slowly rotating stars, this distortion is not visible by eye, but it still causes a shift in the centroid of the line profile, which appears as a Doppler perturbation from the overall orbital motion.

13.3 The Rossiter–McLaughlin Effect in Exoplanets

The R–M effect "hibernated" for almost 80 years until astronomers realized that they could be applied to exoplanets. This was largely possible due to the increase in RV precision over the past few decades.

13.3.1 The Radial Velocity Amplitude

The RV amplitude of the R–M effect can be comparable or larger than the K amplitude due to the planetary companion. The net observed RV variation, $\Delta V(t)$, is a sum of the Keplerian orbital motion, $\Delta V_O(t)$, and the anomalous RV given by the R–M effect, $\Delta V_R(t)$,

$$\Delta V(t) = \Delta V_O(t) + \Delta V_R(t). \tag{13.1}$$

The orbital motion is given by

$$V_O(t) = K_O[\cos(\nu + \omega) + e\cos\omega],$$

where K_O is the RV amplitude of the orbit.

Gaudi & Winn (2007) presented a detailed analytical analysis of the R–M effect (also nicely summarized in Haswell's *Transiting Exoplanets*; Haswell 2010) and showed that the K amplitude due to the R–M effect, K_{RM}, can be expressed as

$$K_{RM}[\mathrm{ms}^{-1}] = 52.8\left(\frac{V_s}{5\ \mathrm{km\ s}^{-1}}\right)\left(\frac{r_p}{R_{Jup}}\right)^2\left(\frac{R_s}{R_\odot}\right)^{-2}, \tag{13.2}$$

where V_s is the projected rotational velocity of the star in km s^{-1} normalized to 5 km s^{-1}, r_p is the radius of the planet, R_{Jup} the radius of Jupiter, and R_s is the stellar radius in solar radii.

It is worth reflecting about several features of this equation:

1. It is truly a rotation effect. For zero stellar rotation, the R–M amplitude is exactly zero.
2. The R–M amplitude depends on the ratio of the projected area of the planet disk to the stellar disk, $\sim(r_p/R_s)^2$.
3. Although it is a Doppler measurement, it cannot give you the mass of the planet.

A visual representation of the R–M *K* amplitude is shown in Figure 13.6 for three stellar radii: 0.2 $R_\odot$ (M dwarf), 1.0 $R_\odot$ (G dwarf), and 2.0 $R_\odot$ (A dwarf), and three planetary radii for Jupiter, Neptune, and Earth. Given a nominal precision of 1 m s^{-1} means that for a Sun-like star, the R–M effect can only be detected down to a Neptune radius planet. If the star shows significant rotation ($v \sin i \approx 20$ km s^{-1}) one could in principle detect the R–M of an Earth-like planet, except for the fact that your RV precision is more like ≈20 m s^{-1}, due to the more rapid rotation of the star.

A-type stars are clearly not amenable for detecting the R–M amplitude. First, they have few spectral lines, and these are usually broadened by rapid rotation of the star. Yes, the R–M effect will be larger due to rapid rotation, but this is countered by the much poorer RV measurement error. As we learned in Chapter 3, a typical RV measurement error on a A-type star is ≈50–100 m s^{-1}. The only hope is detecting the R–M due to giant planets. However, one can instead exploit the rapid rotation of early-type stars to use Doppler tomography to measure the spin–orbit alignment (Figure 13.5).

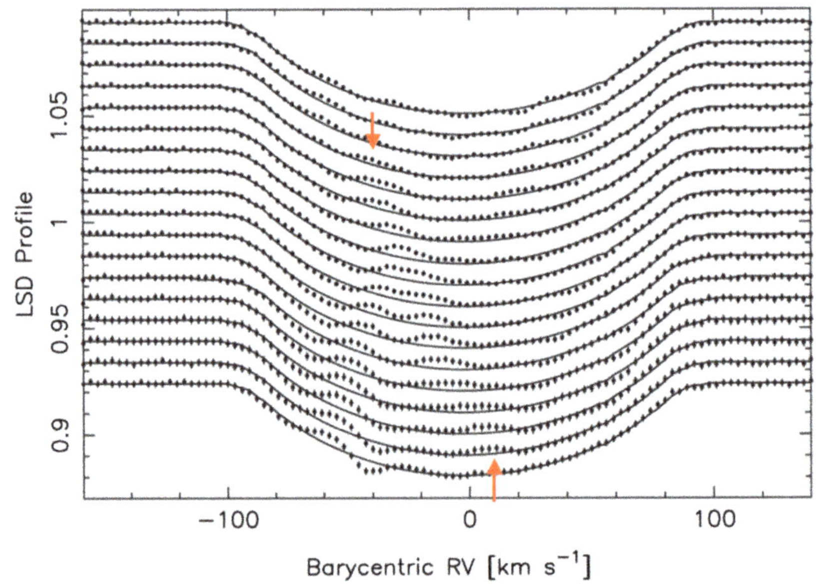

Figure 13.5. Least-squares deconvolved profiles of WASP-33 (HD 15082) during the planetary transit. The stellar rotation profile is shown as a solid line, and time increases vertically. The planet signature is the bump that migrates from $v \approx -5$ km s^{-1} (lower red arrow) in the bottom profile (orbital phase ≈ 0.89) to $v \approx 40$ km s^{-1} (upper red arrow) in the top profile (orbital phase ≈ 1.03). (Reproduced from Collier Cameron et al. 2010. Copyright of OUP Copyright '2010'.)

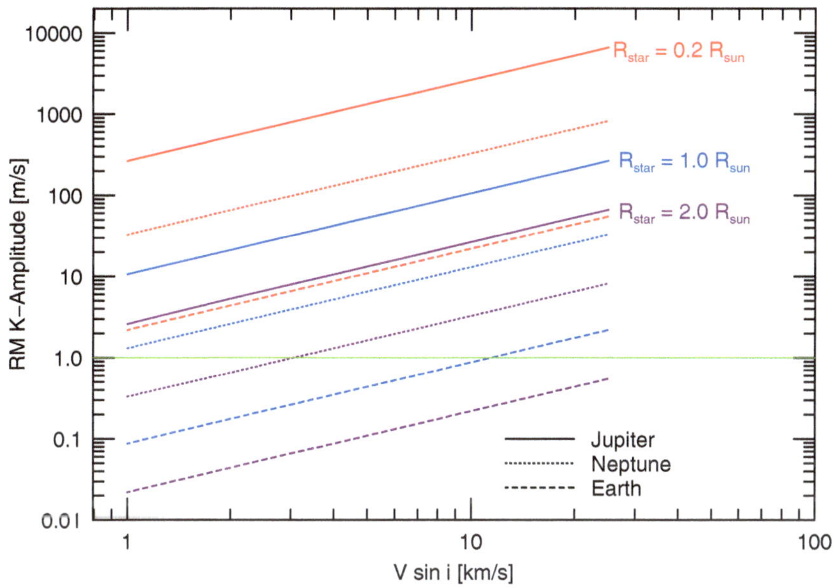

Figure 13.6. The RV amplitude of the R–M effect for three stellar radii, 0.2 $R_{\odot}$ (red), 1.0 $R_{\odot}$ (blue), and 2.0 $R_{\odot}$ (purple), as well as three planets with the radii of Jupiter (solid line), Neptune (dotted line), and Earth (dashed line) radii. The horizontal green line marks the nominal 1 m s^{-1} precision achievable on late-type stars.

Clearly, due to their small radii, M dwarfs are the ideal targets for measuring the R–M effect. Plus, they have a large number of relatively narrow spectral lines. Typically, one can achieve an RV precision of ≈1–2 m s^{-1}, depending of course on the rotational velocity of the star, which means that the R–M effect from an Earth-size planet can be easily measured.

13.3.2 The Spin–Orbit Alignment

The R–M effect is often referred to as "spectroscopic transits" as it gives you the planet radius just as in a photometric transit. So, if the R–M effect only gives you information about the planet radius (ratioed to the stellar radius), why use it? After all, it is simpler to get photometric data to determine the planet radius, and that can be done whether the star rotates or not—which is not the case for the R–M effect. The real power of the R–M effect is that it gives you valuable information on the alignment of the planet's orbit.

Let λ be the angle between the spin axis of the star and the orbital axis ($\lambda = 0°$ means the spin and orbital axes are aligned, $\lambda = 90°$ means the orbit axis is inclined 90° to the spin axis). Figure 13.7 shows the R–M effect for three values of λ and an impact parameter $b = 0$ ($b = 0$ means the planet crosses the stellar equator, $b = 1$ is a grazing transit). A value of $\lambda = 0°$ produces a symmetric R–M distortion. For $\lambda = 30°$, the planet crosses more of the redshifted part of the stellar disk, producing an asymmetric R–M curve. Finally, for $\lambda = 60°$, the planet only blocks light from redshifted regions of the star. The R–M curve always shows a negative velocity displacement.

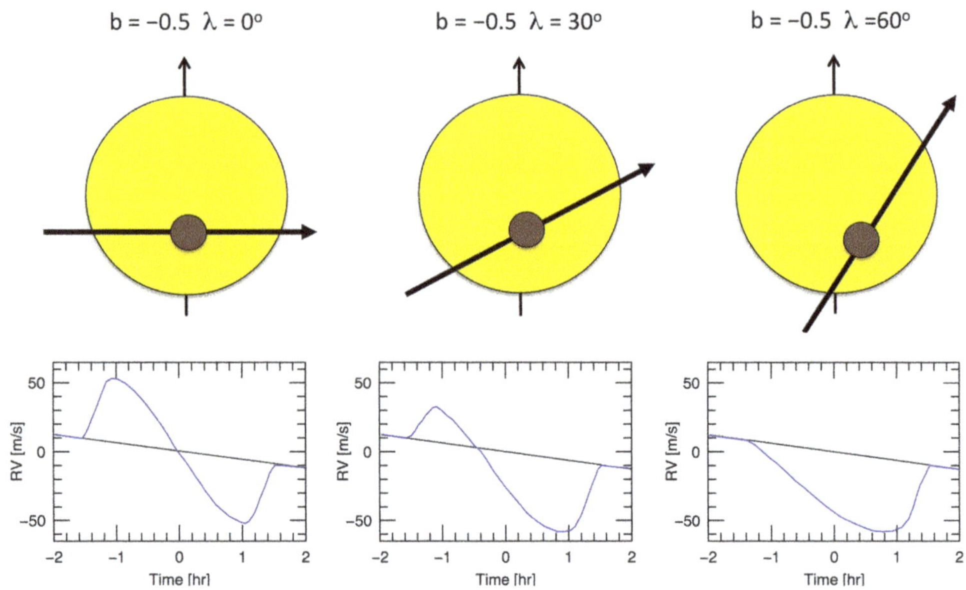

Figure 13.7. The R–M effect for different spin–orbit alignment angles, λ. (Left) Aligned orbits ($\lambda = 0°$) produce an asymmetric R–M effect. For $\lambda = 30°$, the R–M curve starts to become more asymmetric (middle panel). Finally, for $\lambda = 60°$, the R–M curve is purely asymmetric, with no positive values (left panel).

If the planet is in a retrograde orbit ($\lambda = 180°$), then the planet will first block light from the redshifted part of the rotating star and then the blueshifted regions. We therefore expect an "inverse" R–M effect, i.e., an S-curve that first goes down and then up. Figure 13.8 shows a summary of the various R–M curves depending on the track of the planet across the stellar disk.

Figure 13.9 shows the R–M effect measured in four exoplanets. HD 202458b (Queloz et al. 2000) and CoRoT-2b (Bouchy et al. 2008) show the classic symmetric R–M curve. CoRoT-1b shows a purely asymmetric R–M effect (Pont et al. 2010), much like the third panel in Figure 13.7. Finally, HAT-P7 was the first exoplanet to have an R–M effect consistent with a retrograde motion of the planet (Winn et al. 2009; Narita et al. 2009).

13.4 Spin Axis of the Star

The R–M effect does not tell you the true angle between the spin axis of the star and the orbital axis of the planet. Rather, it only tells you the orbital axis and the spin axis of the star projected onto the plane of the sky. You only measure the $v \sin i$ of the star and not i.

To get the true obliquity angle, one needs to measure the inclination of the stellar rotation axis. The best way is to use asteroseismic date (e.g., Corsaro et al. 2017). In the most likely case where asteroseismic data are not available, the only recourse is to estimate i using the rotation period of the star, P_{rot}; its stellar radius, R_{star}; and the true stellar rotational velocity, V_{rot}.

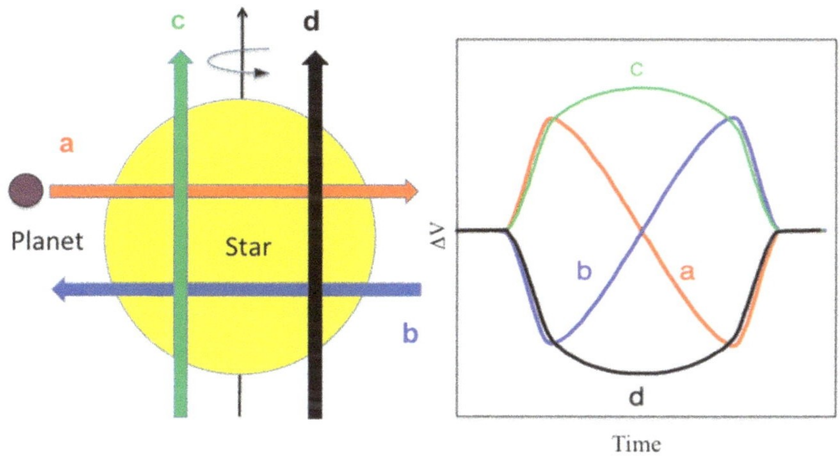

Figure 13.8. (Left) The tracks of a planet across a rotating star. (Right) The observed R–M tracks after removing the orbital motion. (a) A planet in a prograde orbit, (b) a planet in a retrograde orbit, (c) a planet in a polar orbit, but only crossing the blueshifted disk of the star, and (d) a planet in a polar orbit crossing only the redshifted disk of the star.

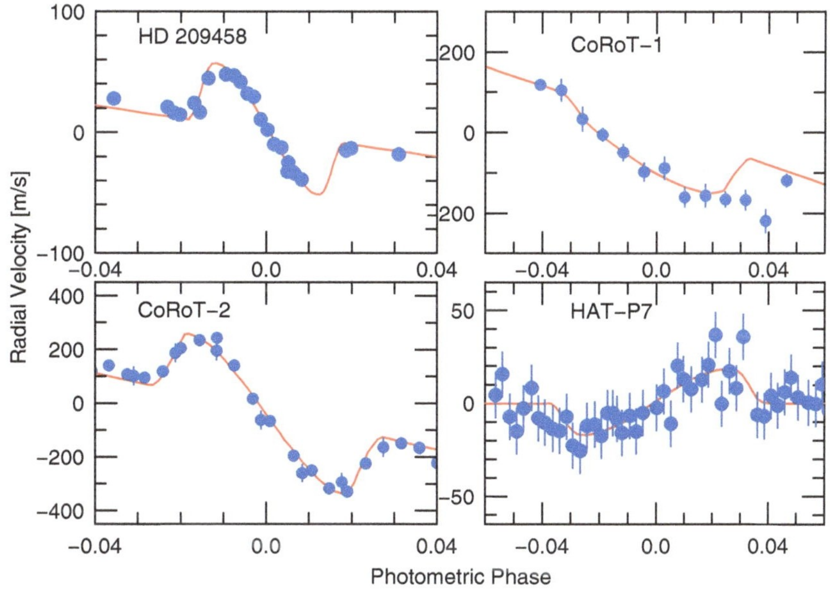

Figure 13.9. The R–M effect in four exoplanets. (Top left) HD 209458b (Queloz et al. 2000), (lower left) CoRoT-2b (Bouchy et al. 2008), (upper right) CoRoT-1b (Pont et al. 2010), and (lower left) HAT-P7 (Winn et al. 2009).

As we have seen, the rotational velocity of the star can be calculated by

$$V_{\rm rot} = \frac{2\pi R_{\rm star}}{P_{\rm rot}}, \tag{13.3a}$$

and the stellar inclination is given by

$$i = \sin^{-1}\left(\frac{v \sin i}{V_{\rm rot}}\right). \tag{13.3b}$$

To calculate the RV of the star requires an accurate stellar radius and rotational period. If the star is active, the rotational period can often be seen in the RV time series. With the advent of high photometric precision space missions (e.g., *CoRoT*, *Kepler*, *TESS*), stellar rotational periods will be available for a large number of stars with accuracies of ≈5%–10%.

Estimating the stellar radius often relies on the spectral type of the star or evolutionary tracks. Fortunately, the *Gaia* mission will have accurate distances for a large number of stars. From the stellar distance, brightness, and effective temperature of the star, one can derive a fairly accurate stellar radius. Asteroseismic data from the *PLATO* mission (Rauer et al. 2014) will also deliver accurate stellar parameters via asteroseismology. One should be able to determine the stellar radius and rotation period to about 10% in each quantity.

The largest uncertainty in determining the stellar inclination will thus fall on the measurement of the projected rotational velocity of the star. For stars where rotation is the dominant broadening mechanism ($v \sin i > 10$ km s^{-1}), this is easily done. However, for slowly rotating stars with projected velocity of a few km s^{-1} (e.g., solar-like stars), the uncertainty can be quite large. The difficulty for these slow rotators is the fact that rotational broadening now has to compete with other broadening mechanisms such as macroturbulence and thermal broadening. In these cases, one is fortunate if one can measure the $v \sin i$ to an accuracy of 0.5–1 km s^{-1}. This means that for a star rotating at 2 km s^{-1}, like our Sun, one can derive an inclination angle via Equation 13.3(b) to no better than about 50°

Here we discuss in more detail how one can measure an accurate $v \sin i$ for a star because this is not only important for trying to estimate the angle of the spin axis of the star, but also for the rotation period if one wants to exclude this as the source of the RV variations (Chapter 9). In this case, we solve Equation 13.3(a) for $P_{\rm rot}$ knowing $R_{\rm star}$ and the rotational velocity.

For slowly rotating stars, the Doppler broadening due to stellar rotation competes with a number of other broadening mechanisms. First, there is the thermal broadening of the line, $V_{\rm th}$, given by

$$V_{\rm th} = 4.3 \times 10^{-7}\,\lambda\left(\frac{T}{\mu}\right)^{1/2}, \tag{13.4}$$

where λ is the wavelength of the line, T the effective temperature of the star, and μ the mass of the element in atomic mass units.

Second, there is the macroturbulent velocity, V_M, which for a solar-type star is about $V_M \approx 5$ km s^{-1} (see Figure 9.12). Finally, there is the broadening of the instrumental profile, V_{IP}, which for an R = 50,000 spectrograph (two-pixel-resolution element) is 6 km s^{-1}. Table 13.1 lists the Doppler widths due to various broadening mechanisms (in velocity units) for a G2 V star with T = 5900 K.

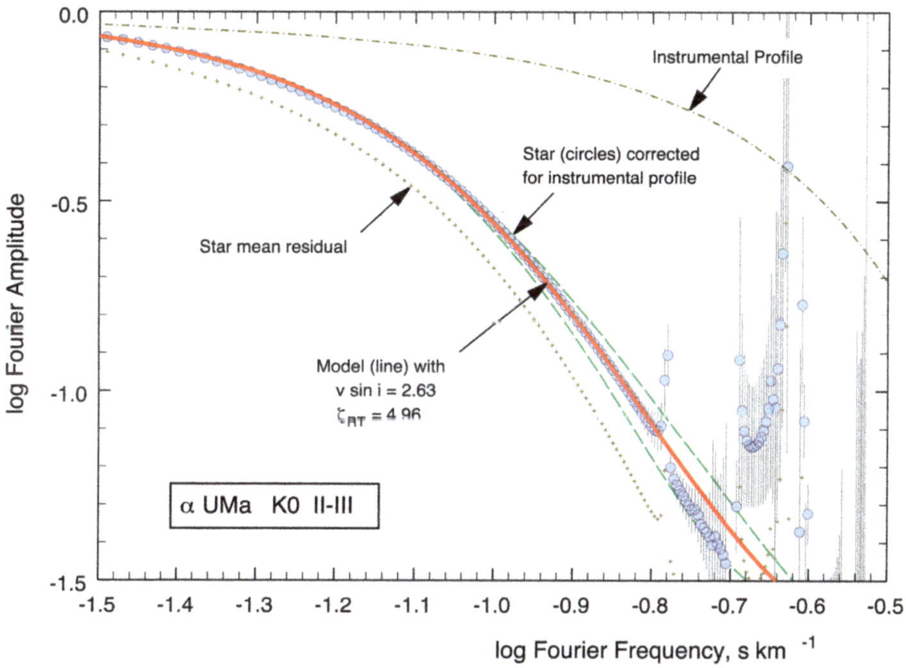

Figure 13.10. The mean residual Fourier transform for nine lines (circles) in α Uma after division by the transform of the instrumental profile (dashed–dot green line). The solid red line is the model with km s^{-1} and ζ_{RT} = 4.97 km s^{-1}. The dashed green lines: show changes in the v sin i altered by ±0.5 km s^{-1}. The star mean residual is after removing the thermal profile (From Gray 2018).

Table 13.1. Broadening Mechanisms for a G2 V Star

Mechanism	Value
Rotation	2.0 km s^{-1}
V_{th} (Fe)	1.3 km s^{-1}
V_M	5.0 km s^{-1}
V_{IP} (R = 50,000)	6.0 km s^{-1}
V_{IP} (R = 100,000)	3.0 km s^{-1}
V_{IP} (R = 200,000)	1.5 km s^{-1}

The total broadening of the spectral line is from a convolution of the individual profiles of the broadening mechanisms,

$$V_{obs} = V_{th} * V_{rot} \sin i * V_M * V_{IP} * V_{int}, \tag{13.5}$$

where V_{obs} is the observed Doppler broadening of the line profile, V_{th} the thermal broadening, $V_{rot} \sin i$ the projected rotational velocity, V_M the macroturbulent broadening, V_{int} the intrinsic line profile with no broadening, and V_{IP} the instrumental broadening. The symbol * represents the convolution process.

All of these roughly have a Gaussian shape, which means that the velocity widths in Table 13.1 add in quadrature. Thus, the broadening due to slow rotation is completely dominated by macroturbulence and the instrumental profile of your instrument.

If one wants to measure a very accurate $v \sin i$ for a slowly rotating star, then this should be done through the individual modeling of a few carefully selected lines. The best procedure is to use

- Spectra lines from species with as high an atomic number so as to minimize the thermal broadening. For instance, a Ca (atomic mass = 40) line in a Sun-like star has a thermal broadening of 1.5 km s^{-1} compared to 1.3 km s^{-1} for Fe (atomic mass = 55). In the unlikely event you can find a Eu line, this would have a thermal broadening of only 0.8 km s^{-1}.
- As high a spectral resolution as possible. With an instrumental broadening of 5 km s^{-1}, $R = 50{,}000$ is inadequate for accurate $v \sin i$. An $R = 200{,}000$ spectrograph is considerably better with an instrumental profile broadening of only 1.5 km s^{-1}.
- High signal-to-noise ratios (S/Ns). This goes without saying, but one should have S/N > 300.
- Perform the analysis in the Fourier domain.

The most accurate measurement of the $v \sin i$ is best done in the Fourier domain (e.g., Gray 1982, 2018) for two reasons. First, we want to measure slight changes in shape in the spectral line profile. We learned in Chapter 7 that a small change in the time or spatial domain results in a large change in the Fourier domain. Subtle differences between the macroturbulent and rotational profiles are thus best seen in the Fourier domain. Second, the observed spectral line profile is a convolution of the individual broadening profiles. This is also easily done in the Fourier domain, where the removal of a profile is done simply by dividing it by the Fourier transform.

Cochran et al. (1991) used Fourier methods to measure the very small rotational broadening of $v \sin i \approx 0$ km s^{-1} for the planet candidate hosting star HD 114762. Since F-type stars should show appreciable rotation this was an early indication that the star and companion orbit, were viewed nearly pole-on. Accurate measurements of such a low $v \sin i$ are only possible with the Fourier method.

Let $d(\sigma)$ represent the Fourier transform of the data (spectral line) profile, $ip(\sigma)$ the Fourier transform of the instrumental profile, and $i_\nu(\sigma)$ the Fourier transform of the intrinsic line profile. The last function is calculated from model atmospheres. The

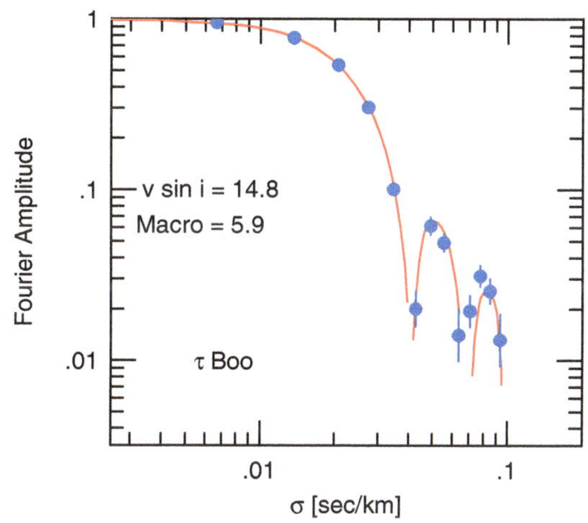

Figure 13.11. The Fourier transform of the observed mean line profiles for τ Boo. Note the zeros and side lobes in the amplitude. The best fit is for a $v \sin i = 14.8 \pm 0.3$ km s^{-1} and macroturbulent velocity = 5.9 ± 0.7 km s^{-1}. (From Gray 1982.)

residual transform, $m(\sigma)$, represents the Fourier transform of the combined rotation and macroturbulent and is obtained from

$$m(\sigma) = \frac{d(\sigma)}{ip(\sigma)i_\nu(\sigma)}. \tag{13.6}$$

Note that because we have divided out the intrinsic spectral line profile (and thus the thermal broadening), we can now average the $m(\sigma)$ of several line profiles. Figure 13.10 shows the fit to the mean values of $m(\sigma)$ for α UMa, resulting in a macroturbulent velocity of 4.96 km s^{-1} and $v \sin i = 2.63 \pm 0.5$ km s^{-1} (Gray 1984). A simple fitting of the line profile in the spatial domain would result in an error of $\approx$1 km s^{-1}.

For more rapidly rotating stars, the differences between the macroturbulent and rotation profiles become more apparent. Macroturbulence has a velocity distribution that has a Gaussian profile. In the Fourier domain, this is also a Gaussian profile with no zeros. On the other hand, the rotational profile has the shape of an ellipse, which in the Fourier domain has zeros as well as side lobes. Once rotation becomes the dominant broadening mechanism, one can fit the location of the zeros and side lobes to get an accurate measurement of $v \sin i$. Of course, one needs data with high S/Ns to ensure that the level of the Fourier noise is below that of the side lobes. Figure 13.11 shows the Fourier transform of the average of six spectral lines in τ Boo. Fitting the Fourier transform results in the measurement of $v \sin i = 14.8 \pm 0.3$ km s^{-1}, an error of only 2%. The macroturbulence is 5.9 ± 0.7 km s^{-1}.

References

Bouchy, F., Queloz, D., Deleuil, M., et al. 2008, A&A, 482, L25
Cochran, W. D., Hatzes, A. P., & Hancock, T. J. 1991, ApJ, 380, L35
Collier Cameron, A., Guenther, E., Smalley, B., et al. 2010, MNRAS, 407, 507
Corsaro, E., Lee, Y.-N., García, R. A., et al. 2017, NatAs, 1, 0064
Gaudi, B. S., & Winn, J. N. 2007, ApJ, 655, 550
Gray, D. F. 1982, ApJ, 258, 201
Gray, D. F. 2018, ApJ, 869, 81
Haswell, C. A. 2010, Transiting Exoplanets (Cambridge: Cambridge Univ. Press)
Lehmann, H., & Mkrtichian, D. E. 2008, A&A, 480, 247
McLaughlin, D. B. 1924, ApJ, 60, 22
Narita, N., Sato, B., Hirano, T., et al. 2009, PASJ, 61, L35
Pont, F., Endl, M., Cochran, W. D., et al. 2010, MNRAS, 402, L1
Queloz, D., Eggenberger, A., Mayor, M., et al. 2000, A&A, 359, L13
Rauer, H., Catala, C., Aerts, C., et al. 2014, ExA, 38, 249
Rossiter, R. A. 1924, ApJ, 60, 15
Vogt, S. S., & Penrod, G. D. 1983, PASP, 95, 565
Winn, J. N., Johnson, J. A., Albrecht, S., et al. 2009, ApJ, 703, L99

www.ingramcontent.com/pod-product-compliance
Ingram Content Group UK Ltd.
Pitfield, Milton Keynes, MK11 3LW, UK
UKHW051903100926
8556IPUK00005B/748

9 780750 316873